RESEARCH LIBRARY

OF

COLONIAL AMERICANA

RESEARCH LIBRARY
OF
COLONIAL AMERICANA

THE NATURAL HISTORY OF *BARBADOS*

GRIFFITH HUGHES

ARNO PRESS
A New York Times Company
New York – 1972

Reprint Edition 1972 by Arno Press Inc.

Reprinted from a microfilm copy furnished by
The University of Illinois Library

LC# 71-141086
ISBN 0-405-03285-4

Research Library of Colonial Americana
ISBN for complete set: 0-405-03270-6
See last pages of this volume for titles.

Manufactured in the United States of America

Publisher's Note: This volume was reprinted
from the best available copy.

THE NATURAL HISTORY OF *BARBADOS.*

IN TEN BOOKS.

By the REVEREND

Mr. *GRIFFITH HUGHES*, A.M.

Rector of St. LUCY's Parish, in the said Island, and F. R. S.

LONDON:

Printed for the AUTHOR;

And sold by most Booksellers in *Great Britain* and *Ireland*. MDCCL.

TO

HIS GRACE

THOMAS,

BY DIVINE PERMISSION,

Lord Archbiſhop of CANTERBURY,

THE FOLLOWING WORK IS,

WITH THE GREATEST DUTY AND RESPECT,

INSCRIBED, BY

HIS GRACE'S MOST OBEDIENT

AND HUMBLE SERVANT,

G. Hughes.

A

L I S T

OF THE

SUBSCRIBERS.

*Those whose Names are marked thus *, are Subscribers for large Paper.*

* *HIS Most Christian* MAJESTY.
* *His Royal Highness the Prince of* WALES.
* *Her Royal Highness the Princess of* WALES.
* *His Royal Highness the Duke of* CUMBERLAND.
* *His Most Serene Highness the Prince of* HESSE.
* *His Most Serene Highness the Prince of* SAXE-GOTHA.
* *His Most Serene Highness the Landgrave of* HESSE-CASSEL.
* *His Most Serene Highness the Duke of* ORLEANS.

A.

* *HIS Grace the Duke of St.* Albans.
* *His Grace the Archbishop of* Armagh, *Lord Primate of all* Ireland.
* *His Excellency the Right Hon. the Earl of* Albemarle.
* *Right Hon. the Earl of* Abingdon.
* *Right Hon. the Earl of* Arran.
* *Right Hon. the Earl of* Ancram.
* *Right Hon. Lord* Anson.
* *Right Rev. the Lord Bishop of St.* Asaph.
Hon. Sir Atwil Lake.
All-Souls College, Oxford.
Mr. Francis Abbyvan.
Thomas Adams, *Esq*;
Mr. Jonathan Adams.
* *Mr.* James Adams.
Mr. James Adamson.
Joseph Addison.
Andrew Afflick, *Esq*;
* Ralph Allen, *Esq*;
Dr. Reynold Alleyne.
Reynold Alleyne, *Esq*;
Mr. James Alport.
* *Mr.* John Alsop.
* *Sir* Stephen Anderson, *Bart.*
Henry Anderson, *Esq*;
* *Mr.* Andrews, *Apothecary.*
Hon. Thomas Applethwaite, *two Books.*
* *Mrs.* Archer.
Mr. Edward Archer, *jun.*
John Archer, *jun.*
William Arderen, *F. R. S.*
George Armstrong, *Esq*;
Mr. Samuel Armstrong.

A LIST of SUBSCRIBERS.

* *Mr.* William Atherly.
* *Sir* John St. Aubyn, *Bart.*
* *Rev. Dr.* Ayscough, *Clerk of the Closet to his Royal Highness the Prince of* Wales.

B.

His Grace the Duke of Bedford.
* *Right Hon. the Earl of* Burlington, *two Books.*
* *Right Hon. the Earl* Brooke.
Right Hon. the Earl of Bute.
* *Right Hon. the Earl of* Blessington.
* *Right Hon. the Lord Viscount* Bolingbroke.
* *Right Hon. the Lord* Vere Beauclerc, *Baron of* Hanworth.
* *Right Rev. Lord Bishop of* Bangor.
* *His Excellency the Chevalier* Bathget, *his Imperial Majesty's Resident at* Venice.
* *Right Rev. the Lord Bishop of* Bristol.
* *Sir* John Barnard, *Knt. Alderman.*
Mr. Robert Babb.
Richard Lovel Badcock.
* *Rev. Mr.* Barnard, *Rector of* Bridgetown *in* Barbados.
Mr. Abraham Barnard.
* Edward Barry, *M. D.*
James Barry, *Esq;*
Mr. Richard Barton, *B. D.*
John Bartrum.
Edward Bartrum.
Hon. William Barwick, *Esq;*
Mr. Samuel Barwick.
Mrs. Susanna Barwick, *Relict of the late President* Barwick.
Mr. Samuel Barrow.
Benjamin Baswain, *Esq;*
Thomas Baxter, *Esq;*
Mr. John Bayly.
* *Rev. Dr.* Bearcroft, *Chaplain in Ordinary to his Majesty.*
* Henry Bellingham, *Esq;*
Mr. George Bell, *Surgeon, F. R. S.*
* Francis Bell, *Esq;*
Mr. Robert Beckles.
Samuel Bedford.
John Best, *Esq;*
Mr. Robert Belgrave.
William Belgrave.
* Slingsby Bethel, *Esq; Alderman.*
* *Rev.* Thomas Berdmore, *D. D.*
* *Mr.* Ambrose Beurer, *Apothecary at* Nuremberg.
Mr. John Berisford.
* *Rev.* Joseph Bewsher, *two Books.*
Mr. Bickham, *Engraver*
* *Rev. Mr.* Thomas Birch, *A. M. F.R.S.*
Henry Bishop, *Esq;*
Rowland Blackman, *Esq;*
Trimmingham Blackwell, *Esq;*
Mrs. Priscilla Blackman.
Jonathan Blenman, *Esq; Attorney General of* Barbados, *three Books.*
* *Mr.* Robert Board.
Mr. Robert Bend.
Charles Bolton.
John Bonnel, *Esq;*
* *Mr.* Booth.
Benjamin Bostock.
* *Hon.* William Boverie, *Esq;*
Robert Bouchier, *Esq;*
Mr. Thomas Bourne.
* *Mr.* Thomas Bowen.
James Bowyer.
John Brace, *Esq;*
Edward Brace, *Esq;*
Mr. Samuel Brace.
* John Brathwaite, *Esq;*
Mr. Robert Brathwaite, *jun.*
* *Mr.* John Brewster.
* *Mr.* Thomas Brewster.
* *Hon.* George Bridges, *Esq;*
* *Mr.* Brindly, *Bookseller.*
* *The Rev. Mr.* Philip Brown, *B. D.*
* *Mr.* Broome.
* John Brougham.
Seigneur de Brummer.
William Bryant, *A. B. two Books.*
Hon. James Bruce, *Esq; two Books.*
Henry Banbrigge Buckeridge, *Esq;*
* *Hon. Sir* Thomas Burnet, *Knt. two Books.*
* *Rev. Dr.* Burton.
Mr. Charles Barton.
* Charles Bush, *Esq;*
Mr. Edward Butcher.
William Butcher.
James Butcher.
* *Mr.* Butterfield.
* *Rev. Mr.* Byam, *Rector of St.* John's, Antigua.

C.

* *His Grace thk Lord Archbishop of* Canterbury, *two Books.*
* *Right Hon. the Lord High Chancellor.*
* *His Grace the Duke of* Chaulnes, *President of the Royal Academy of Sciences in* Paris.
* *Right Hon. the Earl of* Chesterfield.
* *Right Hon. the Lord* Castlemaine.
* *Right Rev. the Lord Bishop of* Carlisle:
* *Sir* William Calvert, *Knt. Alderman.*
* *Hon.* John Campbell, *Esq;*
Walter Caddle, *Esq;*
Thomas Cadogan, *Esq;*
Mr. Joseph Calendar.

Mr.

Mr. Richard Calendar.
Archibald Campbell, *Esq;*
John Campbell, *Esq; of Red-Lion-street, Holborn.*
Archibald Carmichael, *Esq;*
Mr. John Carmady, *jun.*
Hon. James Carter, *Esq;*
Mr. Foster Carter.
Rev Mr. John Carter, *A. M. two Books.*
Mr. William Carter.
Mr. Chilcot, *Organist at* Bath.
Codrington Carington, *Esq;*
Mr. Paul Carington.
John Casson.
William Cattel, *Surgeon, two Books.*
* *Rev. Mr.* Cauthorne, *A. M.*
Rev. Mr. Chamberlaine.
* *Sir* Charles Bagot Chester, *Bart.*
Mr. Charles Child.
Robert Cholmondeley, *Esq;*
Mr. John Clark, *Bookseller.*
Dr. John Clark.
Mr. Samuel Clark.
Thomas Clark, *Esq;*
* *Monsieur* Clement.
Benjamin Clinket, *Esq;*
Mr. Benjamin Cleeve, *Merchant.*
Abraham Chovet, *Surgeon.*
James Cockburne, *Esq;*
John Cobham, *Esq;*
Mr. Thomas Cobham.
Capt. William Cogan.
Dr. Clow.
* *Mr.* Laurence Cole.
* *Hon.* John Colleton, *Esq;*
* James Edward Colleton, *Esq;*
* *Rev. Mr.* Collins.
* *Rev. Mr.* Giles Collins.
* *Rev. Mr.* William Collins.
* *Mr.* Peter Collinson.
Mr. Samuel Cooper.
* *Rev. Dr.* Conner, *Fellow of St. John's College,* Oxford.
Mr. Samuel Cooper, *of* Norwich.
* *Rev. Dr.* Conybeare, *Dean of Christ-Church,* Oxford.
* John Conyers, *Esq; of Corpus-Christi College,* Oxon.
* *Mr.* Mendez de Costa.
Richard Cowlam.
Rev. Mr. James Cox, *Rector of* Suneam *in* Kent.
Dr. William Cox.
Mr. William Coxal.
Mr. William Croft.
Mr. John Crew, *of* Crew-Hall, *Esq;*
Mr. Craten.
* David Crow, *Esq;*
Mr. Thomas Cruise.

D.

* *His Grace the Duke of* Dorset.
* *His Grace the Duke of* Devonshire, *two Books.*
* *Right Rev. the Lord Bishop of St. Davids.*
* *Right Rev. the Lord Bishop of* Derry.
* James Dawkins, *Esq; three Books.*
Mr. Michael Dale.
Mr. John Dapwell.
* *Sir* James Dashwood, *Bart.*
* Henry Dashwood, *Esq;*
* *Rev. Dr.* Derham, *President of St. John's College,* Oxford, *four Books.*
* *Rev. Dr.* Delany.
Theophilus Desbrisay, *Esq;*
Mr. Andrew Derry.
Isaac Del-Peza.
Daniel Devine.
Richard Deveney.
Richard Dilling.
Thomas Dipper.
Hon. John Dottin, *Esq; two Books.*
Dr. John Douglas.
* *Rev. Mr.* Dowding, *B. D.*
William Downes, *Esq;*
Richard Downes, *Esq;*
Mr. Purser Downes.
* William Drake, *Esq;*
Mrs. Elizabeth Drake.
* Thomas Drake, *Esq;*
Mr. Andrew Derry.
Henry Drax, *Esq;*
Mr. Samuel Drayton.
John Drayton,
Francis Drayton.
William Duffy, *Esq;*
Thomas Duke, *Esq;*
Mr. William Duke, *two Books.*
John Dunbar.
Michael Duncan.
Capt. Dunn.
Rev. Robert Dymond.
Mr. Theophilus Dymond.
Rev. Mr. Dyer.
Robert Dynwiddie, *Esq;*

E.

* *Right Rev. the Lord Bishop of* Ely.
* *Right Rev. the Lord Bishop of* Exeter.
Mr. Richard Eaton.

Mr.

A LIST of SUBSCRIBERS.

Mr. Easton, *Bookseller* in Salisbury.
* Matthew Eber, *Esq*;
Mr. Samuel Eburne.
Mr. Simon Edwards.
* Richard Edwards, *Esq*; *Master in Chancery.*
Philip Edwards, *Esq*;
Mr. Edwards *of the College of Physicians, two Books.*
* *Rev. Mr.* Edward Edwards, *A. M.*
Philip Edwards, *Esq*;
* *Mr.* Richard Edwards.
Mr. Simon Edwards.
Rev. Mr. Robert Edwards, *A. B.*
Mr. Robert Edwards.
* *Mr.* G. D. Ehret.
Anthony Eiver, *Esq*;
Richard Estwick, *Esq*;
Mr. Richard Rouse Estwick.
* *Rev. Mr.* Evans, *A. M.*
Rev. Mr. Ewyn.

F.

* *Rev. Dr.* Fanshaw, *Regius Professor of Divinity at* Oxford.
Mrs. Anne Farrell,
Mr. John Farr.
* *Hon. Governor* Fleming.
* Martin Folks, *Esq*; *P. R. S.*
Hon. John Fairchild, *Esq*;
Mr. Joseph Farley, *two Books.*
* Robert Firchurson, *Esq*; *two Books.*
Francis Find, *jun. Esq*;
Thomas Finlay, *Esq*;
Mr. Fitzherbert.
Hon. Richard Fitzpatrick, *Esq*;
Mr. Fletcher.
William Fletcher.
Thomas Forbes.
Rev. Mr. Foote, *L. L. B.*
Francis Ford, *jun. Esq*;
Capt. William Foster, *two Books.*
Rev. Mr. Reynold Foster, *A. M. two Books.*
Mr. Ingham Foster.
Mrs. Hannah Foster.
* *Rev. Mr.* Fothergil, *B. D.*
* —— Fothergil, *M. D.*
* *Rev. Mr.* Fothergil, *A. M.*
* Christopher Freise, *Esq*;
John French, Esq;
* Tobias Frere, *Esq*;
Thomas Fox, *Esb*;
Jonathan Franklin, *Esq*;
Mr. Michael Fortune.
William Fuller.

G.

* *His Grace the Duke of* Grafton.
* *Right Hon. the Lord* Gowran.
* *Right Hon. the Lady* Betty Germaine.
Right Rev. the Lord Bishop of Gloucester.
His Excellency Henry Greenville, *Governor of* Barbados, &c. *six Books.*
Hon. John Gallop, *Esq*;
Mr. Gammon.
Christopher Gall, *Esq*;
Estick Gall, *Esq*;
Mr. William Gall.
Samuel Game.
Dr. Joseph Gamble.
George Gascoigne, *Esq*;
* *Mr.* John Gascoigne.
* *Hon.* John Gibbons, *Esq*; *two Books.*
Hon. William Gibbons, *Esq*; *Speaker of the Assembly at* Barbados, *two Books.*
Hon. Philip Gibbs, *Esq*;
Philip Gibbs, *jun. Esq*;
William Gibbs, *Esq*;
Dr. John Gibbs.
Mr. Joseph Gibbs,
Rowland Gibson.
Rev. Charles Gifford, *A. M.*
Mr. Thomas Gilks.
* Edward Gore, *Esq*;
* *Mr.* Henry Gorsely.
Richard Gosling, *Esq*;
Dr. Charles Goudy, *Esq*;
Mr. William Gough.
Samuel Gould, *Bookseller.*
Capt. Samuel Goulding.
Mr. John Grant.
* Charles Gray, *Esq*;
Gray's *Inn Library.*
George Greame, *Esq*;
* *Rev. Dr.* Greene, *Fellow of* St. John's *College*, Oxford.
Mr. William Green.
* John Grey, *Esq*;
* *Mrs.* Elizabeth Grey.
Mr. Edmund Grey.
Mr. Edward Grey.
Rev. Mr. Walter Griffith, A. M. *four Books.*
Mr. Samuel Griffin.
Robert Grisley, *M. D.*
Edward Grove, *Esq*; *two Books.*
William Gulston, *Esq*;

H.

* *Right Hon. the Earl of* Halifax.
* *Right Hon. the Earl of* Harcourt.

* *Right*

A LIST of SUBSCRIBERS.

* *Right Hon. the Countess of* Huntington.
* *Right Hon. the Lady* Harcourt.
* *Right Hon. the Lady* Howe.
Right Hon. Lord Effingham Howard.
* *Right Hon. and Rev. the Lord Bishop of* Hereford.
* *Hon. Sir* Edward Hawke, *K. B.*
* *Rev. Dr.* Hales,
* *Dr.* Halliar, *Professor of Botany at* Gottingen.
—— Halsey, *Esq*;
* *Hon.* Charles Hamilton, *Esq*;
Mr. Richard Hamilton, *Surgeon.*
* William Hamilton, *Esq*;
* *Capt.* Patrick Hamilton, *of* Jamaica.
* George Hannay, *Esq*;
John Hannington, *Esq*;
* *Mrs.* Hannington.
Mr. William Hall.
Richard Hall, *Esq*;
Mr. Hugh Hall.
Mr. Barnard Hall.
Rev. Mr. Hall, *L. L. B.*
Mr. Richard Halsewood.
Henry Hassel, *Esq*;
Rev. Mr. William Haggat.
Mrs. Susannah Haggat.
Mr. Samuel Hallam.
John Harrison, *Esq*; *two Books.*
Cheney Hart, *M. D.*
Mr. Edward Harrison.
John Hartford
Ward Harris, *Esq*;
Mr. Thomas Harris.
Mrs. Jane Harper.
Mr. John Hartford.
George Hartle.
Mr. Henry Hassel.
John Hothersal, *Esq*;
* John Hay, *L. L. D. Fellow of St.* John's *College*, Oxford.
Capt. Henry.
Mr. William Henry.
Dr. John Herne.
* *Sir* William Heathcote.
* Benjamin Heath, *Esq*;
* James Herbert, *Esq*;
* Thomas Hesketh, *Esq*;
John Hill, *Esq*; *Secretary to the Board of Trade.*
* Thomas Hill, *Esq*;
Mr. Benjamin Hinds.
* *Rev. Dr.* Holden, *President of the* English *Seminary* at Paris.
William Holder, *Esq*;
Mr. Henry Evan Holder.
John Allen Holder.
* Robert Holdford, *Esq*;
Capt. Thomas Holland, *two Books.*
* *Mr.* Hollioake, *Surgeon.*
* Thomas Horne, *Esq*;
Robert Hooper, *Esq*;
Mr. Thomas Hope.
* *Rev. Dr.* Horseman.
John Hothersal, *Esq*;
Rev. Mr. Richard Hotchkiss, *A. M. four Books.*
* Charles Grave Hudson, *Esq*; *Gentleman Commoner of* Edmund-*Hall*, Oxon.
* *Rev.* Obadiah Hughes, *D. D. two Books.*
Rev. Mr. Lewis Hughs, *A. M.*
Mr. Richard Hughes, *Purser of His Majesty's Ship the* Assurance, *three Books.*
Mrs. Susannah Hughes.
Mr. John Hughs, *Printer.*
* —— Hugo *Physician in Ordinary to his Majesty at* Hanover.
* —— Hugo, *Physician to the Houshold.*
* *Mrs.* Hulston.
* *Rev. Mr.* Humphry Humphreys, *A. B.*
* *Rev. Dr.* Hunt, *Professor of* Hebrew *at* Oxford.
Richard Hunt, *Esq*;
* *Rev. Mr.* Hunter, *A. M.*
Hon. Samuel Husbands, *Esq*;
Richard Husbands, *Esq*;
Mr. John Hussey.
Robert Hutchinson.

I.

* *St.* John's *College Library*, Oxford.
St. John's *College Library*, Cambridge.
Mr. Richard Jackman.
William James.
Miles James, *Esq*;
* Robert Janssen, *Esq*;
Rev. Mr. Ibbetson.
Mr. Jeffries *of* Kingston.
* Jeffries, *Engraver to his Royal Highness the Prince of* Wales.
Mr. George Jemmet.
Hon. Edmund Jenkins, *Esq*;
William Jenkins, *Esq*;
Edward Ince, *Esq*;
Mr. Thomas Ince.
Dale Ingram, *Surgeon.*
William Innys, *Merchant.*
* *Dr.* Johnson, *Residentiary of St.* Paul's, London.
* *Rev. Mr.* Rowland Johnson *of* Hempstead, *A. M. two Books.*
* *Mr.* Richard Johnson.
* William Johnson.
Col. John Jones, *Esq*;
Benjamin Jones, *Esq*;

A LIST of SUBSCRIBERS.

* *Rev. Mr.* Lewis Jones, *A. M. Rector of* Mowddwy.
* *Rev. Mr.* Jones.
* *Rev. Mr.* Jones, *A. M. of* Elstree.
* *Rev. Mr.* Jones *of* Jesus *Col.* Oxford. *A. M.*
* *Mr.* Jones *of* Hempstead.
Joseph Jordan, *sen. Esq;*
Mr. Joseph Jordan. *jun.*
Hurdess Jordan, *Esq;*
* Sir William Irby, *Bart.*
Joshua Iremonger, *Esq;*
Edward Ironside, *Esq;*

K.

* *Hon. Admiral* Knowles.
* *Rev. Dr.* Kendrick.
* *Mr.* John Kendrick.
James Keith.
* *Mrs.* Kennons, *two Books.*
Henry Peter King, *Esq;*
* *Rev. Mr.* Arnold King, *A. M.*
Rev. Mr. Richard King, *A. M.*
* Jac. Theod. Klein, *Reip Ged. a secretis Regiæ Societatis Londin. & Academiæ Scientiarum Sodalis.*
Mr. Thomas Knights.
Samuel Knights.
* Francis Knowles, *Esq;*

L.

* *His Grace the Duke of* Leeds.
* *Right Hon. Earl of* Leicester.
* *Right Hon. Lady* Luxborough.
* *Right Rev. Lord Bishop of* London.
* *Right Rev. Lord Bishop of* Lincoln.
* *Right Rev. the late Bishop of* Litchfield.
* *Right Rev. Lord Bishop of* Landaff.
Mr. Lafartie.
* *Sir* Bibye Lake, *Bart.*
* Thomas Lake, *Esq;*
Mr. Samuel Lake.
* *Hon. Lady* Lambert.
Mr. Alexander Lampley.
John Lane, *Esq;*
Rev. Mr. Langford, *A. M.*
Rev. Mr. Larkham, *Fellow of* Corpus Christi *College*, Oxon.
Mr. John Lawson.
Joseph Laycock.
Samuel Leach.
Leach Leach.
John Leary.
Capt. John Leatherhead.
Mr. Matthew Lee.
James Lee.
Henry Leslie, *Esq;*
Mr. Bowman Leslie.
. Thomas Leslie, *Esq;*
George Levine. *Esq;*
Mr. William Levine.
Isaac Levine, *Esq;*
Rev. Mr. James Lewis.
John Lewis, *Esq; Agent Victualler at* Gibraltar.
Mr. Samuel Lewis.
Temple Lewis, *Esq;*
John Liyte, *Esq;*
Mr. Thomas Light.
Anthony Linch.
* David Lisle, *Esq;*
* Charles Long, *Esq;*
Mr. Charles Long.
Philip Lovewell.
Herman Louis.
Charles Lowth.
Samuel Lloyd.
Dr. Nathaniel Lloyd.
* *Rev. Mr.* Lloyd, *B. D.*
Dr. Thomas Lycott.
* *Rev. Dr.* Charles Lyttleton, *Dean of* Exeter.
* *Right Hon.* George Lyttleton, *Esq;*

M.

* *Right Hon. Earl of* Macclesfield.
* *Right Hon. Earl of* Moreton.
* *Right Hon. Lord* Maynard, *two Books.*
Right Hon. Lord Montfort.
* *Right Hon. Lord* James Manners.
* *Right Rev. Lord Bishop of* Meath.
* *His Excellency* ——— Mathews, *Esq; Governor of* Antigua.
Afflick Mackleur, *Esq; two Books.*
——— Mancumara, *M. D. at* Barbados.
Magdalene *College Library.*
* Francis M'Mahon, *Esq;*
Samuel Mapp, *Esq;*
Thomas Mapp, *Esq;*
* *Dr.* Martin, *Dean of* Worcester.
Mr. James Mascol.
Benjamin Massie.
* *Mrs.* Massie.
Mr. John Mashal.
William Massline.
M. Matty, *M. D.*
Mr. John Mayhew.
Hon. William Maynard, *Esq; two Books.*
John Maynard, *Esq;*
Jonas Maynard, *Esq;*
Mr. Mayntown, *of* Gray's *Inn.*
* Richard Mead, *M. D.*
Benjamin Mellowes, *Esq; two Books.*
* Moses Mendez, *Esq;*
Mr. William Meyer, *Bookseller.*
Richard Meyrick, *M. D.*
* Owen Meyrick, *Esq;*

* Hum-

A LIST of SUBSCRIBERS.

* Humphrey Mildmay, *Esq;*
* John Millington, *Esq;*
* *Rev. Dr.* Mills.
Mr. John Middleton.
Francis Miller, *Esq;*
Mr. Joseph Mills,
Timothy Miller.
Elias Mivielle, *Esq;*
* Robert Mitchell, *Esq;*
John Molesworth, *Esq; Gentleman-Commoner of* Baliol *College,* Oxon.
Rev. Mr. Moore, *Curate of St.* Sepulchre's.
Daniel Moor, *Esq;*
Humphry Mostyn, *Esq;*
William Moll, *Esq;*
Mr. Joseph Moll.
John Monroe, *M. D.*
Mr. John Mottley.
Rev. Mr. Edward Morgan, *A. M.*
* *Rev. Mr.* Thomas Morgan, *A. M. of* Cheltenham, *two Books.*
* Samuel Morgan, *Esq;*
Mr. Lewis Morris.
Jasper Morris, *Esq;*
Rev. Mr. Kenneth Morrison.
Mr. Mountford, *Bookseller* at Worcester.
James Mount.

N.

* *His Grace the Duke of* Norfolk.
* *His Grace the Duke of* Newcastle.
* *Right Hon. the Earl of* Northampton.
* *Right Hon. the Earl of* Northumberland.
* *Right Rev Lord Bishop of* Norwich.
* James Nash, *Esq;*
* *Hon.* John Needham, *Esq;*
* *Mr.* Turbeville Needham.
* Samuel Newman.
New *College Library.*
* *Rev. Dr.* Newton.
John Newton, *Esq;*
Mr. Nathaniel Newton.
Mrs. Newton, *Bookseller, at* Manchester.
Mr. William Nibs, *jun.*
Benjamin Nicholls.
Rev. Dr. Nicholls, *Chaplain to the Right Rev. Lord Bishop of* London.
Mr. William Nicholls.
Mordecai Nunes.
William Nut.
Samuel Nurse

O.

* *Right Hon. Earl of* Orrery.
* *Right Rev. Lord Bishop of* Oxford.
* *Hon. Lieutenant-General* James Oglethorpe, *Esq;*
* *Sir* Chaloner Ogle, *Knt.*
* Rowland Oliver, *Esq;*
Oriel *College Library.*
* Robert Osborne, *Esq;*
Samuel Osborne, *Esq;*
Mr. James Osborne, *of* Alban-Hall.
James Osborne, *Esq;*
Mr. Thomas Osborne, *of* Gray's-Inn, *Bookseller, six Books.*
John Ostraham.
* Corbet Owen, *Esq;*
* Hugh Owen, *M. D.*
* Richard Owen, *Esq;*
* Humphrey Owen, *B. D. of* Jesus *College,* Oxon.
* *Rev. Mr.* Owen Owen, *A. M.*
Mr. John Oxley.
Edward Oxnard.

P.

* *His Grace the Duke of* Portland.
* *Her Grace the Dutchess of* Portland.
* *Right Hon. Countess of* Pomfret.
* *Right Rev. Lord Bishop of* Peterborough.
* *Right Hon.* Henry Pelham, *Esq;*
* Francis Paddie, *Esq;*
Mr. Samuel Palmer.
* *Rev. Dr.* Pardo, *Principal of* Jesus *College,* Oxon.
* James Parsons, *M D.*
* Richard Parrot, *Esq;*
Mr. Edward Clark Parish, *Merchant.*
* ——— Parrot, *Surgeon.*
* Samuel Parry, *Attorney at Law.*
* William Paterson, *Esq;*
* Thomas Patridge, *Esq;*
Mr. Eyre Payne.
John Payne, *two Books.*
Thomas Abel Payne.
Thomas Payne.
* *Mrs.* Pearson.
* John Pennant, *Esq;*
* Henry Pennant, *Esq;*
Dr. John Penny.
Rev Mr. Jonathan Perkins.
Rev. Mr. Richard Perry, *A. M. Student of* Christ *Church,* Oxon.
Mr. Samuel Perry.
William Perry, *Esq;*
William Peterson, *Esq;*
Mr. William Peters.

* *Rev.*

A LIST of SUBSCRIBERS.

* *Rev. Dr.* Pickering, *Vicar of St.* Sepulchre's.
John Pickering, *Esq;*
Joseph Pickering, *Esq;*
* *Rev. Mr.* Pearce, *A. M. Vicar of* Tillingham *in* Essex.
Mr. Thomas Pierce, *jun.*
William Pierce.
Conrade Pile, *Esq;*
Mr. Daniel Pindar.
* Philadelphia *Library.*
Mr. George Philips.
John Philips.
* William Phelps, *Esq;*
Mr. Thomas Plumstead.
Mrs. Mary-Anne Popple.
Mr. Powel, *Apothecary.*
Mrs. Anne Poyer.
John Poyer, *Esq;*
* Henry Pratten, *Esq;*
Capt. John Prentice.
Mr. William Prescod.
James Prideaux, *Esq;*
Mr. John Pridee.
Thomas Pridee.
Francis Proines.
Samuel Pyet.

Q.

Queen's *College Library,* Oxon.

R.

* *His Grace the Duke of* Rutland.
* *Right Hon. Earl of* Radnor.
* *Right Hon. Lord* Romney.
* *Right Reverend Lord Bishop of* Rochester.
Dr. John Raven.
Thomas Revell, *Esq;*
Capt. James Rawlins.
Thomas Reeve, *Esq;*
Mr. John Reeves.
Rev. Mr. John Reynolds, *A. M.*
Nicholas Rice, *Esq;*
Mr. Joseph Rider.
George Rider.
Messrs. John *and* James Rivington.
* *Sir* Thomas Robinson, *Knt.*
James Rogers, *M. D.*
Joseph Rolestone, *Esq;*
* *Rev. Mr.* Romaine, *A M. Lecturer of St.* Dunstan.
* *Rev. Mr.* Roman, *B. D.*
Rev. Mr. Patrick Rose, *A. M.*
Mr. William Ross.
Rev. Mr. Thomas Rotherham, *A. M. two Books.*
Rev. Mr. John Rotherham, *A. B.*
Hon. Samuel Rouse, *Esq; four Books.*
Thomas Rouse, *Esq;*
Hillary Rowe, *Esq; two Books.*
Mr. James Roystone, *Merchant.*
Philip Rudder.
David Rudder, *Esq;*

S.

* *Right Hon. Earl of* Sandwich.
* *Right Hon. Countess of* Sandwich.
* *Right Hon. Countess of* Sunderland.
* *Right Hon. Earl of* Shaftsbury.
Right Rev. Lord Bishop of Salisbury.
Hon. Andrew Stone, *Esq; Commissioner of Trade and Plantations.*
Mr. John Salmon.
William Salmon.
* *Hon.* Richard Salter, *Esq;*
Hon. William Sandiford, *Esq; two Books.*
Mr. John Sandiford.
William Sealy, *Esq;*
Mr. James Scott.
Dr. James Sedgwick.
Rev. Mr. Seers, *A. M.*
John Sericold, *Esq;*
John Sharp, *Esq;*
* *Rev. Dr.* Shaw, *Principal of* Edmund-hall Oxon.
James Sheppard, *Esq;*
* *Rev. Mr.* Shipley, *A. M.*
* *Rev. Mr.* Shire, *A. M.*
Mr. Philip Simons.
William Shurland.
Charles Shurland.
Reynold Skeet, *Esq;*
* *Rev. Dr.* Skerret.
* *Sir* Hans Sloane, *Bart.*
Mr. Slaughter, *of Newgate-street.*
* *Rev.* Thomas Smith, *D. D.*
Tennison Smith, *Esq;*
Richard Smith, *Merchant, Esq; two Books.*
Edward Smyth, *M. D.*
Edward Smith, *Esq;*
Mr. Samuel Smith.
Henry Smith.
Henry Snock, *Esq;*
Mr. John Southward.
Mr. Thomas Speck.
Joseph Sponshall.
* Thomas Stack, *M. D.*
Thomas Stevenson, *Esq;*
Mr. Roger Stevens.
* *Rev. Mr.* Stennet.
Mr. James Streaker.
George Stretch.
George Streate.
Demmock Streate.
Benjamin Stoute.
William Sturge, *Esq;*

Rev.

A LIST of SUBSCRIBERS.

Rev. Dr. Stukeley.
John Sutton, *Esq*; of Jamaica.
Samuel Francis Swinden, *Esq*;
* *Mr.* Symons.

T.

* *Right Hon. Earl of* Tilney.
* *His Excellency* Edward Trelawney, *Governor of* Jamaica.
Mr. Fimington Tatem.
James Taylor.
Joseph Taylor.
Joseph Terril, *Esq*; *two Books.*
Mr. William Terril.
Douding Thornhill.
William Thorne.
Thomas Thorneton, *Esq*;
Mr. Samuel Tavers.
* *Dr.* Christophorus Jacobus Trew, *at* Nurenberg.
Mr. John Tucker.
* *Rev. Mr.* Josiah Tucker, *A. M.*
Mr. John Tull.
Thomas Tunckes, *Esq*; *two Books.*
Mr. Nicholas Turner.
Cholmondley Turner, *Esq* ;
* *The Library of St.* Thomas's *Hospital,*
Mr. George Thomas.
Mr. Thomas Thomson.
* Touchitt, *Merchant.*
* Edward Tymewell, *Esq*;

V.

* *Right Hon. the Countess of* Uxbridge.
* William Vaughan, *Esq*;
Thomas Vaughan, *Esq*;
Mr. Stumford Vanhulst.
* Edward Vernon, *Esq*;
Mr. James Virtue.

W.

* *Right Hon. Lord Viscount* Windsor.
Right Rev. Lord Bishop of Winchester.
* *Right Rev. Lord Bishop of* Worcester.
* *Right Hon. Lord Chief Justice* Willes.
* Walter Wade *of* Lisbon, *M. D.*
Robert Wadeson, *Esq*; *two Books.*
John Waite, *Esq*;
Eyre Walcott, *Esq*;
Thomas Walcott, *Esq*;
Mr. Samuel Walcott.
Mr. Charles Waller.
Antony Waldron.
Abel Walker, *Esq*; *two Books.*
Hon. George Walker, *Esq*;
Mr. Joseph Walker.
William Walker, *Esq*;
Mr. Peter Walker.
* *Rev. Mr.* William Warburton, *Chaplain to His Royal Highness the Prince of* Wales.
Aston Warner, *Esq*;
Mr. Richard Warner, *of* Lincoln's *Inn.*
Thomas Warner, *Esq*; *of* Antigua.
William Warren, *Esq*;
Robert Warren, *Esq*;
Thomas Waterman, *Esq*;
Mr. William Wats.
Watson.
William Wayles.
Thomas Weales.
Henry Webb, *Esq*;
Mr. Joshua Webster, *Surgeon.*
Hon. Ralph Weeks, *Esq*;
Roger Weeks, *Esq*;
Mr. Nathaniel Weeks.
Peter Western.
Robert Weston, *Esq*;
Rev. Mr. Joseph Wharton, *A. M. two Books.*
Mr. Dennis Wharton.
John White, *Esq*;
Mr. Samuel Whiteman.
* William Whitaker, *Merchant, in* Dolphin Court, Tower Hill.
Thomas Whitaker, *Esq*;
Mr. William Whitehead.
Wheeler.
John Wheeler.
Thomas Wheeler.
Rev. Mr. Whitford.
Mr. John Wickham.
Nicolas Wilcox, *sen. Esq*; *two Books.*
* *Mr.* Thomas Willbraham.
Rev. Mr. James Williams, *A. M. of* Jesus *College*, Oxon.
* *Rev. Mr.* John Williams, *M. A. of* Twickenham.
Mr. Richard Williams.
Alexander Williams.
David Williams.
Hugh Williams, *Esq*;
Mr. Daniel Williams.
Hugh Williams.
* William Windham, *Esq*;
* *Rev.* Peter Wingfield, *M. A.*
Edward Winslow, *Esq*; *two Books.*
Thomas Withers, *Esq*;
Robert Wrightson, *Esq*; *of* Cuswork *in* Yorkshire.
Rev. Mr. Nathaniel Wright, *A. M.*
William Wollaston, *Esq*;
Rev. Mr. Dudley Woodidge, *two Books.*
Mrs. Ruth Woodhidge.
Mr. Thomas Woolford.

e *Mr.*

A LIST of SUBSCRIBERS.

Mr. John Worrel.
William Worrel.
* Francis Woodward, *M. D.*
* Richard Woodward, *L. B.*
* *Rev.* Hugh Wyne, *D. D. Chaplain in Ordinary to his Majesty.*
* *Sir* John Wynne, *Bart.*
Rev. Mr. William Wynne, *A. M.*

Y.

* *His Grace the Lord Archbishop of* York.
* *Hon.* Philip York, *Esq;*
* *Hon. Col.* York.
* *Hon.* Charles York.
* ——— Young, *Esq; from* Antigua.
* *Rev. Mr.* Richard Younger, *A. M.*
Philip Yeoman, *Esq;*
Mr. Thomas Yard.
* M. Yonstall, *Esq; of* Wycliffe.

SUBSCRIBERS *from VIRGINIA.*

Right *Hon.* Thomas, *Lord* Fairfax.
Hon. William Fairfax.
Hon. John Robinson, *Esq; President of the Council.*
Hon. William Newton.
Hon. and Rev. William Dawson, *D. D one of his Majesty's most Hon. Council, and President of the College of* William *and* Mary.
Col. Charles Carter.
Col. Carter Burwell.
Col. Preffley Thornton.
Col. William Fitz-Hugh.
Col. Joshua Fry.
Col. William Beverley.
Col. Lunden Carter.
Col. Richard Blund.
Col. George Braxton.
Col. Lewis Burwell.
Col. George Lee.
Maj. William Dangerfield.
Maj. Andrew Monroe.
Maj. Peter Wayener.
Thomas Nelson, *Esq;*
James Reid, *Esq;*
Beverley Whiting, *Esq;*
Mr. John Busheron.
George Webb.
Stephen Dewey.
Edward Pendleton.
James Pomer.
Lumford Lomax.
* Techarner Degraffenreid.
John Lee.
——— Curtis, *jun.*
William Kennar.
John Mercer.

E R R A T A.

PAge 1. *l.* 12. *for* greatest Breadth 13. 3-qrs *read* 14 and 5Furlongs. *Page* 52. *for* Swallows *read* Sallows. *Ibid. line* 7. *for* Yeagh *read* Yew. *Page* 136. *dele the Description of* Turk's Head. *Page* 139. *line* 18. *for* expands itself horiontally *read* expanding itself horizontally. *Page* 146. *for* Bee-wood Tree *read* Beef-wood Tree. *Ibid. dele* Fustic Tree. *Page* 164. *for* Buombo *read* Bumbo. *Page* 170. *line* 21. *for* reddish, thick and succulent. *Page* 180. *line* 12. *dele* red *before* Veins. *Page* 182. *line* 22. *for* sort *read* species. *Page* 186. *dele* Cochineal. *Page* 208, *line* 20. *for* brownish *read* brown. *Page* 209. *line* 11. *for* brownish *read* brown. 210 *line* 8. *for* high *read* long. *Ibid l.* 33. *for* high *read* in height. *Ibid. line* 36, *for* brownish *read* brown. *Page* 216, *line* 9. *for* roundish Pods *read* round Pods. *Ibid. line* 10. *for* thoug hsomewhat *read* though somewhat. *Ibid. line* 18. after that Time *add* from their being planted. *Page* 220. *for* Antegoa *read* Antegua. *Page* 221. *dele* Wild Penny-royal. *Ibid. line* 15. *dele* the *after* by. *Ibid. line* 20. *dele* and roundish. *Page* 222. *line* 11. *for* are longish *read* are about five Inches long, narrow and sharp pointed. *Ibid. line* 23. *after* Colt's Foot *add* and the Colour. *Page* 223. *line* 23. *after* leaves *add* somewhat similar. *Idid. lines* 33 *and* 34. *for* three Inches thick *read* two Inches thick near the Earth. *Page* 224. *line* 8. *for* Abundance *read* a great Number. *Page* 226. *line* 16. *for* many leaved Flowers *read* rosaceous Flowers. *Page* 236. *line* 33. *for* Stalk *read* main Stalk. *Page*. 237. *line* 8. for Skill *read* Knowledge. *Page* 253. *line* 15. *for* coral Sea-weed *read the* stiff Sea-weed. *Ibid. line* 20 *dele* Head *after* Brown. *Ibid. instead of* rather Seed, *read* what are here called Seed. *Page* 255, *for* purplish Sea Moss, *read* purple Sea Moss. *Ibid. line* 30. *for* purplish Colour *read* purple. *Page* 258. *line* 3. *after* Resemblance to it *add* in Shape. *Page* 261. *line* 15. *after* reality *add* of. *Page* 262. *line* 3. *for* it often *read* it some Times. *Ibid. line* 14. *for* about the Roots of the great Feelers *read* where the Feelers take their Rise. *Ib. line* 3. *after* Sea *add* this. *Ibid. after* Shoals *add* or such shallow Places. *Ibid. line* 28. *for* which they cast, *read* which theydo. *Page* 266. *line* 8. *for* but oftener *read* and often Graneries near the Sea-side. *Ibid line* 4. *for* twenty Pounds *read* ten Pounds. *Page* 268. *line* 19. for Fancifulness *read* Fancy. *Page* 273, *for* Vergil, *read* Virgil; *for* creditable, *read* credible. *Page* 227. *line* 22. *for* brownish *read* brown. *Page* 278, *line* 5. *for* blackish *read* black Spots. *Ibid. line* 11. *for* Concha Veneris Alba *read the* Rice Shell. *Page* 280, *line* 32. *dele* by which I suppose Nourishment is convey'd to the Fish. *Page* 297. *line* 39. *for* langid *read* languid. *Page* 298 *Plate* 24 *should be marked Plate* 26. Negroes, their Numbers, *Index for Page* 2. *read Page* 14. *For Plate* 11. 352. *read Plate* 11. 152.

ADDENDA. *The Wild* CANE.

THIS grows to about fifteen Feet high. Its Flags resemble to those of the Reeds in *England*, and the Body is generally thicker than a Walking-cane, and strong; the young Canes at about two or three Inches growth are made into Pickles. *Plate* 10 *. *being the Description of the Bay Berry Tree, is to be placed opposite to Page* 145.

ROSE-SHELLS.

THESE are of two Sorts, the white, and those tinged with a Maiden's-blush; the latter are very rare to be met with, but the former are in great Plenty on most of our sandy Bays.

These, with a Species of the small black Buccinum Kind, and another Species of very small Shells of a pale green, and often called Pea-shells, are with great Assiduity, gathered and sent to *England*: With the first Kind the Ladies form a lively Representation of Roses; the second serve to imitate the Seed and Seed-vessels in supposed full-blown Sunflowers, and the third are of no less Service to represent those small Granulations common to the thrummy Discus of some, as well as the Summits of other apetalous Flowers.

THE CONTENTS.

AUTHORS

AUTHORS *Quoted*.

TEXTS *of* SCRIPTURE *Cited or Illustrated*.

THE

PREFACE.

NATURAL Philosophy flourished first in the East. It was in great Perfection among the *Assyrians*, *Chaldæans*, and *Egyptians*; and, if their Knowlege of it had been faithfully conveyed to Posterity, we might have expected, that the next Age of Learning in *Greece* would have been able to have made greater Progress in that noble Science.

* " But, alas! here Philosophy was forced to put " on a poetical Dress, adorned by the Poets, its " Patrons, with Fables, and enlivened with extravagant Fancies. Nor " was *Athens* itself proof against so prevailing a Corruption."

This was the State of every Branch of Philosophy in *Greece*; nor did this Study meet with much better Success in the first Part of the succeeding *Roman* Empire. " For while this was in its Infancy, during the Three " first Ages, which were spent in conquering *Italy*; Strictness of Man- " ners, and the Art of War, were their chief Study, and continued to be " so, till they were Masters of the East." Then Oratory became their Darling. Upon the Declension of the *Roman* Empire, the Darkness of Ignorance, that ensued thro' many successive Ages, spread like *Elijah*'s Cloud, and, in a Darkness, like that of *Egypt*, enveloped the small Remains of Learning in almost a total Eclipse.

Nor did the Philosophical Part of it revive, or shew the least Spark of its native Brightness, till *Galileo* in *Italy*, and the great *Bacon* in *England*, became its Patrons.

A It

* *Sprat*'s History of the *Royal Society*.

It was about this Time that Natural Hiſtory, as well as every Branch of Experimental Philoſophy, became the Study of great Men in different Countries, in *England* eſpecially, and ſome time afterwards in *France*, under the Protection of that great Patron of Literature, *Lewis* the XIVth. whoſe Princely Favours ſought for, and encouraged, Men of Learning, not in *France* only, but in the moſt diſtant Countries.

" But of late Years its greateſt Promoters have been the Royal Society " in *England*, and the Academy of Sciences in *France :* By their means " chiefly, has the World received more uſeful Diſcoveries and Improve- " ments in one Century, than it had done for many Ages before."

" If they go on with the ſame unwearied Diligence and Succeſs for one " Century more, what Depths of Nature will not be explored ? What " Treaſures of Knowlege will not be diſplayed ?"

But tho' many Branches of this excellent Study have, of late Years, been ſo laudably cultivated, to the Glory of God, and the Good of Mankind ; yet we have Cauſe to lament, that our Purſuits of this Nature are ſtill deficient.

The ſeveral ingenious and learned Diſcoveries of *Malpighi*, *Grew*, *Ray*, *Derham*, and *Hales* (among many others that might be added), tho' excellent in their Kind, are yet but ſo many well-proportioned Limbs of an unfiniſhed Piece.

Other Sciences and Arts owe their Perfection, not only to the Capacity of the great Men who ſtudied them, but likewiſe to the confined Nature of the Subjects treated of.

But as to Natural Hiſtory, ſo many are the Species of Animals, Plants, and Foſſils, which are yearly diſcovered, that we may juſtly ſay with *Pliny*, *Multum adhuc reſtat operis, multumq; reſtabit* ; *nec ulli nato poſt mille ſæcula præcluditur occaſio aliquid adjiciendi.* And tho' the Study of ſo extenſive a Subject be attended with ſome Difficulty, yet will it prove no leſs pleaſing than uſeful.

In other Hiſtories we meet with, at leaſt, a great Mixture of Pain with our Pleaſure. If in Biography we are firſt charmed with the great Talents and amiable Character of *Cæſar*, whilſt in the true Intereſt of his Country ; ſhall we not be grieved to find, that his Ambition, at laſt, proved fatal to many Thouſands of his Fellow-Citizens and Countrymen ? If we are pleaſed with the Life and Actions of *Epaminondas* ; as the fatal Cataſtrophe of his laſt unhappy Period draws on, can any Degree of Chearfulneſs check the riſing Sigh, or ſtop the falling Tear ?

When from Biography we purſue Mankind through the general Hiſtory of Nations, we ſhall be ſhocked with monſtrous Examples of Wickedneſs, far overbalancing the few ſucceſsful Inſtances of diſintereſted Virtue.

In

In one Age we ſee *Epictetus* baniſhed, and the venerable *Seneca* doomed to Death, whilſt *Domitian* and *Nero* are covered with Purple. In another Period, injured Majeſty bows the Neck to relentleſs Tyranny.

If, from this unamiable Stream of hiſtoric Truth, we explore its branching Rivulets, and ſeek for Pleaſure in our Reſearches into the Antiquity of Nations; this Study, tho' always harmleſs, and ſometimes uſeful; yet, after an irkſome and tedious Purſuit, thro' Paths rendered obſcure and dark by Length of Time, or Ignorance, or made almoſt impervious by Superſtition, after a Life ſpent in ſuch Inquiries, perhaps the ultimate Reſult will be, how preciſely to determine the Day and Year, that memorable Æra, which gave Birth to a *Nimrod*, who conquered ſome ſmall Province, or built ſome little City, and reigned its Tyrant; or perhaps the Period of our Labours will cloſe with the diminutive Diſcovery how to fix the original Meaning of ſome obſcure Sentence, or even an obſolete Word.

Whereas, in purſuing the Study of Nature, and meditating upon the exact Harmony ſo viſible in the Works of the Creation, we are ſure to meet with untainted Pleaſures; not ſuch as proceed from the Tranſports of an heated Imagination, or a violent Paſſion, but Pleaſures, like that of Health, ſtill and ſerene.

The Accompliſhments we acquire by many other Studies, may, by ſoothing our Vanity, occaſionally miſlead us, and likewiſe byaſs human Nature with a ſtrong Propenſity towards ſome favourite Prepoſſeſſion of the Will.

Thus the Oratory of *Cicero*, indued with every Power to pleaſe ---- to raiſe or calm the various Paſſions in the human Breaſt; ---- (a noble Endowment, when employed in the Cauſe of Virtue!) ---- But (ſuch, alas! is the Inſtability and Imperfection of human Nature) this very Talent, which gave him Pre-eminence above other Men, became ſubſervient to indulge a Weakneſs, which we muſt at once condemn and pity. To living *Cæſar*, the Orator paid the pleaſing Tribute of Adulation and Praiſe: ---But, when dead, loaded him with Reproaches.

If we deſcend to many other Branches of Study, and polite Literature (eſpecially in the preſent Age), we ſhall find, that theſe Embelliſhments too often tend only to inſpire us with a fanſied Superiority over others, and ſerve, at beſt, but to ſet forth and enliven ſome particular Occaſion or Period of Life. Their Amuſements, to make uſe of a great Writer's Simile, "* are like a Fountain, which, on ſome gaudy Days, ſpouts forth a "frothy Stream, but remains dry all the reſt of the Year."

But ſuch are the Subjects of our Inquiries in Natural Philoſophy, that they are as large, and as laſting, as the Univerſe, full of inexhauſtible Variety,

* Lord Viſcount *Bolingbroke.*

Variety, worthy (next the sacred Oracles) of the Attention of him, whom God hath placed at the Head of this lower World.

By contemplating these Subjects, we are gradually led from Things visible, to the Knowlege of him who is invisible. Here we see innumerable Instances of Harmony, Beauty, and Order, not to be imitated by the most laborious Endeavours of any human Art or Contrivance.

Those delighted with the Vegetable Creation may, each returning Spring, not only be pleased with their Profusion of Sweets, and of such beautiful Colours, as excel even a *Solomon* in all his Glory; but likewise our Inquiries may now be as boundless as the Creation; no forbidden Tree checks our Curiosity; but, with *Solomon*, we may explore their Secrets, from the lofty Cedar to the humble Hyssop, springing from the Wall.

When by these, and such-like Inquiries, we find in every thing a wise, good, and useful Design, it will afford us Pleasures, purer and superior to those which the sanguine Glory of Arms, boundless Ambition, or satiated Avarice, can give. It will inspire us with Sentiments most pleasing, as most suitable, to that divine Image the Almighty has pleased to stamp of himself upon the human Mind.

"* What room can there be for low little Things in Minds so usefully "and so nobly employed? What dark and melancholy Passions can over-"shadow his Heart, whose Senses are constantly entertained with so "many various Productions?" The least good Effect attending these Pursuits will be an innocent Joy.

"What Anger, Envy, Hatred, or Revenge, can long torment his Breast, "whom not only the noblest Objects, but even every Insect, every Blade "of Grass, or Grain of Sand, can divert?" To whom the Return of every Season suggests a Circle of the most innocent Amusements.

From these Inquiries, he will learn the infinite Distance between his great Creator and himself. "This will teach him to worship that Wis-"dom by which all Things are so easily sustained, and will lead him to "direct his Praises aright."

There is not the smallest Part of this Globe left without evident Signatures of God's Goodness. If this little Spot we are treating of, produces not the Sapphire, the Carbuncle, Topaz, or Ruby, we are not wanting, however, in the more truly valuable Productions of Nature, which are chiefly subservient to the Necessities and Conveniencies of Life; and even our most barren Hills are not unserviceable: If their high Summits want their grateful Verdure; if out of their Sides we cannot dig Brass; yet it is to these, under Providence, we owe the former and the

* *Sprat's* History of the *Royal Society*.

the latter Rain, by intercepting ſeveral Clouds, and watry Vapours, that would otherwiſe fly over us.

It is from the Sides of theſe Hills and Mountains that the living Streams deſcend, ſo grateful and refreſhing to Men and Beaſts in hot Climates. It is likewiſe to their friendly Shades, that we owe the Growth of ſeveral valuable Plants, whoſe delicate Texture could not bear the long-continued piercing Rays of the Sun.

But, among other providential Bleſſings, the Conſtancy of the Trade-Winds is not the leaſt valuable: Without this, all other Conveniencies, ſuch as Pleaſantneſs of Situation, Richneſs of Soil, and ſeaſonable Weather, would be of no Service; ſince the Return of each Day, now, by the Breezes of this Wind, made cool and grateful, would, without this providential Allotment, be inſufferable through Heat: The tender Plant would loſe its grateful Verdure, fade, wither, and die.

It may perhaps be ſaid by ſome Readers, that theſe, and ſuch Reflections, are too often interſperſed through the following Work; and that it would be ſufficient barely to treat of the Nature and Qualities of the Subjects in a Philoſophical Light, without drawing any Concluſions from them, though they evidently point out the Wiſdom, Power, and Goodneſs of God, in the Make, Nature, and Uſe of each Individual.

The Anſwer to ſuch is eaſy: That the original and true Uſe of Philoſophy was to render Mankind good as well as learned; by raiſing their Thoughts from Things viſible, to the inviſible Author of all Perfection.

Shall *Galen*, a Heathen Philoſopher, break out into Raptures of Admiration, upon the Diſcovery of ſo many Marks of infinite Wiſdom in the Formation of the Eye? And, ſhall Chriſtians be blamed for taking notice of ſuch Inſtances of God's Wiſdom in other Parts of the Creation?

Philoſophy, when rightly applied, is of great Uſe, even in a religious View: It was this that taught *Pythagoras* Integrity of Manners, and Auſterity of Life.

This made *Empedocles* prefer a quiet private Life to the Splendor of a Crown, with all its lucrative Advantages.

This taught *Socrates* to die as free from Haughtineſs as Weakneſs; "and "the moſt beautiful Lineaments, in the Character of *Cato*, are owing to "this excellent Science."

Here the great *Bacon* experienced more exalted Pleaſures than in the Sunſhine of a Court. It was this that not only corrected the paſt Inadvertencies of his Life, unto which an infatuated Attachment to Grandeur had miſled him, but likewiſe recalled his great Talents from meaner Purſuits to their proper Objects---To explore the hidden Secrets of Nature: And, by making Experience and Reaſon go hand in hand, he not only exploded

exploded thoſe groundleſs dogmatical Opinions, which Length of Time had rendered venerable to Ages of Ignorance; but likewiſe, by numberleſs Experiments, both ſought and diſcovered ſeveral valuable Truths; which, like rich Ore, had been long hid under Rubbiſh.

Future Diſcoveries will, no doubt, in ſo inquiſitive an Age as this, bring likewiſe to Light many Secrets of Nature, which even ſtill lie hid in Obſcurity: But, in order to be ſucceſsful in ſuch Attempts; as the Harveſt is large, and the Labourers few, every Aſſiſtance ought to be embraced, even from Men of no extraordinary Talents: Theſe, like Hewers of Wood, and Drawers of Water, may be ſerviceable. It may be, perhaps, neceſſary to premiſe, that we are not to expect the moſt exact Performance of this Kind to be in every Part equally engaging. The hiſtorical Deſcription of a Country, like its natural Appearance, muſt needs be attended with Variety. And as, in travelling over it, we muſt climb high rocky Hills, and paſs through dreary Deſarts, as well as open Lawns, and flowery Meads; ſo the Reader muſt not always expect to be entertained with beautiful Images, and a Loftineſs of Style. In Variety of Subjects, this muſt alter with the Nature of the Things to be deſcribed.

For Inſtance: A pompous Stile and Language would ill fit a bare Narration of Facts, or the Deſcription of a grovelling Plant. Upon the Whole, that Stile or Method of Writing (as a great Orator hath obſerved) is beſt, which repreſents the Subject in Words moſt expreſſive of their Nature and Qualities.

As I have been obliged, in the Courſe of the following Treatiſe, to differ in Opinion from ſeveral Authors, I have endeavoured to do it without ſuperciliouſly condemning them, or, I hope, even leſſening that Benevolence which we owe to one another; and which is the ſureſt Foundation of ſocial Happineſs.

The Miſtakes of many Authors, who have treated of Places not perfectly known to them, are almoſt unavoidable; and will be always ſo, whilſt they receive their Informations from Men, who have neither a Capacity to diſcern, Judgment to repreſent Things in their proper Light; nor are ingenuous enough to relate nothing but Truth.

The Miſtakes, therefore, of ſuch Authors, are by no means the Object of Ridicule: Juſtice, Good-nature, and Candour, ought to be always inſeparable from human Kind; eſpecially when we preſume to ſit as Judges upon other Mens Works; leſt we ſhould raſhly pronounce that a Crime, which, at moſt, was but the Effect of too great a Share of Credulity: A Misfortune, which Men of the greateſt Veracity themſelves have been often led into.

Thus

A MAP OF
the ISLAND of
BARBADOS,
Drawn from an Actual Survey,
and from the Observations of
THE REVD. Mr. Griffith Hughes M.A.F.R.S.
BY Thos. Jefferys, GEOGRAPHER
to His Royal Highness the
Prince of Wales.
To His Excellency
Henry Greenville Esq.r
Governour of Barbados
His Majesty's Hon.ble Members of the Council
The Hon.ble William Gibons Esq.r Speaker,
and to the rest of the
Worthy Members of the Assembly
This Plate is humbly Inscrib'd
&c.
EXPLANATION.
Towns
Churches
Forts or Fortifications
Plantations of greatest Note or
Sugar Works of more Windmills
of two Mills
of one Mill
Plantations of less Note
Rocks under Water
Roads
ST LUCY'S PARISH
ST PETER'S PARISH
ST ANDREWS PARISH
ST JAMES'S PARISH
ST JOSEPHS PARISH
ST THOMAS'S PARISH
ST JOHN'S PARISH
ST MICHAELS PARISH
ST GEORGES PARISH
ST PHILIPS PARISH
CHRIST CHURCH PARISH
LONG BAY
THE BRIDGE TOWN or ST MICHAELS
CARLISLE BAY
HOLE TOWN
OISTIN'S TOWN
OISTIN'S BAY

Thus the great *Boerhaave*, depending upon the Information of others, hath been induced to believe the *Barbados* Green Tar (an evident bituminous Foſſil) to be an Exſudation from a Vegetable.

No Faults, therefore, of this Kind, are, juſtly ſpeaking, proper Objects of Cenſure, but thoſe which Men wilfully commit, and, by pretended uſeful Diſcoveries, impoſe upon the Public: Theſe, like falſe Lights, or erroneous Land-marks, are of an heinous Nature; riſe into Crimes; and become gradually of a deeper Dye, as their Impoſitions upon Mankind are more or leſs of ill Conſequence.

Nor are thoſe Men excuſable, who (tho' in Things indifferent) love to dwell upon the Marvelous; and either ſurpriſingly aggrandize or diminiſh the Subjects they treat of, juſt as it beſt ſuits with their Fancy. Here the Vanity of being thought ſuperior to thoſe who went before them, as well as more learned than their Neighbours, hath a great Share. But the ſhort-lived Portion of all ſuch, is a flaſhy temporary Glare of Applauſe; which, like Lightning, whilſt it ſhines, deſtroys. In purſuing Fame, they fly from Veracity, its beſt and ſureſt Pillar; and, without which, no Character can be truly valuable.

How far I have myſelf obſerved theſe Rules, and avoided theſe Faults, muſt be left to the Public to determine; as well as with what Succeſs I have, upon the Whole, through untrodden Paths, purſued my Subject. And, as this, in its different Branches, required Variety of Reading, as well as Talents peculiarly adapted to each, the learned and candid Reader will look upon any little Inaccuracy as almoſt unavoidable, and therefore excuſable.

This I can with Truth ſay, that I have not repreſented one ſingle Fact, which I did not either ſee myſelf, or had from Perſons of known Veracity. And, as to the Arguments offered to ſupport any ſpeculative Part of the Work, theſe muſt ſtand or fall by their own intrinſic Worth.

ERRATA.

PAGE 6. Line 1. for *inbabited* read *inhabited.* In Note (S) for *bign* r. *high.* p. 36. l. 1. for *Leucopkmatia* r. *Leucophlegmatia.* p. 67. in the Note, for *Barbades* r. *Barbados.* p. 89. l. 26. after *ſometimes* dele *a great deal longer.* p. 116. l. 27. dele *equal* before *Knowlege.* p. 118. l. 22. for *Ten Pints* r. *Ten Pounds.* p. 135. dele *very prickly.* p. 142. l. 10. dele *having never an odd one at the End,* and after *ſmall* add *unequal.* p. 143. l. 32. dele *they,* and add *with.* p. 145. after *Bay-berry* add *Plate X* *. p. 146. l. 14. for *Bee-wood* r. *Beef-wood.* l. 30. *dyeing* dele *e.* p. 159. l. 14. for *its white Flowers* r. *the Flowers which are white.* p. 166. l. 2. r. *its Leaves which are auriculated.* l. 4. dele *as it were into Ears.* p. 170. l. 34. for *Capſulæ* r. *Capſula.* p. 171. l. 33. for *Kitiſonia* r. *Kitiſonia.* p. 176. l. 12. for *Loopholes* r. *Locopholes.* p. 281. l. 3. for *Popo* r. *Papaw.* p. 186. dele *the Deſcription of the Cochineal.* p. 192. l. 13. after *Appearance* r. *hot* inſtead of *ſultry.* l. 14. after *ſultry* add *they are then moſt generally produced.* p. 202. l. 28. for *on the Main* r. *on the Continent of America.* p. 205. l. 32. for *all over the Iſland* r. *in moſt Parts of the Iſland.* p. 212. l. 35. for *Capſula* r. *Capſulæ.* p. 214. l. 4. for *oppoſite to the Stalk* r. *oppoſite to the main Stalk.* p. 218. l. 21. for *red Leaves* read *red Petals.* p. 223. l. 19. for *Popo* r. *Papaw.* p. 224. l. 15. for *ſhort time* r. *ſome time.* p. 242. l. 19. after *with* r. *likewiſe.* p. 249. l. 29. for *Melaſſes* r. *Molaſſes.* l. 36. after *but* add *not ſo.* l. 39. for *Melaſſes* r. *Molaſſes.* p. 250. l. 11. for *Meleſſes* r. *Molaſſes.* p. 252. l. 35. for *capillaricus* r. *capillaceous.* p. 255. l. 11. for *peculiarly* r. *peculiar to it.* p. 260. dele *Plate X. Fig. 4.* p. 263. l. 29. dele *when laſt.* p. 267. l. 30. after *but* add *by.* p. 280. l. 15. after *Toe* add *and Toe Nail.* p. 280. l. 17. for *Echinie* r. *Echini.* p. 285. l. 15. for *Couzy* r. *Coury.* p. 290. l. 27. dele *Plate* XVII. *Fig.* 7. p. 302. l. 24. after *guarded* add *with Sheets.* p. 306. l. 28. for *Slippineſs* r. *Slipperineſs.* l. 36. for *Slippineſs* r. *Slipperineſs.*

The Bay-Berry Tree delineated in Plate 10. marked thus * is to be Paged 145. Plate 24. *for* Page 298. *read* 280. Plate 11. *for* Page 352. *read* Page 152. The Sea-Eggs mentioned in Page 279. and 280. and ſaid to be in Plate 26. Fig. 1. Fig. 2. as well as the Plate Fiſh, or the flat Sea-Egg mentioned in Page 280. and 281. and ſaid to be in Plate 28. Fig. 3. Fig. 4. are to be found in Plate 24. Page 280. Page 280. line 20. *dele* not *after* were. Page 9. line 40. *for* auctiores *read* acutiores. In the Addenda, Page 2. line 27. *for* ελαςε, *read* ελλαςε. *for* πορφυεος, *read* πορφυρεος.

☞ *The Reader is deſired likewiſe to conſult the Errata at the End of the Liſt of Subſcribers.*

The following NAMES did not come in Time to be inſerted in the General LIST,

Henry Bellingham, *Eſq;* *Large Paper*
Edward Barry, *M. D.* *Large Paper.*
Thomas Litton, *Eſq;* *Large Paper.*
Faithful Forteſcue, *Eſq;* *Large Paper.*
Edward Mears, *Merchant,* *Small Paper.*
Edward Meade, *Eſq;* *Small Paper.*
Mr. Cooper, *Small Paper.*

THE NATURAL HISTORY OF THE Island of *BARBADOS*.

BOOK I.

The Extent of the Island.

THE greatest Extent of this Island is from Lat.

- - - - - - - - - - - - - - - - 13 : 10

To Latitude - - - - - - - - - - 13 : 23

Difference - - - - 13 Min.

From Longitude West from *London* - 58 : 49 $\frac{3}{4}$

To Longitude - - - - - - - - - - - 59 : 02 $\frac{1}{2}$

Difference of Longitude 00 : 12 $\frac{3}{4}$

Its greatest Length from *Goulding's Green*, in *St. Lucy's* Parish, to *Ananias Point*, in the Parish of *Christ Church*, is - - - - - - - - - - - - - - 20 : $\frac{3}{4}$ Stat. Miles.

Its greatest Breadth from *Kirtrige's Point*, in *St. Philip's* Parish, to a Point near Mr. *Peyne's*, in *St. James's* Parish, is - - - 13 : $\frac{3}{4}$ Stat. Miles.

The Circumference of the whole Island, from some of the surrounding principal Points, is - - - - - - - - - - - - - 54 : $\frac{3}{4}$ Stat. Miles.

Its first Appearance from the Sea to the Eastward, is somewhat hilly to the South-west, and North-west more level. Its Surface generally appears covered with a grateful Verdure, which, variegated with lofty (1) Trees, and large Buildings, affords many different, and those very beautiful Prospects. The highest Part of the Island is a rocky Cliff, in the Estate of *Francis Vaughan*, Esq; whose perpendicular Altitude is 915 Feet, the horizontal Base being taken from High-water Mark.

B Among

(1) By lofty Trees, we are to understand those chiefly which are planted near Dwelling houses, generally more for Ornament than Use. As for large Timber-trees, so necessary to make and repair the Sugar Mills, and other larger Buildings, we are obliged to bring these, at a great Expence, chiefly from the Islands of *St. Lucia* and *Tobago*.

Among many uncertain Conjectures, thoſe who derive the Name given to this Iſland, from the *Portugueſe* Language, are moſt properly in the Right; for, as they were the firſt Diſcoverers of the *Weſt-India* Iſlands (if not of *America* in general), it is not unreaſonable to ſuppoſe, that they might give this, and the neighbouring Iſlands, a Name analogous to ſomething remarkable in their firſt Appearance; and nothing of this kind could be more ſurpriſing to *Europeans*, than to ſee the Shores ſhaded with a kind of Fig-trees, differing from all other Trees in the manner of their Growth; for, from their Branches hang innumerable ſmall Filaments growing downwards, till they touch the Earth, and then take Root. Theſe Thread-like Reſemblances have been called by the *Engliſh*, from the firſt Settlement of the Iſland to this Time, the Beards of the Fig-trees: And in all Probability the *Portugueſe* might originally have the ſame Notion of their Reſemblance, and from thence called this, and the neighbouring Iſlands, conjunctly, *Las Barbadas*, i. e. *the Bearded Iſlands*; for *Couſa Barbada* in that Language ſignifieth any thing bearded. And, when this Iſland came to be inhabited by the *Engliſh*, it retained the general Name given originally to the Whole, whilſt the neighbouring Iſlands, ſince ſettled, were, for Diſtinction-ſake, obliged to be called by different Names. If we conſult the moſt antient Hiſtories, divine or human, we ſhall find, that the Etymology of the original Names given to Places and Things hath, in general, a relative Meaning, expreſſive of their Nature, Qualities, or Situation.

Why called *Barbadas.*

The original Names of Places expreſſive of their Situation, Nature, or Quality.

Thus, as early as the Time of *Moſes*, when the Spies, who had been to view the Land of *Canaan*, had brought from thence a Bunch of Grapes, ſo large, that they bare it between Two, when the *Iſraelites* got Poſſeſſion of the Country, they called the Valley by the Name of אשכול in Memory of the Bunch of Grapes; for ſo the *Hebrew* Word *Eſhcol* ſignifieth.

An unhealthy Part of *Arabia* was called by the *Hebrews* חצרמות *Hatſarmaveth*, and by the *Arabians* حضرموت, *Hadramaut*, i. e. *the Gate* or *Entrance of Death*. And in *Geneſis* it is ſaid, that the Herdſmen of *Gerar* did ſtrive with *Iſaac*'s Herdſmen, ſaying, *The Water is ours*: And he called the Name of the Well עשק *Ezek*, i. e. *Contention*, becauſe they ſtrove with him. Innumerable are the Inſtances which might be produced from ſeveral Languages, eſpecially from the *Celtic* and *Greek*, to prove that ſeveral Words in thoſe Languages are adapted to repreſent and be expreſſive of the Subjects treated of.

The Works of *Homer* are full of Inſtances of this Kind; and, among many others that might be produced, the original *Celtæ* called the Sun *Titan*; a Name evidently derived from *Ti*, a Houſe, and *Tan*, Fire; which is very ſignificant of the Notion moſt Eaſtern Nations had of that Planet.

If what hath been obſerved of ſimilar Inſtances and Examples have any Weight, we may, from Parity of Reaſon, be ſatisfied with the preſent Etymology of the Name given to this Iſland, and proceed to obſerve, that the Atmoſphere is, in general, ſerene and clear, ſeldom cloudy; and from a total Abſence of Hail, Froſt, or Snow, it is never liable to thoſe many and ſudden Viciſſitudes, no leſs common than prejudicial in variable Climates,

Climates, either to the Southward, or to the Northward, of the Tropics. And ſuch is the Serenity and Clearneſs of the Atmoſphere, that the Air is, in general, very healthy: But what chiefly conduces to its Purity, is the Regularity of the Trade-Winds, which, ſeldom varying throughout the Year, further from the Eaſt, than to the Eaſt-north-eaſt, and conſequently paſſing over a vaſt Tract of Water of about 3127 Miles (for about that Diſtance from us, is the neareſt Point of (2) Land on the oppoſite Shore), or were it only even what lies betwixt the Tropic and our Latitude, it muſt neceſſarily blow upon the Iſland in cool refreſhing Gales. The Clearneſs and Purity of the Air accounted for.

It is likewiſe no ſmall Advantage, or rather Happineſs, that we have neither Bogs nor Marſhes, to ſtagnate our Waters, which, being exhaled into Vapours, might be pernicious; nor large Foreſts of Trees, which not only prevent the Winds in their Paſſage, but likewiſe generate moiſt Air, cauſed by the great Quantity of Vapours which perſpire through their Leaves, as well as from the ſhaded moiſt Soil: By this means the Inhabitants are free from the Fever and Ague, ſo common to the Inhabitants of uncultivated Iſlands.

As it cannot be foreign to my Subject, I beg Leave to offer a few probable Conjectures, to prove that the Current of the Deluge, between the Tropics, ran from Eaſt to Weſt. One Argument that may be offered to prove this, is the ſhattered Condition of the Eaſtward Side of that Chain of Hills and Clifts, which are as Barriers to this Iſland, from *Cuckold's Point*, to a Place called *Conſet's Bay*; for, as they face the Eaſt, their torn State, in that Part alone, and no-where elſe, ſhews, that they not only, by their Situation, firſt ſtemmed, but, as they were higher than any other Part of the Iſland, they wholly bore the repeated Percuſſions of the Current in the gradual Aſcent of the Deluge, as well as, during the Continuance thereof, the Force of the Tides, until they were intirely overwhelmed; and even then, the higher they were, the nearer their Summits would be to the Surface; therefore the greater Arch of a Circle they would deſcribe; and conſequently, they would be more violently preſſed and torn by the Water, than Places nearer to the Centre. Arguments offered to prove, that the Current of the Deluge, between the Tropics, ran from Eaſt to Weſt.

This is evidently ſeen from the boiſterous Surface of the Sea in a Storm, whoſe Violence often waſhes, or rather tears off large Pieces of Rocks, whilſt, at the ſame time, the tendereſt Sea Plant, at any conſiderable Depth, is not diſturbed.

From hence we may obſerve, that by how much the lower the level Land, and the Valleys were, than the Tops of the Hills, by ſo much leſs they were diſturbed; and conſequently, upon the Decreaſe of the Deluge, the ſooner, and in greater Quantity, would the Dregs and Sediment (which trailed gently along an even Bottom) ſubſide and ſettle: And from hence, in part, proceed the different Degrees of Fertility in the Soils of Hills and Valleys (3). Another Argument, that the Current of the Inundation,

(2) *Portaventura* in *Africa*.

(3) The very deep Soil of moſt fertile low Countries, ſuch as *Egypt*, may be, in all Probability, primarily owing to the ſubſided *Fæces* of the Deluge.

undation, in the Time of the Deluge, ran, as above deſcribed, is the coping Figure of the Iſland, from Eaſt to Weſt; for, if we narrowly and attentively view the ſeveral gradual Deſcents of ſo many continued Ridges of Rocks, like Caſcades, declining precipitately to the Weſtward (for Inſtance, that long Chain of Hills, from *Mount Gilboa*, in *St. Lucia's* Pariſh, to the *Black Rock*, in *St. Michael's* Pariſh), we cannot well otherwiſe conclude, from the deep Soil, on the Eaſtward of theſe, where the Land is level; and from the ragged, and bare-waſh'd Surface, to the Weſt; but that the latter was thus torn by the Violence of the Waters falling over them; and the former, the Effect of the ſubſided Sediment, upon the Decreaſe of the Deluge. The Want of ſuch a Bed of Rocks, from *Black Rock* to *St. Anne's Caſtle*, cauſed that Chaſm which opens to the Sea thro' *Bridge-Town*, oppoſite to the Valley of *St. George's*. What further confirms me in my Opinion, and brings what I have firſt offer'd, as conjectural, to a greater Degree of Certainty, is the Obſervation I made upon the ruinous Tops of moſt of the lofty Mountains in *North America*, as well as in *England*. In thoſe Northern variable Climates, Nature felt the univerſal Shock promiſcuouſly on every Point of the Compaſs; whereas our Hills and Clifts, which extend along the Eaſt Side of the Iſland, diſcover no Indications, on the Weſtern Sides, of any Violence receiv'd, but what was naturally occaſion'd, by the falling off of the Waters, upon the Decreaſe of the Deluge. The Courſe of ſo many deep Chaſms, or abrupt Fiſſures in the Earth (call'd here *Gullies*) always running from Eaſt to Weſt, is likewiſe an additional, and even an unanſwerable Argument in favour of what hath been urg'd upon this Head: And, that theſe were originally the Effect of the Deluge, tho' ſince much inlarg'd by repeated Torrents, will evidently appear by viewing the oppoſite Sides of thoſe deep Chaſms in the Earth, where are to be ſeen large Pieces of Rocks, that appear to have been once intire, but afterwards forcibly torn aſunder: And as theſe divided Rocks are often to be found near the Heads or Beginnings of theſe Gullies, where the greateſt Stream of collected Rain-water, even in the greateſt Flood, is too ſmall and weak to be the Cauſe of ſo violent a Separation; therefore we may, with great Reaſon, conclude, that this Diſunion, and torn State, was effected by the Current of the (4) Deluge; and that its regular Courſe, to the Weſtward, between the Tropics, was but the natural Conſequence of an (5) Eaſterly Trade-Wind, which guided and byaſs'd its Courſe to that Point. Former

The Diſcovery of the *Weſt-India* Iſlands by the *Portugueſe*, and inhabited by the *Engliſh*.

Deſcriptions of this Iſland begin with barely mentioning its Diſcovery by the *Portugueſe*, and the Settlement of the *Engliſh* there, in the Reign of King *James* the Firſt, in the Year 1625, without the leaſt Inquiry, whether it had been ever before inhabited, and by whom. It is, indeed, ſaid, that ſome of the firſt Diſcoverers of this Iſland

(4) This Opinion was confirm'd to me by the concurrent Conjecture of Dr. *Warren*, a very ingenious Phyſician, late of this Iſland.

(5) The Cauſe and Nature of Trade-Winds are excellently explained by the great Dr. *Halley*.

Iſland, found no Inhabitants upon their Arrival. However, we ought not to conclude too haſtily, that there never were any, until what is offered to prove the contrary, be fully conſidered. I was, indeed, once partly inclined to that Opinion, induced to it from the Remoteneſs of this, from the whole Cluſter of Iſlands, *viz. St. Lucia*, *Dominico*, *Mary-galant*, *Guardaloup*, *St. Chriſtopher's*, *Antigua*, and *St. Vincent*; which laſt, tho' neareſt, is about one hundred Miles diſtant. And as moſt of theſe are in Sight of one another, they are more convenient for mutual Commerce in Time of Peace, and Embarkation in Time of War. And what gives ſeveral of theſe the Advantage over this Iſland, with regard to an *Indian* Settlement, is their far greater Number of open Bays and Rivers, ſtored with Tortoiſes, and almoſt an incredible Plenty of Fiſh. But as we have late Inſtances of their coming hither from *St. Vincent's*, in their ſmall Canoes, or Perriawgers, even for their Pleaſure, I concluded, that they might formerly, more probably, come for their Intereſt; eſpecially at certain Seaſons of the Year, when the Fiſhing, or Game, in the other Iſlands, grew either ſcanty or ſhy, by being too often diſturbed.

Theſe probable Conjectures, upon a farther Inquiry, were corroborated by the Suffrage of many aged Perſons; ſeveral of whom were between eighty and ninety Years old, who, not only agreed in their received Tradition, that there were *Indians* formerly in this Iſland, but likewiſe ſome of them added farther, that their frequent Arrival to, or Departure from it, was always in the Wane of the Moon, for the Benefit of light Nights; and that when a Difference aroſe between them and the *Engliſh*, the *Indians* retired to their Faſtneſſes in the Woods; and that in their Way down to their Canoes, they would artfully hide themſelves with Coverings of green Boughs, to elude the Search of the *Engliſh*.

Now, ſince the Parents of theſe aged Perſons, who give this Account, might be old enough to be Eye-witneſſes of theſe Things; ſuch, and ſo early a Teſtimony, where they had no apparent Inducement to deviate from the Truth, muſt, at leaſt, be allow'd to carry with it the uſual Weight and Credit in ſuch Caſes, till theſe Conjectures, ſtrengthened by Tradition, and confirm'd by apparent Facts, grow to ſuch a Degree of Certainty, as to leave every Doubt inexcuſable.

The Method I ſhall take to prove their former (6) Reſidence in this Iſland (which to Poſterity may be ſerviceable) will be to make it appear, *Firſt*, That there are ſeveral Places, in this Iſland, called, to this Day, after their Names. *Secondly*, That in theſe very Places, there are daily dug up ſuch Marks of their former Reſidence, as were peculiar to *Indians*. *Thirdly*, I ſhall compare theſe ſeveral Evidences with thoſe confeſſedly ſatisfactory ones, in almoſt parallel Caſes. I ſhall begin, by obſerving, that

 the

(6) I am countenanced, in this ſeeming Digreſſion from the Subject, by many *Greek* and *Roman* Authors; eſpecially by the Example of *Julius Cæſar*, who, in his Commentaries, carefully traces the Origin, as well as deſcribes the Manners and Cuſtoms, of the different Nations whoſe Countries he treats of.

the *Indians*, who inhabited this Iſland, could, in all Probability (by reaſon of the great Diſtance from the Continent), be no others than Colonies from ſome of the *Leeward Iſlands*, moſt probably from *St. Vincent's*, *St. Lucia*, or *Tobago*. From either of theſe, according to their Situation, with reſpect to *Barbados*, as well as with regard to a ſafe Harbour, they muſt probably land to the Weſt or Weſt-ſouth-weſt of this Iſland. Accordingly, as *Carliſle* (7) *Bay* is the largeſt and moſt commodious Harbour, it is natural to ſuppoſe, that they landed here, and made the adjacent Part of the Iſland their Place of Reſidence. This is evidently confirmed by the Buttings and Boundings of ſeveral Tenements near this Bay; which, in very old Deeds, are ſaid to terminate at or within a certain Diſtance from the *Indian* Bridge; from which the great Number of Houſes built about this convenient Place, came ſoon afterwards to be called *Bridge Town*. The above-mention'd Bridge was placed over that Part of the Creek, or narrow Neck of the Bay, which divides Major *Gidney Clerk*'s Houſe from Colonel *John Fairchild*'s. The Neceſſity that the *Indians*, reſiding on the North ſide of the Creek, were under to make a Bridge (which in all Likelihood conſiſted of no more than a few Trees, fell'd down, and laid acroſs this Creek), will appear, when we conſider, that their beſt, and almoſt only Supply of freſh Water, was on the South ſide of it, at a Place now call'd *Hannington's Spring*.

The former Reſidence of *Indians* in this Iſland proved.

Bridge Town, why ſo called.

The next ſmall Rivulet running to the Sea to the Leeward of *Bridge Town* is at preſent, as well as in many old Deeds, call'd *Indian River*. It was likewiſe at this Place that ſome *Indians*, from *St. Vincent's*, landed, in or about the Year 1738. And as *Indians* are remarkably nice in chuſing a dry healthy Situation, this, with ſo few or no Footſteps of their long Abode here, gave me room to believe, that they kept moving along the Sea Shore ſo long as they met with convenient fiſhing (8) Bays: And as the laſt of theſe, of any Note, to the Leeward of the Iſland, is *Six Mens Bay*, and *Rider's Bay*, let us but allow theſe to be then, as they are at preſent, as well ſtored with Fiſh as any other, and we ſhall ſoon fix their Reſidence, for a while at leaſt, in this convenient Place; for theſe *Indians* (as ſhall be preſently ſhewed) were but ill provided with Tools to fell Timber. This, join'd to their great Indolence, cauſed them to ſearch (eſpecially in wet Seaſons) for their natural ſheltering Place, the firſt convenient Cave; and, as there is a very commodious one in the Side of a neighbouring Hill, called to this Day the *Indian Caſtle*, and almoſt in a direct Line from *Six Mens Bay*, and not above a Mile and an half off, in a pleaſant Part of the Country, it is more than probable (eſpecially as there was no

The Reſidence of *Indians* further proved.

(7) This Harbour is call'd *Carliſle Bay*, from a Royal Grant of this Iſland to the Earl of *Carliſle*. The adjacent Town, call'd *Bridge Town*, conſiſts of about twelve hundred Houſes, all built of either Stone, Brick, or Timber. The other chief Towns are *Speight's Town*, *Auſtin's Town*, and the *Hole Town*.

(8) As the Bays, near the *Hole Town*, were well ſtored with Fiſh, this ſeemed to be their ſecond Settlement from *Indian River*: But as it was neceſſary, in ſtormy Weather, to provide a Shelter higher up in the Country, I found ſeveral Remains of their Abode, under the Shelter of an high Cliff, in the Eſtate of *Samuel Barwick*, Eſq; deceaſed. We are likewiſe informed by Tradition, that five *Indian* Women, upon Promiſe of good Uſage from the *Engliſh*, upon the Deſertion of the reſt of the *Indians*, lived and died in that Place; and, about two Miles from hence, there is a Tract of Land called *Indian Wood*, or *Indian Town*.

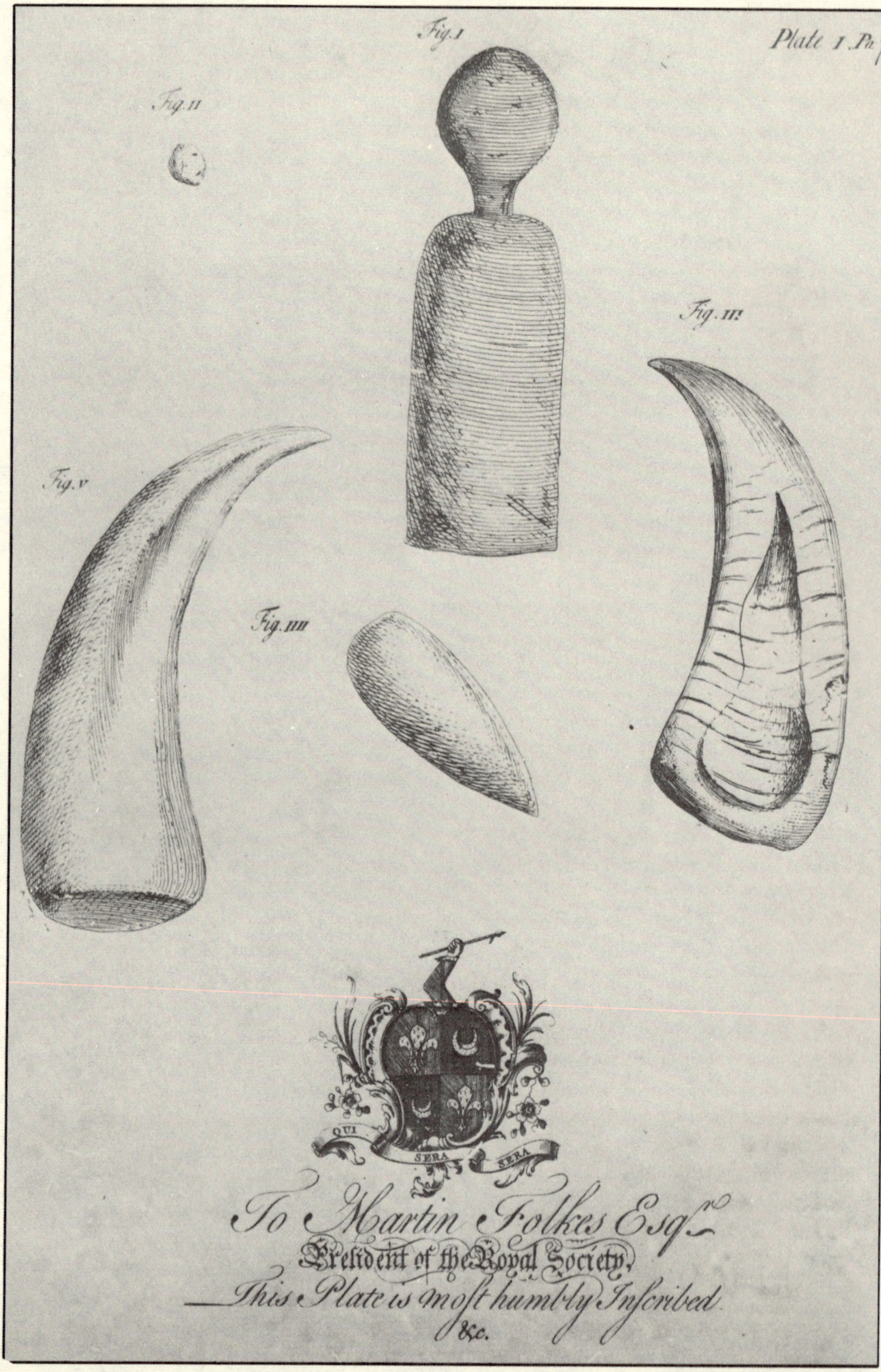
Plate I Pa 7
Fig. I
Fig. II
Fig. III
Fig. IIII
Fig. V
QUI SERA SERA
To Martin Folkes Esqr
President of the Royal Society,
This Plate is most humbly Inscribed
&c.

no other so near, and so convenient), that they should pitch upon this, being upon several Accounts very commodious; for, as the Mouth of it faced the West, and, being under the Shelter of an Hill, was secured from the Wind and Rain, and even from Danger by Hurricanes, and as the Entrance to it is so steep and narrow, that, upon Occasion, one Man may defend himself against an hundred, it may be justly called their Castle. But what made this Place more complete (and affords, I think, an undoubted Proof of their Residence here), is an adjoining clayey Bottom, where they dug a Pond or Reservoir to hold Rain-water; which Place is, and hath been, since the Memory of the oldest Neighbours alive, call'd the *Indian Pond*. With Part of the Clay (9), which they dug out, they made their Earthen-ware, such as Pots and Pans; and, like the Idolaters of old, out of the same Materials they made to themselves Gods, and worshiped them. Among several broken Fragments of Idols, said to be dug up in this Place, I saw the Head of one, which alone weighed above sixty Pounds Weight. This, before it was broken off, stood upon an oval Pedestal above three Feet in Height; and is describ'd in *Plate* 1. *Fig*. 1. The Heads of all others that came within my Observation, were very small: One of these (which is described in *Plate* 1. *Fig*. 2.) exceeds not in Weight fifteen Ounces; and all, that I have hitherto seen, are of Clay burnt. These lesser Idols were, in all Probability, their *Penates*; made small for the Ease and Conveniency of being carried with them, in their several Journeys; as the larger Sort were perhaps designed for some stated Places of Worship.

Indian Idols discovered.

The last Proof of their Residence in this Place, is a great Number of their Stone Hatchets and Chissels, that are here dug up: These are represented in *Plate* 1. *Fig*. 3. 4. and 5. The Use of these (10) Hatchets and Chissels was in all Likelihood to cut down Timber, to make Huts, where they had not the Conveniency of Caves; as well as with the Help of Fire to fell some of the largest Kind, to make Canoes. These, I imagine, after they were roughly squared, were half-burnt with live Coals; and then, with their scooping Chissels, they, by Degrees, made them hollow.

Indian Hatchets and Chissels discovered.

Having proved their Residence in this Place, it will not be surprising to those, who know the roving Disposition of *Indians*, to find, that they moved under the Shelter of the Hills, and almost always at equal Distance from the Sea, till they came to a large convenient Cave, under an Hill, called *Mount Gilboa*, in the Estate of Colonel *John Pickering*, where I found several of their broken Images, Pipes, Hatchets, and Chissels. And, though this, and the before-mentioned Cave, were for several

(9) By the great Number of the Remains of *Indian* Pot-kilns, in this Island, and the known Scarcity of Clay in many, if not in all of the *Leeward Islands*, I make no doubt, but they were supplied with Earthen-ware from hence.

(10) Some of the Hatchets, and a great Number of the Chissels, especially the scooping ones, are made of the most substantial Inside Part of a Conch-shell; others, of the hardest common Stones; and a few, of the cutting Sort of a Stone of an opake greenish Colour, variegated with several white Veins, somewhat resembling the Spleen-stone. *Vide* Plate 1. Fig. 5.

ſeveral Reaſons very convenient; yet there was one eſſential Happineſs wanting; for, upon very heavy Rains, the Water in theſe artificial Ponds, being diſturbed, became muddy: And, as their greateſt native Luxury (if it may now-a-days be ſo called) was to drink of the pureſt Stream, it was natural and cuſtomary for them to ramble in Search of Spring-Water, which at a little Diſtance from hence they found in the Eſtate of *Edward Bonnet*, Eſq; Here they ſettled, near a ſmall perennial Spring, which is the only one for ſeveral Miles round it. The Situation of this Place afforded another additional Conveniency, being, by the Goodneſs of the Soil, very proper to produce *Yams*, *Plantain*, and *Banana* Trees; but as there was no Cave to afford Shelter, they were obliged to cut down Timber to build Huts; and from thence, in all Probability, it comes, that there are found, about this Place of their Reſidence, ſo many Fragments of broken Hatchets, Chiſſels, and Pots. There are not only near this Place, but likewiſe at *Maycock's Bay* (11), *Joan's Hole*, *Cluff's Bay*, and near *Scotland* Church, as well as in other Places, many Tokens of their former Abode. To this may be added, that there is a Tradition in a Family of Negroes, belonging to *Thomas Tunckes*, Eſq; the Anceſtors of which Family came over with the firſt Negroes that ever came hither from *Guiney*, that before the Country was cleared from Woods, there was an *Indian* Town near a Pond, in his Eſtate, in the Pariſh of *St. Michael's*, which Reſervoir to this Day is called the *Indian Pond*; and, when theſe *Indians* could not for a long time be brought into Subjection by the *Whites*, the laſt Attempt was ſo vigorous, that it obliged all the *Indian* Inhabitants of the Town to make their Eſcape in their Canoes to the neighbouring Iſlands; which they all did, except one Woman, and her Son a young Lad. The latter ſoon afterwards making his Eſcape alſo, his Mother, in a ſhort time, pined to Death. Several Earthen Veſſels, of different Sizes, have been dug up near the abovementioned Pond, within theſe thirty Years laſt paſt: Theſe were generally of a globular Figure, of a Slate Colour, but very brittle otherwiſe, far ſurpaſſing the Earthen-ware made here by Negroes, in Thinneſs, Smoothneſs, and Beauty.

An *Indian* Town.

Indian Earthen-ware.

Let us now ſum up theſe Teſtimonies, and put the Certainty of what I have offered on this Head, upon the Authority of theſe Proofs. As for the firſt Proof, that there are ſeveral Places in this Iſland called after their Names; this, tho' the leaſt, carries with it a Degree of Conviction, which cannot be well gainſaid; for, who will pretend to diſpute (without being led into the ſeveral Particulars of the Hiſtory of the Times) whether the *Celtæ* did not formerly poſſeſs and inhabit that Part of *France*, called *Armorica*, when he finds the moſt antient Churches, and notable Places, retain their *Celtic* Names to this Day? Or, what Sceptic would be hardy enough to diſpute, whether the Urns dug up in *England* were not originally depoſited there by the *Romans*? Such a moral

(11) There is near this Bay a very convenient Cave for Shelter in ſtormy Weather.

a moral Certainty commands the Aſſent of every unprejudiced Perſon. The ſeveral Places in this Iſland called and ſuppoſed to be *Indian* Settlements, confirmed by early Tradition, and further corroborated by the Diſcovery of ſeveral rude Idols, Hatchets, and other Tools, dug up in theſe Places, different in Shape and Subſtance from thoſe uſed by every civilized Nation, and peculiar (after the Uſe of Iron was found by the more civilized) to *Indians* only; theſe Things (I ſay) being conſidered, and impartially ſuffered to have their proper Weight; there will, I believe, be very little occaſion to multiply or add any further Arguments, to prove that *Indians* formerly reſided (for at leaſt certain Seaſons of the Year) in this Iſland.

The Reſidence of *Indians* being proved, it may be expected, that we inquire into the Diſpoſition of the preſent Inhabitants. Of theſe it may be juſtly ſaid, that they are generous and hoſpitable, firmly attached in their Principles to the preſent happy Eſtabliſhment, in Church and State. The Men have a natural, as well as, by the frequent Uſe of Arms, acquired Bravery; with no ſmall Share of uſeful Learning, and Knowlege in Trade. The Women, in general, are very agreeable; and ſeveral of them might any-where paſs for Beauties. There are many Inſtances of their prudent Behaviour and Oeconomy, greatly aſſiſting to improve a moderate, and retrieve a broken Fortune.

The preſent Inhabitants of *Barbados*.

Their Character.

I muſt here beg leave to endeavour (a Thing hitherto unattempted) to aſcertain ſome reaſonable Cauſe of that general Obſervation, that the Inhabitants of hot Countries are of a more volatile and lively Diſpoſition, and more iraſcible in general, than the Inhabitants of the Northern Part of the World. The former is evidently apparent from the more ſublime Compoſitions of almoſt all Eaſtern Authors, their lively Imaginations (as a learned (12) Critic hath obſerved) tranſporting them, with incredible Warmth and Activity, from one Thing to another; and thereby cauſing them to overlook thoſe Rules of Method and Connexion, that are obſerved by *Europeans* of a cooler and more regular Fancy. Inſtances of this we have in the Works of *Pindar*, and throughout the whole Book of *Job*. Nor is it with me a Doubt, whether different Climates may not cauſe a conſtitutional Difference. *Herodotus*, as well as *Cicero*, was of this Opinion. The latter ſays, *Videmus quam varia ſunt terrarum genera .... aliæ quæ acuta ingenia gignant, aliæ quæ retuſa: quæ omnia fiunt ex cœli varietate, & ex diſparili adſpiratione terrarum.* CICERO de Div. No. 79.

Idem de Fato, No. 7. *Inter locorum naturas quantum interſit videmus; alios eſſe ſalubres, alios peſtilentes ....*

Athenis tenue cœlum, ex quo auctiores etiam putantur Attici: craſſum Thebis; itaque pingues Thebani.

And *Horace*, ſpeaking of the Works of a dull Author, ſays,

Bæotum in craſſo jurares aere natum.

HOR. Lib. II. Ep. I. 244.

D But,

(12) *Vide* Doctor *Hunt*'s Explanation of *Job*, Chap. vii. Ver. 22. 23.

But, without laying any Streſs upon the above Inſtances, it will not be very difficult to explain, in ſome meaſure, the Reaſon of ſo viſible a conſtitutional Difference in the Inhabitants of hot and cold Climates.

The conſtitutional Difference of the Inhabitants of hot and cold Climates examined in a new and phyſical Manner.

As (13) Heat in the former rarefies and increaſes the Velocity of the Fluids, conſequently the Particles of Blood, thus expanded, meeting in their Circulation (even in the minuteſt Veſſels) no Obſtacle from the external Preſſure of Cold, nor any Languidneſs by immoderate Heat, it follows, that as Health conſiſts in an equal Motion of the Fluids, and Reſiſtance of the Solids, ſuch an even Temperament of the Air muſt be more friendly to, and productive of, Health and Chearfulneſs, than in Climates where often, by ſevere cold Weather, the Globules of Blood loſe in a great meaſure their Motion, eſpecially towards the Extremities of the Body, where at ſuch times they cohere in Maſſes too large to paſs freely through the minute capillary Veſſels. When this Harmony between the Solids and Fluids is diſconcerted, then ariſes, as Experience teaches us, a ſudden Senſe of Pain, which cannot be removed till the Blood is, as it were, thawed by Heat, and ſo recovers its uninterrupted Motion: And tho' we know not certainly how to define what the Animal Spirits are; yet, on Suppoſition that they are the moſt refined and active Particles of the Blood, as they are with great Reaſon thought to be, their pleaſing, or painful, Influence on the Body muſt, in a great meaſure, depend upon the regular or irregular Motion of their original conſtituent Fluid, the Blood: Now, when this is checked in its Circulation, either by a ſudden Preſſure of an heavy Atmoſphere, or the Variableneſs of the Winds from warm to cold; the Animal Spirits, which before were agreeably diffuſed through the whole Body, by theſe ſudden Interruptions in the Blood, their original Source, likewiſe become unable to perform their Functions; for, when Nature is impeded in any of her Operations neceſſary to our Well-being, whether natural, vital, or animal, that mutual and inſeparable Connexion between the Mind and the Body will appear ſo viſibly, that if the latter be affected with Pain, the neceſſary Conſequence of Obſtructions in the Fluids, the former ſoon loſes that Chearfulneſs which is naturally leſs interrupted, and may be better preſerved, in Climates obnoxious to no ſudden Viciſſitudes of Weather, from moiſt to dry, from hot to cold; for an even Temperature of the Air is beſt adapted to ſupport the (14) Body in that State which is requiſite

(13) It may perhaps be ſaid, that as Heat increaſes ſenſible, and diminiſhes inſenſible Perſpiration, it will render the Blood, in hot Climates, more viſcid, and conſequently its Circulation more languid: This, in general, may be true, eſpecially of thoſe who are daily expoſed to hard Labour, and conſequently ſweat much: But, with regard to the *white* Inhabitants of this Iſland, this ſuppoſed Viſcidity is not ſo great; or, if it were, it is more than counterbalanced by the daily Uſe of a great Quantity of Diluters of every kind, *viz.* Water, and weak Punch made very acid with Lime or Lemon Juice. The ſame Argument will be of Force, with regard to the Inhabitants of any other hot Climate, under equal Circumſtances.

(14) This is evidently proved by the great Number of decayed Conſtitutions in variable Climates, who, whilſt expoſed to the Viciſſitudes of the different Seaſons of the Year, are in daily Expectation of making their Exit; but, when once bed-ridden, have lived many Years: And will any one ſay, that their Lives, under Providence, were not prolonged by being always kept in an equal Degree of Heat and Cold, in a cloſe Room, and a warm Bed? However, it muſt be owned, that warm Climates have their Inconveniencies; they are very improper for thoſe Conſtitutions which require much bodily Exerciſe; for it

requiſite to the due Exertion of the Animal Oeconomy, and conſequently, in ſome meaſure, of the rational Faculty.

Nor is the ſudden Alteration, upon the Change of Wind and Weather, peculiar to the human Conſtitution alone; for other Animals feel the Viciſſitudes of the Weather, eſpecially in variable Climates. This is a Remark as early as the Time of *Virgil*, which appears by the following Lines:

> . . . *Ubi tempeſtas & cœli mobilis humor*
> *Mutavere vias, & Jupiter humidus Auſtris*
> *Denſat, erant quæ rara modo; & quæ denſa, relaxat.*
> *Vertuntur ſpecies animorum, & pectora motus*
> *Nunc alios, alios, dum nubila ventus agebat,*
> *Concipiunt. Hinc ille avium concentus in agris,*
> *Et lætæ pecudes, & ovantes gutture corvi.* Geo. I. 416.

> But with the changeful Temper of the Skies,
> As Rains condenſe, and Sunſhine rarefies;
> So turns the Species of their alter'd Minds,
> Compos'd by Calms, and diſcompos'd by Winds.
> From hence proceeds the Birds harmonious Voice:
> From hence the Crow exults, and friſking Lambs rejoice.

Thus far have we briefly endeavoured to account, from the Nature of the Climate, and the Mechaniſm of the human Body, for that volatile Diſpoſition, ſo peculiar to the Inhabitants of hot Climates: That Iraſcibility of Temper, likewiſe, which is aſcribed to them, is, in a great meaſure, the natural Conſequence of the above-mentioned Diſpoſition; for, as Water that is already hot, will, with a little additional Heat, boil over, ſo when the Animal Spirits are in a high Flow, and the Will, by the Propenſity of long-rooted Habits, unhappily aſſiſts, and is bent to gratify ſome favourite Paſſion, the Tranſition from a Degree of Sprightlineſs to Iraſcibility is natural and eaſy.

If we purſue this Argument a little further, it will appear more concluſive by reaſoning upon, and experimentally comparing the Correſpondency between artificial and natural Heat, and their ſimilar Effects upon human Bodies.

This will, in ſome meaſure, anſwer to two different and oppoſite Climates: For Inſtance, If, in very cold Weather, the Inhabitants of far Northern Countries drink ſuch a Quantity of ſtrong ſpirituous Liquors, as will cauſe a free Circulation to the Blood, before almoſt congealed with Cold, they will aſſume a ſudden lively Chearfulneſs: Should they ſtill drink

it cannot be taken without Danger of being ſurfeited; nor is it leſs improper to Perſons who labour under any Degree of Phrenſy or Madneſs, whoſe periodical Fits, at the Full and Change of the Moon, return here with greater Violence than in a cold Climate: Yet, tho' we enjoy, or may enjoy here, in general, by a regular Conduct, a more chearful Scene of Life, free from the unequal Changes of Heat and Cold; yet it muſt be owned, that we are, in a great meaſure, Strangers to thoſe invigorating Starts of Livelineſs, which the Animal Spirits diffuſe thro' the whole Body in moderate froſty Weather, and a clear Air, in *England*.

drink on, they then hurry on the Blood, from a free to a rapid Motion: And if these are the natural Effects of spirituous Liquors in a very cold Climate, what and how great must they be in the opposite Extreme? For the Velocity of the Blood, occasioned by the Warmth of the Climate, naturally attenuates and breaks the Cohesion of its Parts, hereby accelerating its Motion to such a Degree, as to raise a more than ordinary Flow of Animal Spirits: These, in such Cases (as Experience teaches us), are too often, in most Constitutions, accompanied with an uncontroulable Flow of the irascible Passions of the Soul.

If we allow *Horace* to be a Judge of Nature, such frequent Observations upon the Effect of strong Drink might give him the Hint to caution us against the Excess of it:

Tres prohibet supra
Rixarum metuens tangere Gratia.

Hor. Lib. III. Ode 19.

It might likewise give Birth to that frequent Custom of raising Courage in common Soldiers, who often want nobler Motives to heroic Deeds, by giving them strong Liquors to heat their Blood immediately before an Engagement, that, by the Assistance of such a borrowed Flow of Animal Spirits, they may be the more resolute, and thoughtlesly rush into the Heat of Battle, and there act with an Intrepidity suitable to such dangerous Circumstances; for, as the same Author elsewhere observes,

Quis post vina gravem militiam, aut pauperiem crepat?

Hor. Lib. I. Ode 18.

Who in his Cups can feel the Weight
Of Arms, or of a pinching State?

Should the grave Behaviour of the *Spaniards*, compared with the Sprightliness of the *French*, who live in a colder Climate, be brought as an Argument against what I have said, it is easily answered; for perhaps the too grave Behaviour of the one, and the too volatile Disposition of the other, may not be intirely constitutional to either; for the one may be partly the Effect of cultivated Levity, as the rigid Deportment of the other in Part the Improvement of an affected Gravity.

But, however this may be, another very natural and very convincing Reason why the Inhabitants of some of the very hot Parts of *Spain* are more liable to be pensive, melancholy, and revengeful, than others, is, that the scorching Heat of the Sun is there of long Continuance: And as Heat, in general, produces that Disposition which is called *bilious*, it increases the oily Parts of the Blood, and renders them lighter, and more moveable, by lessening their Tenacity: Then the other constituent Parts become more piquant, and the Salts they contain are more active, in proportion as the blunting Oil is attenuated.

The

The greater the Heat, the longer it is continued, the more obvious will its Effects be; Livelineſs and Activity will be remarkable where this Diſpoſition is moderate, as it is in the *Weſt-India* Iſlands, which enjoy the Benefit of the Trade-Winds, and moiſt Vapours; but where ſuch Breezes are wanting, long-continued Heat, ſuch as they feel in *Spain*, is capable of abſorbing every thin Fluid; and, conſequently, ſuch a Diſpoſition of the Juices inclines to Penſiveneſs, Reſerve, and often deliberate Revenge.

Add to this what I before hinted at, with regard to the *Weſt-Indies*, that Habits, eſpecially ſuch as have been early indulged to our depraved Nature, have a great Share in forming our future Conduct in Life.

This leads me to obſerve, that Children, in theſe *Weſt-India* Iſlands, are, from their Infancy, waited upon by Numbers of Slaves, who, in their moſt unwarrantable Conduct (unhappy for both!), are obliged to pay them unlimited Obedience; and as, in theſe tender Years, their natural Appetites are ſtronger than their Reaſon, when they have thus their favourite Paſſions nouriſhed with ſuch indulgent Care, it is no Wonder, that by Degrees they acquire (unleſs happily prevented, or corrected, by the good Examples of Parents, or Education) an overfond and ſelf-ſufficient Opinion of their own Abilities, and ſo become impatient, as well as regardleſs, of the Advice of others.

And, as this is a Matter of Fact, unhappily verified by numberleſs Inſtances, it is not ſurpriſing, if, in Minds thus early indulged in the Gratification of their Appetites, and too often undiſciplined by the Reſtraints of Education, we ſometimes find the iraſcible Paſſion domineering, and inſolently triumphing, over Reaſon: Nor can ſuch an Influence be worthy of Surprize, when we conſider, that Habits and Cuſtoms leave deep Traces, and laſting Impreſſions, upon the more ſolid Structure of the human Frame.

A notorious Inſtance of this kind is mentioned by *Herodotus*, who ſays, that, paſſing by *Peluſium*, where there had been fought, many Years paſt, a bloody Battle between the *Perſians* and the *Egyptians*, the Skulls of the Slain, on each Side, being ſtill in different Heaps; he found, upon Trial, that a ſtrong Blow could not eaſily break thoſe of the *Egyptians*, whereas thoſe of the *Perſians*, by the conſtant Warmth of their Turbants, ſcarce ever feeling the Variety of the Seaſons, were ſo very tender, that they did not reſiſt the leaſt Blow.

From hence *Herodotus* (15) juſtly attributes the Hardneſs of the *Egyptians* Skulls to the Habit of that Nation, whoſe Cuſtom it was to ſhave

E their

(15) Θῶυμα δὲ μέγα εἶδον πυθόμενος παρὰ τῶν ἐπιχωρίων· τῶν γὰρ ὀστέων περικεχυμένων χωρὶς ἑκατέρων τῶν ἐν τῇ μάχῃ ταύτῃ πεσόντων (χωρὶς μὲν γὰρ τῶν Περσέων ἔκειτο τὰ ὀστέα, ἃ ἐχωρίσθη κατ' ἀρχὰς, ἑτέρωθι δὲ τῶν Αἰγυπτίων) αἱ μὲν τῶν Περσέων κεφαλαὶ εἰσι ἀσθενέες οὕτω, ὥστε, εἰ θέλεις ψήφῳ μούνῃ βαλέειν, διατετρανέεις· αἱ δὲ τῶν Αἰγυπτίων οὕτω δή τι ἰσχυραὶ, ὡς μόγις ἂν λίθῳ παίσας διαρρήξειας· αἴτιον δὲ τούτου τόδε ἔλεγον (καὶ ἐμέ γ' εὐπετέως ἔπειθον) ὅτι Αἰγύπτιοι μὲν αὐτίκα ἀπὸ παιδίων ἀρξάμενοι, ξυρεῦνται τὰς κεφαλὰς, καὶ πρὸς τὸν ἥλιον παχύνεται τὸ ὀστέον· τὠυτὸ δὲ τοῦτο καὶ τοῦ μὴ φαλακροῦσθαι αἴτιόν ἐστι· Αἰγυπτίων γὰρ ἄν τις ἐλαχίστους ἴδοιτο φαλακροὺς πάντων ἀνθρώπων· τούτοισι μὲν δὴ τοῦτό ἐστι αἴτιον ἰσχυρὰς φορέειν τὰς κεφαλάς· τοῖσι δὲ Πέρσῃσι ὅτι ἀσθενέας φορέουσι, αἴτιον τόδε· σκιητροφέουσι ἐξ ἀρχῆς πίλους τιάρας τε φορέοντες. Herod. Thalia, Cap. XII. Edit. Gale.

their Heads often; and, by exposing them to the different Vicissitudes of Wind and Weather, their Skulls grew to an extreme Hardness.

The former *Indians*, and the present *English* Inhabitants, being considered, it will not be here improper to take some Notice of the Nature and Disposition of our Negroes, or black Inhabitants, employed in cultivating our Land: These are between Sixty-five and Seventy thousand, tho' formerly we had a greater Number: Yet we are obliged, in order to keep up a necessary Number, to have a yearly Supply from *Africa*.

The Number of Negroes in the Island.

The hard Labour, and often the Want of Necessaries, which these unhappy Creatures are obliged to undergo, destroy a greater Number than are bred up here: These new Slaves are chiefly brought from the Kingdoms of (16) *Coromantee*, *Angola*, (17) *Whiddaw*, *Ebo*, and *Anamabw*. The first of these, in general, are looked upon to be the best for Labour, being, in some measure, inured to it in their own Country.

Their different native Countries.

We have had likewise formerly some Slaves brought hither from the Island of *Madagascar:* These differ something from the *Africans* in the Colour of their Skins, being not of a Coal-black, but somewhat inclined to the Tawny; yet still a Degree blacker than the *Indians*.

It hath perplexed the Learned to find out some natural Cause of the Negroes Complexion, so remarkably differing from the rest of Mankind.

The black Colour of Negroes proved to be not owing to the Heat of the Sun.

Some have endeavoured to account for it, from the intense Heat of the Sun in these and such-like hot Climates: But this is so far from being true, that I have always observed, that the Hair of those who are exposed to the Sun's Heat, turns from a true black, to a brown reddish Colour. As to the Blackness of the Negroes Skin, this reaches no deeper than the outward *Cutis*; for, when this peels off by being scalded, or by any other Accident, the Part ever after remains white.

Neither can the extraordinary Curling of their Hair be owing to the Heat of the Sun; for the (18) *Indians* have always lank Hair, tho' generally exposed to its Heat.

As to the Stature and Make of Negroes, excepting that a greater Number of them have their Noses shorter, and Lips thicker, than the *Whites*, I never could find out any extraordinary Difference: They are generally strait-limb'd, which is occasioned, in some measure, by their not lacing with Bandages their Children when young, according to the

Shape and Stature of Negroes.

too

(16) This Nation of Negroes, above all others, deprecate the divine Vengeance when it thunders.

(17) The *Whiddaw* and *Angola* Negroes are generally circumcised.

(18) The *Indians*, tho' they resemble *Mulattoes* in Colour, yet they are a distinct Race, and far from deriving their Origin from a Mixture between a *white* Man and a *black* Woman; for such a Mixture, which are called *Mulattoes*, hath always very curled Hair, the *Indians* always lank and strong. A *Mulatto* Woman cohabiting with a *black* Man, and the Children of these with *Blacks* likewise, their Descendents, in three or four Generations, will gradually lose their Copper-colour Complexion, and Length of Hair; the former being almost absorbed in the natural Blackness of the Negro Hue, and the latter loses its borrowed Length, becoming weak, short, curled, and woolly: Likewise, on the contrary, if a *Mulatto* Woman is married to a *white* Man, and their Children continue to marry *white* People, their Complexion will be so far bleached in about six or seven Generations, as to terminate in what we call in *England* a Nut-brown Complexion. A few Generations further will lose all Distinctions in Colour, and the Quality of their Hair; so that the *Whites*, the *Indians*, and the *Blacks*, differ not accidentally, but originally and really: And therefore, as those Men who pretend to account for the Origin of the *Blacks*, or *Indians*, bring neither Proof nor Authority to strengthen their Opinions, these must be looked upon at best but uncertain Conjectures.

too usual Custom of a few of the *white* Inhabitants here, as well as the almost universal Custom in most Northern Countries, which not only prevents the free Circulation of the Blood, but is often the Cause of unnatural Distortions of the Body.

The Negroes in general are very tenaciously addicted to the Rites, Ceremonies, and Superstitions of their own Countries, particularly in their Plays, Dances, Music, Marriages, and Burials (19). And even such as are born and bred up here, cannot be intirely weaned from these Customs: They stand much in Awe of such as pass for *Obeah* (20) Negroes, these being a sort of Physicians and Conjurers, who can, as they believe, not

Negroes strongly addicted to the Rites and Ceremonies of their native Countries.

(19) There are but few Negroes who believe that they die a natural Death, but rather that they are fascinated, or bewitched. The Bearers, in carrying the Corpse of such a one to the Grave, when they come opposite to, or in Sight of the House of the Person who is supposed to have bewitched the Deceased, pretend to stagger, and say, that the Corpse is unwilling, and will not permit them to carry it to the Grave, until it is suffered to stop near, or opposite to, that House: After this is complied with for a few Minutes, the Corpse is, as they think, appeased, and then the Bearers, without Difficulty, carry it to the Grave. If likewise, in digging a Grave, they find a Stone which they cannot easily get out, they immediately conclude, that the Deceased is unwilling to be buried there; therefore they dig elsewhere, until they find a Place more propitious to the supposed Inclination of the Dead. Most young People sing and dance, and make a loud Noise with Rattles, as they attend the Corpse to its Interment: Some Days after, especially on their Feasts, they strew at Night some of the dressed Victuals upon the Graves of their deceased Parents, Relations, or Friends. Something like this Custom was observed by the *Romans*, in their Feasts called *Silicernia*, at which Times there was a Repast prepared for the Dead, at least in Honour of them, and laid on their Graves. This appears from the following Words of *Ovid de Fastis*, Lib. II. 533, &c.

Est honor & tumulis. Animas placate paternas,
Parvaq; in extinctas munera ferte pyras.
Parva petunt manes. Pietas pro divite grata est
Munere. Non avidos Styx habet ima Deos.
Tegula projectis satis est velata coronis;
Et sparsæ fruges, parcaq; mica salis.

Tombs have their Honours too; our Parents crave
Some slender Present to adorn the Grave.
Slender the Present, which to Ghosts we owe:
These Pow'rs observe not what we give, but how:
No greedy Souls disturb the Seats below.
They only ask a Tile with Garlands crown'd,
And Fruit, and Salt to sprinkle on the Ground.

(20) These *Obeah* Negroes get a good Livelihood by the Folly and Ignorance of the rest of the Negroes. I shall here insert one Instance of their pretended Method of curing the Sick, which was performed upon a Negro Woman; who, being troubled with Rheumatic Pains, was persuaded by one of these *Obeah* Doctors, that she was bewitched, and that these Pains were owing to several Pieces of Glass, rusty Nails, and Splinters of sharp Stones, that were lodged in the different Parts of her Body; adding, that it was in his Power, if paid for it, to cure her, by extracting these from her through her Navel. Upon the Payment of the stipulated Præmium, he produced his Magical Apparatus, being two Earthen Basons, a Handful of different Kinds of Leaves, and a Piece of Soap. In one of these Basons he made a strong Lather, in the other he put the bruised Herbs; then clapping these with one Hand to the Navel, and pouring the Suds by Degrees upon them, he stroked the Parts most affected with the other Hand, always ending towards the Navel: In a short time after, thrusting his Finger and Thumb into the Cataplasm of Herbs, he produced several Pieces of broken Glass, Nails, and Splinters of Stones (which he had before artfully conveyed among the bruised Herbs). As such a great Number extracted, was looked upon as an extraordinary Instance of the Doctor's great Skill, he unluckily demanded a farther Reward than what was stipulated: But as the Woman's Husband was one of those very few, who had no Faith in such pretended Cures, being accidentally knowing in some of their Secrets, instead of an additional Reward, he made him by Threats refund the Money he had already received, bidding him, if he was a Conjurer, find out by his Art some Means of getting it again restored to him.

Their Method of clearing themselves from imputed Crimes hath a near Affinity to the bitter Water used among the *Jews*. In the latter Case, the Priest took some of the Dust of the Floor of the Tabernacle; and, mixing it with Water, he gave it to the Woman suspected of defiling her Husband's Bed, saying unto her, If thou hast gone aside to Uncleanness with another, instead of thy Husband; then this Water, that causeth the Curse, shall go into thy Bowels to make thy Belly to swell, and thy Thigh to rot: And the Woman shall say, *Amen, Amen*. In like manner, the Negroes take a Piece of Earth from the Grave of their nearest Relations, or Parents, if it can be had; if not, from any other Grave. This being mingled with Water, not only ſaſcinate them, but cure them when they are bewitched by others. And if once a Negro believes, that he is bewitched, the Notion is ſo ſtrongly riveted in his Mind, that, Medicines ſeldom availing, he uſually lingers till (21) Death puts an End to his Fears.

Their Superſtition.

The Capacities of their Minds in the common Affairs of Life are but little inferior, if at all, to thoſe of the *Europeans*. If they fail in ſome Arts, it may be owing more to their Want of Education, and the Depreſſion of their Spirits by Slavery, than to any Want of natural Abilities; for an higher Degree of improved Knowlege in any Occupation would not much alter their Condition for the better.

That Slavery not only depreſſes, but almoſt brutalizes human Nature, is evident from the low and abject State of the preſent *Grecians*, when compared with their learned and glorious Anceſtors.

Our Slaves, in their Mirth and Diverſions, differ according to the ſeveral Cuſtoms of ſo many Nations intermixed: However, all agree in this one univerſal Cuſtom of adorning their Bodies, by wearing Strings of Beads of various Colours, intermixed ſometimes by the richer ſort of Houſe Negroes with Pieces of Money. Theſe Beads are in great Numbers twined round their Arms, Necks, and Legs. This Cuſtom is not peculiar to the Inhabitants of *Africa*; for we find it by the moſt antient Account to be common in all Eaſtern Nations, eſpecially among the *Jewiſh* Women, as early as the Patriarchal Age: And their Fondneſs for theſe Ornaments came at laſt to ſo great Exceſs, that we find it among the Crimes reproved by the Prophet *Iſaiah*.

Their Cuſtom of wearing Strings of Beads round their Legs and Arms.

That they were likewiſe in Uſe among the *Greeks*, is evident from a Piece of Painting of above Seventeen hundred Years old, now in the curious Collection of Dr. *Mead*, where the *Graces* are repreſented in a Dance, having their Legs adorned with Bracelets.

Theſe Slaves, in ſome of their rude Dances to (22) Muſic ſtill ruder, uſe Geſticulations very unſeemly and wanton; at other times, they have a ſort

Water, they drink it, imprecating the divine Vengenace to inflict an immediate Puniſhment upon them; but in particular, that the Water and mingled Grave-duſt which they have drank (if they are guilty of the Crime) may cauſe them to ſwell, and their Bellies to burſt. Moſt of them are ſo firmly perſuaded, that it will have this Effect upon the Guilty, that few, if any (provided they are conſcious of the imputed Crime), will put the Proof of their Innocency upon the Experiment. This Cuſtom, with the Uſe of Bracelets, and the Circumciſion of ſome Nations of Negroes, are the almoſt only Inſtances of this kind, wherein I find that theſe Inhabitants of *Africa* agree with the former *Jews*, except that they likewiſe, in ſaluting their Friends and Acquaintance, often call them by the Name of Brothers: This ſeems to be analogous to the Cuſtom of the antient *Jews*, who included Conſanguinity, as well as Fraternity, in that Relation. We are Brethren, ſays *Abraham* to *Lot*; whereas he was only his Nephew. So *Jacob* told *Rachel*, that he was her Father's Brother. The wearing of Ear-rings among the richeſt of the Negroes, is likewiſe agreeable to the Cuſtoms of ſome of the antient *Jews* and *Carthaginians*: The former we find breaking their Ear-rings to make the molten Calf: and *Plautus*, in his Play called *Pœnulum*, taking notice of ſome *Carthaginian* Slaves, ſays, that their Hands ſhould be without Fingers; one would think, becauſe they wore their Rings in their Ears

(21) When theſe Negroes die a natural Death, or eſpecially when they deſtroy themſelves, they believe that they ſhall return to their own Country. It would be too groſs to believe, that they expect to be reſtored and to live there in their mortal Bodies: Therefore we muſt conclude, that they have ſome Notion of the Immortality of the Soul; and what they mean by their own Country is, that they ſhall, after this Life, enjoy the Company of their Friends and Relations in another World.

(22) The Inſtruments they generally make uſe of in their Dances, are a Banjau, a Gambay, and a Drum, which they likewiſe call a Pump. The latter is made of an hollow Trunk of a Tree, about two Feet high, and about a Foot in Diameter, the Dimenſions of the Whole, more or leſs. This is covered over

a ſort of *Pyrrhic*, or a Martial Dance, in which their Bodies are ſtrongly agitated by skipping, leaping, and turning round.

Since I have made this Digreſſion to treat of the Manners and Cuſtoms of theſe Negroes, it may perhaps be expected, that I ſhould conſider the ſeveral Arguments for and againſt making our Fellow-creatures Slaves.

But, without engaging in a Controverſy foreign to my Deſign; If we agree with the Civilians of the moſt poliſhed and learned Nations, they are of Opinion, that the Power of making Slaves is, and hath been, a natural Conſequence of Captivity in War. As to the Slavery of theſe *African* Negroes, this Hardſhip is not ſo unſupportable to them; for they are very little better than Slaves in their own Country. Slavery conſidered.

However, I will not preſume to determine how defenſible this Cuſtom may be under a Chriſtian Diſpenſation; but thus far may be ſaid, in favour of it, that by the beſt Accounts we have from the Coaſt of *Guiney*, antecedent to our purchaſing any Slaves, the ſeveral different Nations were ſo very ſavage and barbarous, that they were at continual Wars with one another, and the greateſt Mercy that was ſhewn to the Conquered was to be put to ſudden Death. Hence the Trees along the Sea-Coaſts were daily to be ſeen horribly adorned with the Limbs and Heads of their vanquiſhed Enemies.

Upon the Truth of this Suppoſition, proved by repeated credible Atteſtations, that theſe *African* Nations were, and are, ſo inexorable to their Captives taken in War, will it not, with the ſevereſt Caſuiſt againſt the Purchaſe of Slaves, be of ſome Weight to conſider, that if they are bought, and tranſported into Chriſtian Settlements (without laying any Streſs upon taking away by this Means the Guilt of Murder from their Conquerors, or the Benefit that ariſes to Mankind from their Labour), at leaſt a few, among many Thouſands imported, may probably come to a better Knowlege of their Duty to God and Man.

I barely mention this Argument, and leave it to ſtand or fall by its own Weight: Though to bring theſe in general to the Knowlege of the Chriſtian Religion, is undoubtedly a great and good Deſign, in Intention laudable, and in Speculation eaſy; yet I believe, for Reaſons too tedious to be mentioned, that the Difficulties attending it are, and I am afraid ever will be, unſurmountable.

The only Happineſs, even in Temporals, that theſe poor Creatures meet with, is when they fall into the Hands of Maſters influenced by the Principles of Humanity, and the Fear of God: By theſe they are treated (though often their ill Behaviour deſerves the contrary) with great Lenity. So true is the Saying of the Poet, in the moſt literal Senſe:

over with a Goat's or Sheep's Skin. *Iſidore*, in his *Origines*, deſcribes the *Symphonia* to be hollow like a Drum, and covered with Leather, which was beaten, or played upon, with a Stick or Quill.

From hence we may conclude, that the laſt-mentioned Inſtrument, in Uſe among our Negroes, hath a great Reſemblance, if it is not the ſame, with the *Symphonia* of the Antients; for this *Weſt-India* Inſtrument is always played upon with a Stick, or the Fingers.

------*Ingenuas didicisse fideliter artes*
Emollit mores, nec sinit esse feros.
Ovid. de Ponto, II. 9. 47.

T'imbibe soft Arts, and yield to Learning's Sway,
Soon wears the Edge of Savageness away.

On the other hand, when they are so unfortunate as to have Owners unpolished in their Manners, and insatiable of Riches (who, like the *Egyptian* Tyrants, require Brick without Straw, or, more literally, exact severe Labour from an hungry Belly, or a naked Back), they have scarce a Gleam of Rest from hard Toils, and heavy Stripes, but whilst Sleep, with its short Interval, eludes the painful Scene, which must again be renewed with the returning Dawn.

Such Task-masters (23) consider not, that Compassion to the Miserable is a just Debt; and that this humane Disposition, as *Cicero* justly observes, is not the Institution of worldly Policy, is not the bare Effect of any particular Custom, but the universal Voice of Nature, whose Dictates the Brave, the Generous, and the Good, in every Age, and in every Nation, hear and obey: For, although God suffered the Children of *Israel* to be made Slaves in *Egypt*, till such wholsome Severity recalled them to their Duty; yet he brought Plagues upon the *Egyptians* for their Cruelty and Inhumanity towards them.

The obvious and natural Inference that we may and ought to draw from so many complicated Scenes of Misery, in the Condition of so many Thousands of our Fellow-creatures (who are continually liable to the Insults of the Imperious, and the Lusts of the Debauched, and whose own and their Childrens Childrens incessant Labour will never be at an End, but with their Lives), is gratefully to acknowlege the Happiness of living where neither our Lives nor Fortunes are at the Mercy of any tyrannical Oppressor.

Their Ignorance in Religion compared with some other Nations.

How happy, I say, are we, when compared with the several Nations above-mentioned, whose Ignorance in all Knowlege about Religion is so great, that, in their original native Country, their Adoration is often paid to Crocodiles, Rivers, Snakes, and certain coloured Fowls!

Howsoever ridiculous the Custom of the *Egyptians* may appear in worshiping Beetles and Crocodiles, yet there is some Shadow of Reason, at least of Excuse, to be given for this their Extravagancy; for, as the Passion of Fear is almost universally predominant, as these Beetles came some Years in great and numerous Swarms from *Assyria* into *Egypt*, and almost

(23) It seems by the following Passage from *Horace*, that the Treatment of the *Romans* of their Slaves was not less severe, than that of the present Age; for the Poet, putting *Mena*, *Pompey*'s Freed-man, in mind of his former Slavery, describes him thus:

Ibericis peruste funibus latus,
Et crura dura compede. Hor. Epode 4.

almoſt covered the Face of the whole Land, and, like their kindred Locuſts, devoured every green Thing; and as the Crocodiles proved often very deſtructive to their Herds; the *Egyptians* therefore built Temples and Altars to theſe, and worſhiped them, in Hopes, that either they, or the ſuppoſed Deity, who preſided over them, would be more propitious, and leſs hurtful, to them for the future.

It was upon the ſame Principle that the polite *Romans* deified *Fevers*, and dedicated Temples to them. But it is hard to conceive, that there was any Good to be hoped, or Evil be dreaded, from an harmleſs black Snake, which was one of the Objects of Worſhip among theſe Slaves, or from a parti-coloured Fowl, which was never thought by any other Nation to be even ominous, much leſs to have Qualities that would render it, to a reaſonable Creature, the Object of divine Worſhip.

Here we ſee the Effect of unaſſiſted Reaſon; and that it cannot, to the Bulk of Mankind, under ſuch Circumſtances, be a ſufficient Guide to a reaſonable Service towards God.

Indeed, ſuch a Religion is ſo void of improved Reaſon, and ſpeculative Thinking, that it ſurpaſſeth the groſſeſt Apoſtaſies of the *Jews*, in their Imitation of the *Egyptians*, *Syrians*, and (24) *Phœnicians*, ſo much complained of by the Prophets *Iſaiah* and *Ezekiel.*

The Nature of the Soil.

From the Diſpoſition and Manners of the Inhabitants, the next Thing that will naturally fall under our Conſideration, will be the Nature of the Soil. This chiefly is black, in low deep Lands, in ſhallower Parts ſomewhat reddiſh, on the Hills frequently of a whitiſh chalky marly Nature, and, near the Sea, generally ſandy.

By ſuch Variety, Providence hath wiſely adapted different Soils to the different Nature of the ſeveral Kinds of Trees, Shrubs, and Plants.

Where

(24) Theſe, among the many Apoſtaſies of the *Iſraelites*, mentioned by the Prophets, are elegantly deſcribed by the great and ingenious Mr. *Milton*, who, ſpeaking of the *Egyptian* and *Syrian* Idols, ſays,

For thoſe the Race of Iſrael *oft forſook*
Their living Strength, and unfrequented left
His righteous Altar, bowing lowly down
To beſtial Gods; for which their Heads as low
Bow'd down in Battle, ſunk before the Spear
Of deſpicable Foes. With theſe in Troop
Came Aſhtoreth, *whom the* Phœnicians *call'd*
Aſtarte, *Queen of Heav'n, with creſcent Horns;*
To whoſe bright Image nightly by the Moon
Sidonian *Virgins paid their Vows and Songs:*
In Sion, *alſo, not unſung, where ſtood*
Her Temple on th'offenſive Mountain built
By that uxorious King, whoſe Heart, tho' large,
Beguil'd by fair Idolatreſſes, fell
To Idols foul. Thammuz *came next behind:*
Whoſe annual Wound in Lebanon *allur'd*
The Syrian *Damſels to lament his Fate*
In am'rous Ditties all a Summer's Day;
While ſmooth Adonis *from his native Rock*
Ran purple to the Sea, ſuppos'd with Blood
Of Thammuz *yearly wounded: The Love-Tale*
Infected Sion's *Daughters with like Heat;*
Whoſe wanton Paſſions in the ſacred Porch
Ezekiel *ſaw, when, by the Viſion led,*
His Eyes ſurvey'd the dark Idolatries
Of alienated Judah. —— Paradiſe Loſt, Lib. I. Ver. 432.

Where the Soil is black, as it imbibes the Rays of the Sun, and reflects few or none, there the circumambient Air is not near ſo hot as where the Soil is ſandy, or gravelly; for moſt, if not all Bodies, reflect the ſolar Rays, in proportion to their reſpective Denſity and Smoothneſs; ſo that by how much the more ſolid and poliſhed the Particles of a ſandy or gravelly Soil are, than black Mould, by ſo much the more intenſe will the Heat of their Reflexion be.

It is owing to theſe reflected Rays, and the Want of Moiſture, that the Blades of Canes, as well as other Plants, are more apt to ſcorch in ſandy, or gravelly, than in a black Soil: And, as the Fertility of this, as well as others of the *Weſt-India* Iſlands, depends upon ſeaſonable Rains, the long Chains of (25) Hills, interſperſed with deep Valleys, are therefore providentially ſituated to the Eaſtward of ſome Parts of the Iſland, to intercept the Clouds and Vapours: For, if the Wind, at Eaſt, drives before it even a thin ſcattered Cloud towards the Weſt, this, as long as it is not augmented by adventitious Exhalations, will hover in the Air; but as ſoon as it comes oppoſite to, and as low as theſe Hills, the impelling Force of the Wind at Eaſt, and the Reſiſtance of the now cloſe oppoſite Hills will compel theſe thin watry Veſicles to coaleſce and aſſociate into Drops; by which means this thick Collection of Water becomes ſo denſe, that its Gravity is greater than the Strength of the Air in that Situation can uphold, and it deſcends therefore in refreſhing Rain; but, when the Collection of Vapours is not ſo great as to form Clouds, and if theſe are not raiſed before the Cold of the Evening above the Height of the Hills and Clifts, being then deſtitute of their chiefeſt Cauſe of Expanſion, I mean the Heat of the Sun, they then are compreſſed, and deſcend in ſoft Miſts upon the Earth; yet in far greater Quantity, for the Reaſons above-mentioned, upon the Sides of theſe reſiſting Hills to the Eaſtward. It is partly from hence, and not intirely, as ſome imagine, from the ſuperior Excellency of the Soil, that particular Parts of *St. Joſeph's* and *St. Andrew's* Pariſhes yield a more certain, and often better Increaſe, than moſt other Parts of the Iſland.

This is ſo remarkably true in general, that it hath been obſerved, that as long as the Wind blows oppoſite to one Side of the high Mountains, near the *Ganges*, they have no Rain on the other Side, the Clouds being intirely broken by the Reſiſtance of thoſe very high Hills. And, as the Winds in and near *England* blow near half the Year from the Weſtward Points, it muſt follow, that the Clouds and Vapours are, in a great meaſure, carried that Way: If theſe, as in *England*, are in their Courſe intercepted

(25) That a long Chain of Hills is very ſerviceable to intercept the Clouds and Vapours, was known as early as the Time of *Moſes*; for, in his Deſcription of the Land of *Canaan*, he ſays to the Children of *Iſrael*, *The Land whither thou goeſt in to poſſeſs it, is not as the Land of* Egypt, *from whence ye came out, where thou wateredſt thy Seed with thy Feet, as a Garden of Herbs*; i. e. In *Egypt* (as the learned Dr. *Shaw*, in his Travels, obſerves) the *Egyptians*, for want of ſeaſonable Rains, were obliged to water their Grounds by artificial Canals, which received the Waters upon the Overflowing of the *Nile*. *But the Land whither ye go to poſſeſs it, is a Land of Hills and Valleys, and drinketh Water of the Rain of Heaven; a Land which the Lord thy God careth for: The Eyes of the Lord thy God are always upon it from the Beginning of the Year even unto the End of the Year.* Deut. xi. 10, 11, 12.

cepted by high Hills, they become more denſe, and conſequently break into Rain or Miſt. If they meet with little or no Reſiſtance, many of theſe thin ſcattering Clouds paſs over an open champaign Country. It is, I believe, owing to this chiefly, that *Picardy* in *France* is leſs ſupplied with Moiſture, and conſequently wants that grateful Verdure ſo remarkable in *England*.

That theſe Hills are ſerviceable in ſtopping the flying Clouds from paſſing over the Iſland, is what is daily confirmed by Experience; for that Part of the Iſland called *The Thickets*, in *St. Philip's* Pariſh, being low Land, and having no Hills, nor high Clifts to the Eaſtward, to make any Reſiſtance to theſe Clouds, is often ſcorched with great Drought, when the middle and more hilly Parts of the Iſland are repleniſhed with Rain. **High Lands ſerviceable to intercept Clouds.**

It muſt not from hence be concluded, that ſuch Countries are intirely deſtitute of Rain, becauſe they have no Hills or Clifts to the Eaſtward. I mention theſe only as Helps, to prevent ſome low Clouds and Vapours from paſſing over them; for, if the general Cauſe of the Deſcent of Rain was the Reſiſtance made by the Hills and Mountains, there would be little or no Rain at a great Diſtance from Land, which we are ſure there is.

It will likewiſe perhaps be ſurpriſing to thoſe unacquainted with the Nature of the Soil of ſome Parts of this Iſland, to hear, that in the moſt hilly, which is called *Scotland*, conſiderable Quantities of Land ſhould run away (as it is here termed), and become Part of a neighbouring Eſtate.

Sometimes large Pieces of Ground planted with Canes, and even Land with Plantain and Banana Trees growing upon it, have ſlid down to the Valleys from the Sides of the Hills. This happens in very rainy Seaſons; for, as the Soil upon theſe Hills is commonly not above Eight or Nine Inches deep, and of an oozy and ſoapy Nature underneath, it eaſily ſeparates from the next immediate *Subſtratum*, which is of a ſlippery Chalk, flat Stones, or looſe Gravel. **The Soil in rainy Seaſons often ſlides from the Sides of the Hills to the Valleys below.**

When the Soil ſlides in large Pieces, its Motion is leſs violent, than when it is confined in narrow Chaſms in the Meeting of two Hills, eſpecially if the Situation be very ſteep; for there the Collection of Water being conſiderable and heavy, inſtead of gliding ſoftly between the two *Strata*, it breaks out in different Places at once, and then runs down the Precipice a mingled Torrent of Earth and Water.

A remarkable Inſtance of this kind happened ſome Years ago in *St. Andrew's* Pariſh, where a large Garden, the Soil of a Potato-Garden, with its growing Produce, ſlid from the Side of a neighbouring Hill, and richly covered the next Neighbour's Land, ſituated in the Vale below.

Another Inſtance, ſimilar to this, happened at an Eſtate, now in the Poſſeſſion of the Reverend Mr. *Reynold Foſter*, where the greateſt Part of a poor Tenant's Land, with all its Produce growing upon it, firſt cracked, and then ſlid over the Clifts into the Sea; but during the long Continuance of the ſame Rain, in a few Hours afterwards, the adjoining Land of

G Mr.

Mr. *Foſter*, being then planted with Canes, came tumbling down, and richly ſupplied the Place of the Soil ſo lately waſhed off, and there remained.

The Violence with which the Land moves at ſuch times, may be gueſſed at likewiſe from the following Inſtance : A Cabbage-tree, of about Thirty Feet high, moved Roots, Soil, and all, ſeveral Feet from the Place where it grew ; it moved in an upright Poſture, and remained ſo upon the Settlement of the Soil, and continues to this Day to grow and flouriſh. This Tree is to be ſeen at *Banana Hole*, an Eſtate in *St. Joſeph's* Pariſh, belonging to Mrs. *Warren*, at the *Black Rock*.

However fertile theſe hot Climates are, if refreſhed with ſeaſonable Dews and Rains, and with the warm enlivening Rays of the Sun ; yet it will ſcarce appear credible, at leaſt to the Northern Inhabitants of the World (where the Vegetable Creation languiſhes for near half the Year ; and, when the Sun ſheds its benign Influence, is comparatively weak and faint), that in an Iſland, containing not above 106470 Acres of Land, there ſhould ever have been near 70,000 *black*, and 30,000 *white* Inhabitants ; though at this Time, indeed, as I have obſerved before, the Number is not quite ſo conſiderable.

The great Fertility of the Land.

What I ſhall add will ſtill appear more incredible (though upon good Grounds I affirm it) ; That if all the Valleys, and other Places ſhaded from the Wind, were to be planted with Plantain, Banana, and Caſſado Trees, and the open level Land with Yams, Potatoes, Eddoes, Corn and Pulſe (eſpecially Pigeon-peas), and a ſufficient Number of Cattle was kept to provide Manure, this ſmall Iſland alone, without any foreign Aſſiſtance, would, in ſeaſonable Years, produce a Sufficiency of ſuch Food to maintain more than the (26) Number of its preſent Inhabitants.

The numerous Armies of the Children of *Iſrael* not incredible.

From hence we may learn not to wonder at, nor disbelieve what is ſaid in Scripture of the Fertility of the *Holy Land*, and the numerous Armies brought into the Field by the Children of *Iſrael*. Though their ſettled Kingdom from *Dan* to *Kadeſh*, upon the Northern Boundaries of *Arabia Petræa*, was not above One hundred and Twenty Miles long, and in Breadth, from

(26) *Romulus* allowed every *Roman* but Two Acres of Land (which were much leſs than the *Engliſh* Meaſure) in even a Winter Country. How much more fruitful an Acre in this Iſland is, than Two about *Rome*, will appear by the following Calculation : An Acre of Land deſigned for Yams hath generally three thouſand one hundred Holes dug in it. This planted in the Beginning of *July*, with Five hundred Weight of ſmall Yams, will come to their Maturity, and be fit to be dug up the latter End of *December*, or the Beginning of *January*, and will yield Two Pounds and Three Quarters Weight of Yams from every Hole. This is a Computation ſo very ſmall, that the Inhabitants often dig Four, Five, or Six Pounds, more than what I have mentioned : However, by this very low Calculation, an Acre thus planted (beſides the ſmall Yams or Seedlings, equalling the Weight of what was planted) will produce a Sufficiency for Five Men a whole Year, allowing each Man Four Pounds of this kind of Food a Day ; ſo that Twenty-ſeven thouſand Acres of the preſent beſt Cane Land, planted in Yams, would produce yearly a Sufficiency of this kind, at the above Allowance, for One hundred and Thirty thouſand Men, which is above the Number of its preſent Inhabitants. After the above-mentioned Quantity is ſubtracted from the Whole, there will remain Seventy-nine thouſand Four hundred Acres, which (except what the Buildings, Orchards, and other Conveniencies take up) may be planted, ſome in Corn and Pulſe, and the reſt left for Paſturage, which Quantity will be more than ſufficient to raiſe Cattle for Manure, and other Uſes : Beſides, there might be raiſed upon the moſt barren Part a great Number of Sheep and Goats ; and the whole Land, that was planted in Yams, would be open for other Proviſions, ſuch as Potatoes and Pulſe, for above Six Months in the Year.

from the *Mediterranean Sea* to the Deſart on the Eaſt, about .ignty ; yet, when *Joab* numbered the Children of *Iſrael*, they were found to be Thirteen hundred thouſand fighting Men, beſides Women and Children.

The great Fertility, and prodigious Growth, of Vegetables in warm Climates, when compared with the Northern Parts of the World, is almoſt incredible ; at leaſt, it will appear to be ſo to thoſe who are unacquainted with the Nature of Vegetation.

This will be made more evident, when we come to deſcribe the Plantain-tree, and the great *American* Aloe ; nor will it be leſs agreeably ſurpriſing to find, that theſe hot Climates are far from being uninhabitable ; for the Heat is daily cooled by the conſtant Breezes of Wind, which at the Sun's firſt Riſing gently fan the Air, and increaſe in Strength in proportion to the active Influence of the Sun's Heat, till towards Evening, when the ſolar Heat is abated, theſe refreſhing Gales, generally ſpeaking, die away.

The Conſtancy, Regularity, and Gentleneſs of the Wind, eſpecially in what we call the Winter Months, the Serenity and Clearneſs of the Air, the continued Verdure of Trees, diverſified with Variety of Objects of Art and Nature, all terminating in and ſurrounded with the Sea, may be compared, with ſome ſmall Allowance, to *Homer*'s beautiful Deſcription of the Plains of *Memphis*, near the *Acheruſian* Lake.

Οὐ νιφετὸς, ὐτ' ἄρ χειμὼν πολὺς, ὔτέ ποτ' ὄμβρος·
'Αλλ' αἰεὶ Ζεφύροιο λιγυπνείοντας ἀήτας
'Ωκεανὸς ἀνίησιν ἀναψύχειν ἀνθρώπους.

Stern Winter ſmiles on that auſpicious Clime :
The Fields are florid with unfading Prime.
From the bleak Pole no Winds inclement blow ;
Mold the round Hail, nor flake the fleecy Snow :
But from the breezy Deep the Bleſt inhale
The fragrant Murmurs of the Weſtern Gale. POPE.

The beautiful Proſpects from ſeveral Hills to the Vales below, eſpecially from the Top of a ſmall Hill, near the Honourable *John Dottin*'s Eſtate, from another in *Batten*'s Eſtate, and from *Brigs*'s Hill, are not perhaps inferior to that ſo celebrated Proſpect from the Top of *Mount Tabor* towards the ſpacious Valley of *Eſdraelon*, and the *Mediterranean Sea*.

Nor ought we here to neglect the Deſcription of *Hackleton's Clift*, where Nature at one View diſplays a great Variety of ſurpriſing Proſpects. Here the high impending Rocks yield a dreary rueful Appearance : The ſeveral deep Chaſms below, over which they project, are imbrowned with the thick Foliage of lofty Trees. The adjacent ſteep Declivity is crouded with irregular Precipices, and broken Rocks ; the whole View terminating, in the tempeſtuous Sea, over whoſe craggy Shores the foaming Waves inceſſantly break.---All ſolemnly awful, if not horriſying Scenes ! except when the Eye is relieved by a Glimpſe, or ſometimes a full Sight, of the neighbouring

neighbouring Plantations. To complete this uncommon Contrast, a deep rapid River runs through the Valleys at the Bottom of these Precipices; which, tho' in the dry Seasons it is almost without Water; yet in the rainy Months of *June*, *July*, and *August*, it often overflows its highest Banks; and its then irresistible Course may justly be said to answer *Horace*'s Description of the River *Tiber* in a Storm.

- - - - - - - - - - Cætera Fluminis
Ritu feruntur, nunc medio alveo
Cum pace delabentis Etruscum
In mare; nunc lapides adesos,
Stirpesq; raptas, & pecus, & domos
Volventis una, non sine montium
Clamore, vicinæq; silvæ,
Cum fera diluvies quietos
Irritat amnes.

Like *Tiber*'s smooth and glassy Face,
When gently rolling to the Main;
Life sometimes glides with a soft Pace,
Unvanquish'd with the Edge of Pain.
Like *Tiber*, oft 'tis in a Roar,
When his rough Billows lash the Shore;
Rocks, Trees, and Cattle, down are borne,
And Domes from their Foundation torn;
Huge Mountains sound, as well as Woods;
And all around the Plains are Floods.

As the above-mentioned stupendous Clifts are intirely impassable for many Miles, except by Three narrow Chasms hewed through the Rocks, where Ten Men can resist a Hundred; this, with their Vicinity to the Sea, reminds me of Mr. *Glover*'s beautiful Description of the Streights of *Thermopylæ*.

- - - - - There the lofty Clifts
Of woody Oeta *overlook the Pass;*
And far beyond, o'er half the Surge below,
Their horrid Umbrage cast. - - - - LEONIDAS.

This Place is still more melancholy, if we cast our Eyes upon the several Catacombs, which on our first Settlement in this Island were dug out of the Rocks in the Side of this Clift, where lie the Remains of those, who, like the Patriarchs of old, procured to themselves Places of Rest.

The Antients, especially *Aristotle*, were of Opinion, that the Torrid Zone was uninhabitable, by reason of its supposed great Heat; but when we consider the almost constant Regularity of the Trade-Winds and

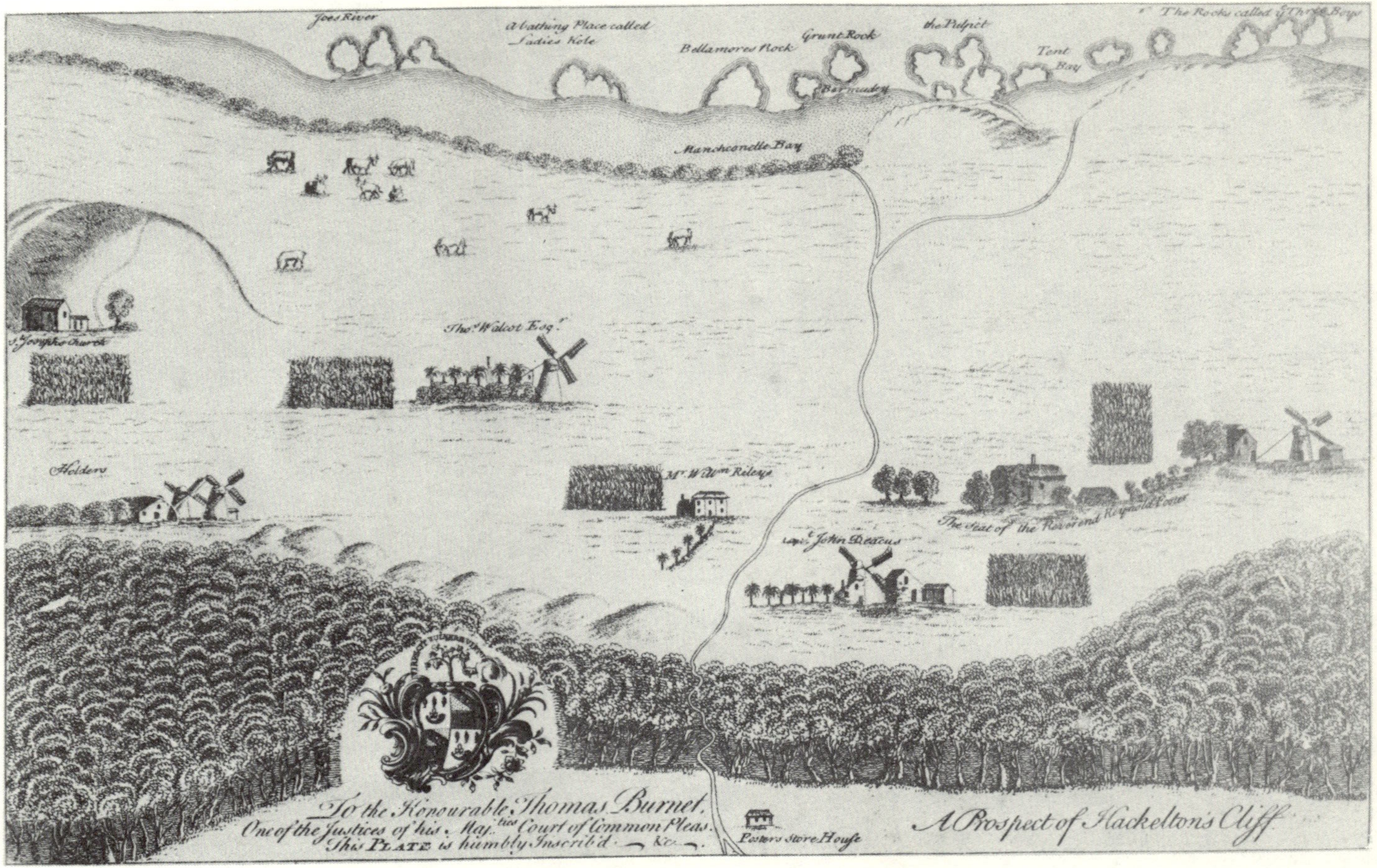
Joes River
A bathing Place called Ladies Hole
Bellamores Rock
Grunt Rock
Bermudas
the Pulpit
Tent Bay
The Rocks called ye Three Boys
Manchaneelle Bay
Thos Walcot Esqr
Holders
Mr Wm Rileys
Capt John Deacus
The Seat of the Reverend Reynolds Estate
To the Honourable Thomas Burnet,
One of the Justices of his Majesties Court of Common Pleas.
This PLATE is humbly Inscrib'd &c.
Fosters Store House
A Prospect of Hackelton's Cliff

and Weather, and the quick Return of Night, with its moist Vapours, we cannot be at a Loss to conceive how these otherwise hot and sultry Regions, near the Equator, are by this Means rendered not only habitable, but pleasant; and, as the Inhabitants are not liable to sudden Changes from Heat to Cold, they not only enjoy a great Share of Health, but likewise live to an advanced old Age.

The Island proved to be healthy from the Longevity of some of the Inhabitants.

There are now alive (and most of them in a tolerable good State of Health) within Three Miles round, in a Country-place, Part in *St. Lucy's* Parish, and Part in *St. Peter's*, Eleven Women, and One Man, whose Age, added together, make a Thousand and Five Years. We have likewise, in the Island, Three Persons, whose Ages, being added together, amount, by the best Computation that can be made, to near Three hundred Years.

Piso, in his Account of *Brasil*, says, that several of the Natives there live, some to an Hundred, others to an Hundred and Fifty Years: Yet I beg Leave to dissent, with regard to the Certainty of it, from that great Man, however a Man of Veracity in other respects; for neither the *Indians* upon the Continent of *America* (at least the different Nations that I knew), nor the native Negroes of *Africa* (as far as I could learn by the most strict Inquiry from such of them as have been naturalized in these Colonies), have any certain Method of reckoning their Years; nor of the latter is there one of an Hundred that can give you any tolerable room to guess at their Age, unless something very remarkable happened when they were young, to leave a lasting Impression upon their Minds.

When I examined a very old Person about her Age, all that she could remember was, that she was a very lusty Girl when the great Hurricane happened.

Hurricanes, their Force and Effect.

This indeed was an Æra remarkable enough; for of all Storms called Hurricanes, which, in Truth, are inexpressibly violent (though not so frequent in their Returns here, as it is imagined in *Europe*), that which happened in *Barbados*, on the 31st of *August* 1675. was not inferior, in its deplorable Consequence, to the Earthquake which happened at *Jamaica* (27); for it left neither House nor Tree standing, except the few that were sheltered by some neighbouring Hill or Clift.

Some Hours before the Storm began, the Heaven was overcast with thick Clouds, of a black reddish Hue; the Air calm, but sultry; and the more it lost its expansive Force (which before buoyed up the watry Particles in a dispersed State), the closer the Clouds condensed, and the blacker they appeared.

H The

(27) In the terrible Earthquake at *Jamaica*, which happened on the 7th of *June* 1692. the Sand in the Streets moved in Heaps, like the Waves of the Sea; then on a sudden the Earth opened in several Places, and swallowed in whole Rows of Houses with their Inhabitants. The remaining few knew not whither to fly; for Destruction was on every Side. As the Earth opened and closed with a very quick Motion, several Hundreds of the Inhabitants were absorbed alive; others jammed in the close Fissures, some up to their Necks, some to their Waists; and others had their Legs and Thighs broken in these Yawnings of the Earth, which were so numerous, that a Gentleman reckoned above Two hundred open in a very short time. The Sea was in such strong convulsive Motions, that in almost an Instant it receded back near a Mile, and as quick returned to its former Bounds. A large Mountain, not far from *Port Morant*, was quite swallowed up, and the Place where it stood is now become a large Lake of about Four Leagues over. *Vide* Phil. Trans. Vol. II. p. 411.

The Afternoon, when the Violence of the Storm began, the Wind was high and varying, almoſt in an Inſtant, to every (28) Point of the Compaſs; but ſettled chiefly at North, being attended with dreadful Rain, Thunder, and Lightning.

The Sea, where not guarded with high Clifts, overflowed its Banks, in ſome Places, above an Hundred Yards.

The Day-light, while it laſted, as *Milton* expreſſes it,

Serv'd only to diſcover Sights of Woe;

for there was nothing to be ſeen but one rueful Spectacle of almoſt univerſal Ruin.

If one looked in to the Land, *Behold, Darkneſs, and Sorrow, and the Light was darkened in the Heavens thereof.*

The Night, as it came on, was uſhered in with an almoſt continued rumbling Noiſe in the Air, with the Increaſe of Wind, Rain, Thunder, and Lightning; eſpecially the latter, which now with redoubled Force, darted not with its uſual ſhort-lived Flaſhes, but in rapid Flames, skimming over the Surface of the Earth, as well as mounting to the upper Regions.

The next (29) Morning, when the Storm was abated, the whole Iſland afforded a lively, but terrible, Idea of the Tenth *Egyptian* Plague of old; for there was ſcarce a Houſe but lamented one dead in it, or in general ſomething equal, or worſe.

Several Families were intirely buried in the Ruins of their Houſes; and there were few that eſcaped, but with the Loſs either of ſome Relation, Friend, or Acquaintance.

Theſe,

(28) There was not an Houſe left in *Speights Town*, whoſe Roof was not blown off. Major *Streate* lived above the Town, at an Eſtate, now in the Poſſeſſion of Mr. *Andrew Derry*, whoſe Roof being firſt blowed off, the Violence of the Wind was ſo great, that he and his Bride, up one Pair of Stairs (being married but that Evening), were carried from their Bed by the Force of the Wind, and thrown into a prickly Pimploe-Hedge; in which uncomfortable Situation, they were both found next Morning, unable to aſſiſt themſelves. Likewiſe one *Humphry Waterman*, then an Infant, was found as ſoon as the Storm was over, with his Arm broken, and in that Condition ſucking his dead Mother, killed by the Lightning or Storm.

There are many Inſtances that might be given of the great and deſtructive Force of Storms; eſpecially that which happened in the Reign of *Cambyſes*, which was ſo violent, that it raiſed the Sand in the Deſart between *Egypt* and *Ethiopia*, ſo as to cover and deſtroy a whole Army, conſiſting of Fifty thouſand Men, which were intended againſt the *Ethiopians*.

In the above-mentioned Hurricane in this Iſland, there were Seventeen Perſons killed in one Houſe, in that Part of the Iſland called *Scotland*. I have theſe Informations from creditable Perſons now alive in the Iſland, and who were, at the time of the Hurricane, of ſufficient Age both to remember, and make their Remarks on, the Violence of the Storm, and the Occurrences that then happened.

(29) It is here reported, that, during this Hurricane, one Mrs. *Groome*, a Midwife, in St. *Philip's* Pariſh, living upon the Eſtate of Mr. *Thomas Hardin*, endeavouring to go from her Dwelling to an Out-houſe, was ſnatched up by a Whirlwind, and carried through the Air to a great Diſtance from Home, where ſhe was found many Hours afterwards graſping the Roots of a large Tree newly blown up.

Thoſe here, who believe this Relation, urge, in favour of their Belief, that there are many well-atteſted Inſtances of as great Storms, which have been frequent at *Rome*, at *Capua*, at *Lavinium*, and ſeveral other Places of *Italy*. See *Diodorus Siculus*, Vol. III. Lib. XII.

However, I leave the Reader at Liberty to judge for himſelf, whether ſuch an Inſtance is conſiſtent with Probability, and whether any ſimilar Inſtances, from good Authority, be brought to countenance, if not confirm it?

Theſe, and ſuch-like Storms, when compared with the moſt remarkable ones, deſcribed by Heathen Authors, either in Proſe or Verſe (30), are infinitely more ſurpriſing and dreadful; for, when the Almighty, in theſe terrible Viſitations, *rideth upon the Wings of the Winds*, when he *maketh the Darkneſs his Pavilion*, then it may be ſaid without a Figure, that he *taketh up the Mountains in a Balance, and the great Deep as the Drops of a Bucket:* Not the Tops of *Rhodope*, or the *Ceraunian* Hills, alone are torn; it is not theſe alone that skip like Lambs, nor is it the River *Jordan* only that flies back, but Hills upon Hills ſink into Valleys, and the Bottom of the great Deep uncovers, and ſtarts into Land, at his angry Nod. The Foundations of the Earth ſhake, and the Pillars of Heaven are aſtoniſhed, at his Rebuke. *Lo! theſe are but Part of his Ways. How little a Portion is heard of him!*

The ſudden ſhifting of the Wind, in theſe and ſuch-like Hurricanes, to the ſeveral Points of the Compaſs, may be juſtly compared to *Virgil*'s beautiful Deſcription of a Storm, in the following Words:

- - - - - - - Ac venti, velut agmine facto,
Qua data porta, ruunt, & terras turbine perſlant.
Incubuere mari, totumq; a ſedibus imis
Una Euruſq; Notuſq; ruunt, creberq; procellis
Africus; & vaſtos volvunt ad littora fluctus. Æn. I. 86.

The huddled and tumultuous Band of Winds
Sweeps o'er the Earth, and a wide Portal finds:
Then broods upon the Deep; and in their Turns
South, Eaſt, and Weſt, with eddying Fury burns: They

(30) As the moſt celebrated Deſcription of a Storm is that of *Virgil*'s, it will not perhaps be improper to inſert it in his own Words: G. I. 322.

Sæpe etiam immenſum cœlo venit agmen aquarum,
Et fœdam glomerant tempeſtatem imbribus atris
Collectæ ex alto nubes: ruit arduus æther,
Et pluvià ingenti ſata læta, boumq; labores
Diluit: implentur foſſæ, & cava flumina creſcunt
Cum ſonitu, fervetq; fretis ſpirantibus æquor.
Ipſe Pater, media nimborum in nocte, coruſca
Fulmina molitur dextra; quo maxima motu
Terra tremit; fugère feræ, & mortalia corda
Per gentes humilis ſtravit pavor: Ille flagranti
Aut Atho, aut Rhodopen, aut alta Ceraunia telo
Dejicit: ingeminant Auſtri, & denſiſſimus imber:
Nunc nemora ingenti vento, nunc littora plangunt.

Oft from above deſcend a Troop of Floods:
Oft gather from the Deep the thick'ning Clouds;
Down ruſh the Skies, and with impetuous Rain
Waſh out the Ox's Toil, and ſweep away the Grain.
The Dikes are fill'd; no Bounds the Torrent keep;
And with the breathing Surges boils the Deep.
Amidſt a Night of Clouds, his glitt'ring Fire,
And rattling Thunder, hurls th'eternal Sire.
Far ſhakes the Earth; Beaſts fly; and mortal Hearts
Pale Fear dejects: He, with refulgent Darts,
Or *Rhodope*, or *Atho*'s lofty Crown,
Or ſteep *Ceraunia*'s Clifts, ſtrikes headlong down.
The Rains condenſe; more furious *Auſter* roars:
Now with vaſt Winds the Woods, now laſhes he the Shores.

They rouse the lowest Waves, that slept before,
And heave huge Billows tow'rds the lab'ring Shore.

As these quick Transitions of the Wind to the several cardinal Points, is a notorious Matter of Fact, with respect to Hurricanes, how inexcuseable must those be, who, without any Inquiry, ridicule the Description of a Storm in the Book of *Job*; because the Author makes the Storm to affect the Four Corners of the House, as if it were at once! which indeed is no less a true, than an elegant Description, of such a Storm.

It is observed, that the Glare of Lightnings is, in general, more expanded in these hot, than in Northern Climates. This is undoubtedly occasioned by the greater Rarefaction of the Air near the Equator. However, in Times of violent Hurricanes (if of any long Continuance), the Rapidity of the Wind and Rain presses on, and forces a prodigious Quantity of heavy Air in the room of the more rarefied, which is obliged to fly upwards, and to give way: Then the Atmosphere becomes as much condensed, during the Storm, as in any Northern Climate: And therefore the Lightning, being more closely pressed by so thick a Medium, would appear, and in reality be, more compact, than when less resisted by the common more rarefied Air. It is chiefly owing to this, with a small Allowance for the Consternation and Fear that the Inhabitants were then possessed with, that we can in some measure reasonably account for their describing the then Lightnings to be so very uncommon to them as to appear like so many Wedges or Bolts of Fire in the Air.

I am of Opinion, that all Hurricanes begin in variable Climates; otherwise we should not in all Probability have here any such Storms at all; for as the Air is very much rarefied near the Equator, its Particles are at a great Distance from one another, and consequently act with a less Force and Struggle, than when nearly compressed between a thicker Medium. This *(cæteris paribus)* may be evidently seen by the great Power of Air pent up in a Wind-Gun, which will force Leaden Bullets through an Half-inch Deal-board; so that where-ever the Particles of Air are most compressed together, there the greater will their Struggle be to expand, or dilate. Hence all such variable Climates will be more liable to Storms. We may likewise add, that the nearer any of our *West-India* Islands is to the main Land, whose Mountains near the Shore are very high, if the Storm blows upon the Island from the opposite Point; for Instance, If it comes from the North-east; and the adjacent Continent, opposite to that Point, rises with very high Lands (the Storm meeting with this Resistance, and being still violently pressed against the Sides of these Hills); the remaining Particles of rarefied Air will, according to their Nature, mount upward, leaving the lowermost still more condensed; and as every Particle of Air hath an elastic Quality, the greater the Number is of these that are compressed together, their united Force will

will with greater Power recoil; ſo that the Storm, being thus ſtopped in its Courſe, will more violently blow, as well as laſt longer.

Some ſlight Earthquakes have been likewiſe felt in this Iſland; but probably not ſo much here as in the neighbouring Iſlands of *St. Vincent's* and *St. Lucy's*, where there are burning Mountains, that have at times vomited Fire (tho' not of late Years); and where there is a great Quantity of crude Brimſtone to be found in the Neighbourhood of their Volcano's. The Hurricanes are much leſs frequent of late Years; nor hath there appeared any thing like one in this Iſland, ſince the Year (31) 1731. which blew from the Eaſt and North-eaſt.

All the Ships were driven out of the Harbour; but there was not much Damage received, except on Shore, where ſeveral Houſes were thrown down, and ſome large Trees torn up by the Roots.

From theſe loud ones, let us take a View of the more ſilent Strokes of the Almighty's Hand, whereby he nevertheleſs aſſerts his Kingdom over all the World, I mean the Diſeaſes peculiar to this and ſome neighbouring Iſlands.

(31) The Violence of this Storm was ſuch, that it covered, near *Bridge Town*, a Shoal at a ſmall Diſtance from the Land, oppoſite to Colonel *Hilary Rowe*'s and Mr. *Waldron*'s Houſes, of near Two hundred Yards long, Twenty broad, with a Bank of Stones Four Feet thick: Theſe adventitious Stones have been ſince partly waſhed off by high Tides, and partly carried away for Ballaſt, or to be burnt into Lime by the Inhabitants.

The End of the FIRST BOOK.

THE

NATURAL HISTORY

OF THE

Iſland of *BARBADOS*.

BOOK II.

Of the Diſeaſes peculiar to this and the neighbouring Iſlands.

IT is the Opinion of many great Men, that the Almighty never ſhews his Power by inflicting Diſeaſes, or even great Inconveniencies, peculiar to any certain Part of the World, but at the ſame time he diſplays forth his Goodneſs, in providing ſome extraordinary Remedies. If *Egypt* is often denied refreſhing (1) Rains, its Dews deſcend not only, like thoſe of *Hermon*, upon the thirſty Plains, but the wide-extended *Nile* alſo is commanded more plenteouſly to ſupply that Bleſſing from his copious Urn. When he hath thus fertilized the Earth, his Waters recede within their former Bounds, and return not again till the ſtated Periods, to cover the Land; ſo that if this Part of the World is ſometimes viſited with Fevers far more acute than any known in the Northern Parts (the Plague excepted), I make no doubt but indulgent Providence

(1) This muſt be underſtood chiefly of *Upper Egypt*.

Providence hath provided Plants, Minerals, &c. in the same Climate, which, upon a thorough Search into their Qualities, are capable of affording not only great Relief, but also most effectual and specific Remedies.

That they are not already found, is rather an Argument, that we have not been sufficiently inquisitive, than that there are no such Plants endued with these Virtues.

The Heat of the Climates sublimates the Juices of Plants.

The several Discoveries that have enriched the *Materia Medica* plainly shew, that the Virtues of Plants and Minerals are inexhaustible. And we may likewise observe, that the Heat of the Sun in these hot Climates is so intense (especially upon those Plants which grow sheltered from the Wind), that it sublimates their Juices, Salts, and Spirits, to a far greater Degree of Perfection, than a chemical Fire, by its inconstant Heat, can possibly effect.

Hence it is, that the most valuable Gums and Balsams are brought from hot Climates. Witness that of *Gilead* and *Peru*.

It is observed by the ablest Practitioners in Physic of this Island, that Peripneumonies and Pleurisies are almost the constant Attendants of the Change of Wind, if of any Continuance, from the true Trade-Wind, especially to a sultry South-west Wind.

The Difference of Purity between this and our Trade-Wind, is easily discovered; for the South-west Wind blows from a very hot moist Part of the Continent of *America*, which is not above 254 Miles distant from this Island: And as this is by far hotter than our usual Air, it adds too great a Relaxation to the Animal System.

The Trade-Winds constitutional to the Inhabitants.

This, with the different Degrees of Circulation in all the Juices (which such a Difference in the Air must cause), gives Birth to various Diseases: Whereas the Trade-Wind, by its Frequency, or rather by its Constancy, is not only constitutional to the Inhabitants, but it is in itself purer than the other, because it blows upon the Island at East-north-east; and as the nearest Part of the Continent, upon that Point, is 3127 Miles from us, the Air must be far colder by passing over so much more Water, than the South-west Wind, and consequently more wholsome.

The learned Dr. *Mead*, in his Treatise *de Peste*, observes, that it was the Opinion of *Hippocrates*, that the Constitution of the Air that preceded pestilential Fevers, was mixed with great Heats, much Rains, and Southerly Winds. And *Galen* takes notice, "that no other than a moist "Air brings the Plague." *Lucretius* is of the same Opinion, in his admirable Description of the Plague of *Athens*. "These Diseases, says he, either "come from the Air, or rise from the Earth."

> - - - *Ubi putrorem humida nacta est,*
> *Intempestivis pluviisq; & solibus icta.*
>
> Lucret. Lib. VI. Ver. 1098.

In

In or about the Year 1741. a great Number of Dogs ran mad; and I observed, that the Temperament of the Air, for many Months before, was very hot. In this Season, a great Number of Cattle, being bit by these Dogs, ran mad likewise.

I cannot here omit inserting a very lucky unexpected Cure of a mad Cow, belonging to *Hurdiss Jordan*, Esq; in *St. Lucy's* Parish. When found to be mad, she was confined; but as the usual Symptoms, attending that terrible Distemper, appeared more and more, and she growing to extreme Madness, the Owner determined to destroy her; but as she was a favourite Beast, her Doom was respited till he had tried the following Experiment.

Having thrown her down upon a Dunghil, he directed his Slaves to keep, by Force, her Mouth open, whilst he poured by Degrees down her Throat a large Pailful of cold Water. In a short time after, she began to feed, and in about Twenty-four Hours drank Water as usual, fully recovered, and remained so.

As this is a Matter of Fact, sufficiently attested by that Gentleman, and his Family, I make no Apology for inserting it: Indeed I should have been inexcusable, if I had omitted so extraordinary and surprising a Circumstance; for who knows but indulgent Providence may at last point out some Remedy against so dreadful a Calamity, and that the same Method may be of Service towards the Cure of the human Species?

It appears to me, at least, probable; for even Nature, in Instances not intirely dissimilar, points out something like this: Thus, when the Stomach is overloaded, it generally makes many Efforts to relieve itself by Vomits, which, when assisted by Art, prove often an effectual Remedy: And as we find, that the miserable Patient, when afflicted with this Disorder, hath an ardent Desire of Water, though the Nature of the Distemper is such, that it will not permit him to drink; therefore, if, in this Case, a considerable Quantity of Water were forcibly poured down his Throat, perhaps it would be of great Service; for as the Poison is of a very hot Nature, which appears by the violent Thirst it causes, it is more than probable, that such a Quantity of cold Water, mixing with the *Virus*, would, at least, abate its Force, till Sudorifics, or other Medicines, had time to expel and throw off the Poison by Perspiration, or otherwise: That some Poisons, of an hot Nature, act less vigorously in Cold, than when assisted with Heat, is evident, from the more dangerous Consequence of the Bite of Scorpions and Vipers in *Italy*, and elsewhere, in an hot Summer Season, than in moderate cold Weather.

As the *English* Physicians are no less remarkable for their Humanity, than Skill and Judgment, it is to be wished, that these ingenious Gentlemen would (in Pity even to the dumb Creation, as well as to their Fellow-Creatures) try this Experiment first on Beasts; which if successful, it might likewise be of Service to the human Species, under such deplorable Circumstances.

From this ſeeming Digreſſion, let us return to obſerve, that the Dry Belly-ach (ſo called from its affecting that Part of the Body with great Coſtiveneſs and Pain) was formerly much more frequent and fatal, than it hath been of late. Diſtillers of Rum, and Boilers of Sugar, and Overſeers, were chiefly ſubject to it; the firſt (who are generally of the poorer Sort), from immoderately drinking new hot Rum; the ſecond and third, from taking Cold, after ſweating in hot Boiling-houſes, and drinking very ſtrong Punch, or almoſt as ſtrong (2) Cowow, or (3) Mobby.

This Diſtemper is likewiſe ſaid to ariſe from Grief and Trouble of Mind; and ſometimes it hath been thought to be epidemical.

The Belly-ach, its Symptoms.

After it hath for ſome Days afflicted the Patient with intolerable Pains in the Belly, Coſtiveneſs, with voiding little Urine, and that like Mum, the Breath ſtinking very offenſively, the Pains are frequently tranſlated to the Limbs, attended ſometimes with an intolerable Pricking and Burning in the Feet; at other times, great Pains in the Stomach ariſe; and then the Patient hath Epileptic Fits, which often relieve the Pains, but weaken and impair the Uſe of the Limbs.

When the Diſeaſe becomes inveterate, the Patients grow hoarſe, and loſe their Voice, coveting Anodynes, and ſtrong Drink, to that Degree, that the common Rum is much too weak; and they drink even *Hungary* Water, or put red Pepper into Rum, to augment its Heat. When they loſe the Uſe of their Limbs, they commonly have profuſe Sweats.

The former Method of Cure.

The Method of Cure formerly was to purge the Patients both with Cathartics and Clyſters: From this Practice the unhappy Creatures were ſubject to loſe the Uſe of their Limbs.

However, this prevailed until about the Year 1700. or ſomewhat before, when the Phyſicians began to make uſe of Anodynes with better Succeſs; which hath been their Way of proceeding ever ſince.

The preſent Method of Cure.

Their preſent Method of Cure is univerſally followed, being both ſafe and eaſy, and the Patients run no Danger of either Life or Limbs: For, as ſoon as Dr. *Warren* found, that it was a convulſive Conſtriction, chiefly in the Ilion, he judged that Purges and ſharp Clyſters muſt be attended with bad Succeſs, and Emetics, at leaſt, uſeleſs, if not hurtful; from whence he concluded, that Anodynes mixed with Anti-hyſterics would be the likelieſt Means of Relief: He therefore purſued that Method, as neceſſary to be followed, for a few Days, till the Pains of the Belly intirely ceaſed for the Space of Twenty-four Hours: Then he concluded, that the ſpaſmodic Conſtrictions were over; at which time he ordered ſome emollient Clyſter, and then gave a gentle Purge of ſome of the bitter Pills, which generally perfected the Cure, only taking ſome Anodyne for a few Nights ſucceſſively afterwards.

Dr. *Warren*'s Method of Cure.

The bitter Pills are thought propereſt, becauſe the Bile in this Diſtemper ſhews itſelf to be deficient in its due Secretion, or its Quality.

There

(2) A Drink made with the Scummings of the boiled Cane Juice, mixed with Water, and fermented; and then drank.

(3) Mobby is a Drink made with pounded Potatoes, and Water fermented with Sugar or Molaſſes.

There are other Kinds of Colics, which often afflict the Inhabitants; likewise the Yellow-Jaundice, and Hysteric Disorders: But as these are not peculiar to this Climate alone, it cannot be expected, that I should enter so far into the Province of the Physician, as to treat medicinally of them.

As Diarrhœas, however, and Dysenteries, make no small Havock among the poorer Inhabitants of the Island; and as the Cause and Method of Cure, in some measure, differ from the same Distemper in Northern Climates, being partly constitutional; they deserve to be mentioned: Nor can this be better done, than in the Words of Dr. *Towne*.

Fluxes are very common in the *West-Indies*, but more especially in rainy Seasons; and may be imputed chiefly to the Negligence of those who either too unwarily or unavoidably expose themselves to the Injuries of the wet Weather; by which means Perspiration being interrupted, the thin Part of the Blood, which should have been exhaled through the Pores of the Skin, is thrown upon the Bowels, and thence discharged in loose Stools. This plainly appears from the great Number of Negroes, and the poorer Sort of White People, who in these Seasons are much more afflicted with this Distemper, than such whose Conduct of Life does not subject them to the like Inconveniencies. Fluxes very common in rainy Seasons.

Besides catching Cold, there are other antecedent Causes of a Diarrhœa; the principal of which are, "An immoderate Use of crude fugacious Fruits, " unwholsome Food, and Meats of difficult Digestion; all which, by stimu- " lating the Guts, will likewise occasion a Diarrhœa.

" When these last-mentioned Causes occur, with a damp rainy Season, " the Bowels will not only be loaded with the thin Juices, which ought " to pass off by Perspiration, but they will also, by reason of the Stimulus " lodged in them, be continually solicited to expel their Contents more " frequently, and of a thinner Consistence, than usual. A diminishing " Perspiration will likewise contribute towards inlarging the Orifices of the " Hepatic and Pancreatic Ducts; and, on this Account, the Secretions of " their respective Juices will be more plentifully made into the Intestines: " And hence we have an additional Cause of Looseness."

These Circumstances, I think, are sufficient to account for every Species of the Diarrhœa; and, when we are once fully ascertained of the Cause, we need not be much at a Loss what Method of Cure ought to be pursued in each Species.

Fluxes have been here very often neglected in the Beginning, from an Opinion, that they are salubrious, and of Service to the Constitution, by affording an Outlet to some offending Matter, which, if retained, would have proved prejudicial.

This Remark may in some Cases be very true; but it is not to be confided in without great Caution in the *West-Indies*, where a simple Flux frequently rises up into an obstinate Dysentery in Three or Four Days; and, when the Diarrhœa is suffered to continue any time, it too commonly terminates

terminates in a Leucophlematia, or Dropſy; to which Diſeaſes People in theſe Parts of the World are exceedingly diſpoſed.

But as a Diarrhœa is ſometimes truly critical, and contributes a great deal towards the Cure of other Diſtempers, ſuch a Diarrhœa ought by no means to be ſtopped, ſo long as the Strength of the Patient can ſupport him under it.

The Matter of a Diarrhœa in theſe Parts is frequently ſo ſharp and corroſive, as not only to carry off the mucous Subſtance of the Inteſtines, but alſo in a few Days to abrade and tear away their villous Coat, and excoriate the Mouths of the Blood-veſſels. The ſtimulating Matter ſtill remaining, the Flux is continued; and, the Orifices of the Blood-veſſels being opened at the ſame time, the Flux muſt unavoidably become a Bloody one.

Likewiſe ſultry Weather, and hot ſpirituous Liquors, as well as acrid pungent Food, are capable of producing a Diarrhœa; and as they greatly rareſy the Blood, this Rarefaction, ſuperadded to the Looſeneſs, gives us a ſatisfactory Idea of their being the Cauſe of a Dyſentery.

This is confirmed, by obſerving how much this Diſeaſe rages among the White Servants, as well as Negroes, in our Plantations; which ſort of People are much addicted to debauch in Spirits, and Punch made exceedingly ſtrong with new Rum, very acid with the Juice of Limes, fermented with coarſe Sugar.

The almoſt unanimous Conſent of Phyſicians, as well as Proofs drawn from Practice, have ſo fully demonſtrated the Benefit received from Blood-letting, that it is ſufficient only to mention it, to remind Practitioners of the Neceſſity there is not to omit it in the Beginning of a Dyſentery.

The *Indian* Root, or (4) Ipecacuanha, hath juſtly eſtabliſhed its Reputation in the Cure of this Diſtemper, both in thoſe Parts of the World from whence it was originally brought, where Fluxes are (5) endemic, and in all other Places where it hath been uſed on thoſe Occaſions.

But perhaps the Reaſon why it is found to be ſuperior to the reſt of the Emetic Tribe, was not at all conſidered before the learned and judicious Dr. *Freind* (6) gave us an Inſight into that Matter, unleſs we except one Paſſage in *Piſo* (7).

If the Ipecacuanha does not only exert its emetic Faculty, but likewiſe paſſeth through the Bowels in ſuch manner as to promote a Stool or two,

(4) Si quando evacuationibus locus, radicem vomitivam Ipecacuanha, exquiſitiſſimum naturæ munus, cæteris remediis præferre conducit. *Piſo*, p. 39.

(5) Affectus dyſentericus nimirum his terris eſt familiaris, ita ut perpetuo nobis ſporadica, & popularis. *Piſo* ibid.

(6) Radix Ipecacuanha præter vim vomitoriam, quam obtinet, uberrimum ſudorem excitare ſolet. Atq; in hoc, quantum ego conjectura aſſequi poſſum, præcipue conſiſtit egregia illa in dyſentericis affectibus virtus, quam ſibi præ aliis vomendi inſtrumentis vendicat. *Freind Comment. de Feb. p.* 40.

(7) Ad radicem Ipecacuanha confugiendum, tanquam ad ſacram ancoram, qua nullum præſtantius aut tutius, tum in hoc, tum in pleriſq; aliis, cum vel ſine ſanguine, fluxibus compeſcendis, natura excogitavit remedium. Quippe, præterquam quod tuto & efficaciter tenaciſſimos quoſq; humores per ipſam alvum, ſæpiſſime autem per vomitum ejiciat, & a parte affecta derivet, vim quoq; aſtrictivam poſt ſe reliquit. *Piſo*.

two, which often happens, it will be ſufficient that Night to give the Patient a Bolus made with Rhubarb and Diaſcordium; but if the Phyſician obſerves no ſuch Effect from the Vomit, then he generally gives half a Dram of Rhubarb in any convenient Vehicle, in order to carry off any Remains of the ſtimulating Matter which may adhere to the Guts; for many are the Occaſions which confirm the ill Conſequences of giving aſtringent Medicines in the Infancy of this Diſeaſe, when the Flux of Blood hath not been immoderate.

After the Bowels have been prepared in this manner, the Phyſicians have recourſe to Balſamics, Agglutinants, Aſtringents, and Opiates, which muſt be contrived in proportion to the Neceſſity of the Patient, the Violence of the Flux, and the Length of its Duration.

If the Sick be much debilitated by the Continuance of the Diſtemper, and the great Effuſion of Blood, a Cordial Draught is generally allowed him to ſupport and invigorate the languid Spirits, ſuch as Claret, or *Madera* Wine, burnt with Spices, or ſtrong Cinnamon-Water, diluted with the white Decoction.

The Food in this Caſe muſt be cooling, mucilaginous, and aſtringent. Panada, made with Caſada Bread, is with very good Reaſon commended by *Piſo*, who adviſeth it not only as a cooling healing Diet, but as an uſeful Ingredient in the Compoſition of Clyſters.

The ſame Author recommends the Jelly of white and red Guavas, Granadilla, Hog-plumb, *&c.* as proper Nouriſhment. Theſe are mentioned, becauſe they are always at hand in *Barbados*.

Unripe Plantains and Bananaes, roaſted, are alſo proper for theſe Patients; but, above all, a Milk Diet.

This Iſland is likewiſe ſubject to a very malignant Fever (though I believe in common with other Countries between the Tropics) now called the Yellow Fever. **The Yellow Fever.**

Dr. *Warren*, in his ingenious Treatiſe upon this Diſtemper, concludes it to be a Species of the Plague, and that the Infection was unhappily brought to *Martinico* in Bales of Goods from *Marſeilles* in the Year 1721. though others, who have reſided much longer in the Iſland, are of a different Opinion, eſpecially Dr. *Gamble*, who remembers that it was very fatal here in the Year 1691. and that it was then called the New Diſtemper, and afterwards *Kendal's* Fever, the Peſtilential Fever, and the Bilious Fever.

The ſame Symptoms did not always appear in all Patients, nor alike in every Year when it viſited us.

It is moſt commonly rife and fatal in *May*, *June*, *July*, and *Auguſt*; and then moſtly among Strangers, though a great many of the Inhabitants, in the Year 1690. died of it, and a great many at different Periods ſince.

Symptoms of the Fever.

The Patient is commonly ſeized with a ſhivering Fit, as in an Ague, which laſts an Hour or two, more or leſs; and the Danger is gueſſed at, according to the Severity and Continuance of the Ague.

After the ſhivering Fit, a violent Fever comes on, with exceſſive Pains in the Head, Back, and Limbs, Loſs of Strength and Spirits, with great Dejection of Mind, inſatiable Thirſt and Reſtleſneſs, and ſometimes too with a Vomiting, attended with Pains in the Head, the Eyes being red, and that Redneſs in a few Days turning to a Yellowneſs.

If the Patient turns yellow ſoon, he hath ſcarce a Chance for Life; and the ſooner he does ſo, the worſe.

The Pain in the Head is often very great, when firſt ſeized with this Fever.

After ſome Days are paſt, this Pain abates, as well as the Fever; and the Patient falls into a breathing Sweat, and a temperate Heat, ſo that he appears to be better; but, on a narrow View, a Yellowneſs appears in his Eyes and Skin, and he is viſibly worſe.

A Yellowneſs appearing in the Eyes, a bad Symptom.

About this Time he ſometimes ſpits Blood, and that by Mouthfuls; as this continues, he grows cold, and his Pulſe abates, till at laſt it is quite gone; and the Patient becomes almoſt as cold as a Stone, and continues in that State with a compoſed ſedate Mind.

In this Condition he may perhaps live Twelve Hours without any ſenſible Pulſe or Heat, and then expire.

Such were the Symptoms and Progreſs of this Fever in the Year 1715.

Sometimes likewiſe the Patients burſt out with bleeding at the *Anus*, and ſoon after die; and ſometimes likewiſe at the Noſe, by which means they have been relieved; but when the Blood iſſues from thence but in few Drops, it is a bad Prognoſtic, and is generally the Harbinger of Death.

The Texture of the Blood.

In moſt of theſe Caſes, the Patients are generally hot and dry; the Blood taken from them very red, and ſcarce will coagulate; the Grume ſwimming upon the Surface of the Serum in a thin Leaf, having ſcarce any Conſiſtence.

The Patients have likewiſe often intolerable Pains in and about the Stomach: Sometimes with thoſe Pains they ſhall have a Livor, and the plain Marks of a Sphacelus ſhall poſſeſs the greateſt Part of the Abdomen before they die, particularly the Region of the Stomach and Liver.

It often alſo happens, that the ſick Perſon ſhall lie almoſt ſtupid; and, being asked how he does, ſay, He is very well; at other times he labours under the greateſt Agonies, and Fits of Groanings.

A looſe Tooth being drawn from a Perſon who had the Fever very ſeverely, there iſſued out from the Hole a great Quantity of black ſtinking Blood, which ſtill kept oozing till the third Day, on which the Patient died in great Agonies and Convulſions.

After Death the Corps of ſuch appear livid in ſome Parts or other; or elſe marked with peſtilential Spots, Carbuncles, or Buboes.

I am

I am of Opinion, that the Blood is from the Beginning full of putrid Alcaline Salts.

If this be the Case, I leave it (with Submission) to the Learned in Physic to consider, whether diluting Acids would not be of great Service (especially in the Beginning of this Disorder) to raise a Conflict, by their mingling with the Alcaline Salts; by which means a Stagnation in the Fluids would be, in a great measure, prevented, until Nature, with other seasonable Assistances, might have time to try her Efforts, and so recover at last.

Without this, or some Help like this, from the Physician, the Patient often dies in three Days time.

Such are the Symptoms, Progress, and Conclusion of this malignant Distemper: And though Dr. *Warren*, in his learned Treatise, is of Opinion, that it can be cured by Diaphoretics and Sudorifics; yet, as this Disease appears in so many various Shapes, these have very often failed.

The next Disease worthy our Notice is the Small-pox; for we are seldom free from it in some Part of the Island or other: However, since Inoculation hath been introduced among us, our Loss by that Distemper hath hitherto been very small. This Island seldom free from the Small-pox.

It hath been observed, that this Distemper proves, in general, very fatal to such of the Inhabitants of the *West-Indies* as receive the Infection in *England*, or in any other Northern Climate, especially if they are taken ill soon after their Arrival at it.

This, I believe, will not be difficult to be accounted for, when we consider, that the Inhabitants of warm Climates do almost continually strongly perspire, by which means their Pores are more relaxed and dilated; and consequently, if this Infection (as surely it may) can be communicated by the Pores, the more open these are, the more capable they are of receiving a deeper Degree of Infection, than those Persons, who every Winter have their external Vessels compressed and contracted by the cold Air. This Distemper often fatal to *West-Indians*, if taken ill in cold Climates.

However, we are not from hence to conclude, that the Pores of such are never to be reconciled to cold Climates; for those of the Skin, as well as the Juices of the Body, will in time become adapted to the different Climates we are in.

It is remarkable, that the Chicken-pox visits this Island at set periodical Times; for it hath been observed by Dr. *Gamble*, a very old and ingenious Physician, as well as a Person of undoubted Veracity, that the following were its periodical Returns within his Observation; *viz.* in the Year 1692-3. 1711. 1728. 1746-7. which is at the Distance of about Eighteen Years. The Measles a periodical Distemper.

The Leprosy is another Distemper, which some unhappy Persons are afflicted with; nor is it less surprising to Strangers, than a Concern to the most thinking Part of the Inhabitants, that public Care hath not been taken to keep the Clean from the Unclean. The Leprosy.

Among

Among the *Jews*, the Law concerning Leprosy was so strictly observed, that they deposed *Uzziah*, tho' a Prince, and thrust him out of the Temple, and confined him to a separate Apartment during Life.

We cannot trace this Distemper up to its Origin, as far as it relates to the *West-India* Islands; but its first Appearance here was about Sixty Years ago. It hath spread very much within these Twenty Years, and more lately among White Inhabitants, as well as Black. Children have been often seized with it, without any known Cause, either in themselves, or their Parents: However, it is, in general, thought to be hereditary, at least, in the Male Line; which the following Instance will, in some measure, evince.

The Leprosy propagated chiefly in the Male Line.

A certain Negro Woman, in *St. James's* Parish, cohabiting with a leprous Negro Man, had Two Children by him, both leprous, though she herself did not receive the Infection. After some Years she turned him off, and cohabited with another Black, who was free from this Disorder; by him likewise she had Children, but neither of them in the least infected with the Leprosy. After a Quarrel with this last pretended Husband, she returned to her former leprous Husband, and had this Second time by him several Children; who, as they grew up, proved leprous also, tho' she herself received not the Infection this time.

From hence we may perhaps learn to know how emphatically, and with what great Propriety of Expression, the inspired Prophet passed Sentence upon *Gehazi*, saying, *The Leprosy of* Naaman *the* Syrian *shall cleave to thee, and thy Seed for ever.*

Its Symptoms.

The first Symptom of this loathsome Disease, in this Part of the World, is a permanent Swelling on the Tips of the Ears, and the Falling off of the Hairs of the Eye-brows; then the Face appears unctuous, shining, and full of protuberant superficial Spots of a brown Copper Colour; the Lips and Nose generally much swelled, the Fingers and Toes distorted, and at last ulcerated; the Infection creeping from Joint to Joint, till it hath corroded all the Fingers and Toes. These Ulcers never kindly digest; however, there issues from them a thin corroding Ichor.

Many are the miserable Objects, that are daily seen afflicted with, and labouring under, this great Misfortune.

The Elephantiasis.

The Elephantiasis is likewise a Disease very common among the Blacks, and hitherto incurable.

The Seat of this Distemper is the Legs and Feet. It happens most commonly after long Illnesses, especially severe acute Fevers, or long Intermittents, or other tedious Distempers, or Surfeits, occasioned by too violent Exercise.

Its Symptoms.

In the Beginning the Patient appears cachectical and emaciated, and soon after the vitiated Humours subside, generally in one Leg, sometimes in both Legs and Feet: These become tumefied, and the Veins distended with varicose Swellings, which are very apparent, from the Knees to the Extremities of the Toes; "then the Skin begins to grow rugged and "unequal,

" unequal; its vascular and glandulous Compages is inlarged; and a scaly " Substance, with a sort of Chaps and Fissures in the Interstices, appears " upon its Surface. These seeming Scales do not dry up and fall off; but " are daily protruded forward, and stretched in their Dimensions, till the " Leg is inlarged to an enormous Bulk; so that in the Size, Shape, and " all other external Appearance, it minutely represents the Leg of an " Elephant, from whence the Disease receives its Denomination."

But, notwithstanding that this scaly Coat appears to be hard, callous, and insensible; yet if it be touched ever so superficially with a Lancet, the Blood will freely ouze out; and, if the Epidermis, which affords this monstrous Appearance, be pared off to the Thickness of the Scarf-skin in those Parts, an Infinity of Orifices of the Blood-vessels will present themselves to the Eye, when assisted with a Microscope.

" Though the Limb continues to proceed to this inordinate Magnitude, " yet the Appetite of the Negro remains good, his Digestion strong, and " his Secretions regular; nor is he sensible of any other Inconveniency, " than the Burden of carrying such a Load of a Leg along with him."

In this Condition, several have been known to live Twenty Years, and have performed chearfully all the Duties of their Servitude, which were consistent with such disproportionate Limbs.

Amputation of the diseased Leg hath been performed many times, but has always failed of a Cure; for the Distemper constantly takes Possession of the remaining Leg.

As for other cutaneous Disorders, we are not more afflicted with them than *Europeans*, unless such as are occasioned by the *Guiney* Worms and Ciegoes, called here *Chiggers*. The *Guiney* Worms.

The former are seldom or never known among the White Inhabitants; nor often among our native Slaves; but chiefly among new Negroes brought hither from the Coasts of *Guiney*.

These Worms are generally bred in Ponds of stagnant fresh Waters; and it is supposed, that they enter oftener in by the Pores of the Skin, than by drinking the Water: For those who most frequently bathe in such Ponds, are oftenest troubled with them. They are exceeding long, in respect to their great Exility and Thinness; and are easily observed, when some of that Water is mingled in a Glass with some other more transparent Water.

The first Day or two they cannot be clearly seen, being as pellucid as the Water itself; but soon afterwards they grow so opaque as to be discernible, even by the naked Eye.

Dr. *Gamble* had one of these, which measured an Ell long; it exactly resembled a waxed white Thread. Their Shape.

They move very quick under the Skin, along the *Membrana adiposa*; and what is one Day seen in the Breast, or Belly, shall often, in a Day or two, be found in a distant Part of the Body, perhaps the Thigh, the Leg, or under the Ham.

However, theſe are eaſily cured, by a Method ſometime ſince found out.

The Method of deſtroying them.

Take, for Example, one Ounce of Garlick, one of Black Pepper, pulverized, and one Ounce of the Flour of Brimſtone; mix theſe well together in a Quart of Rum; and for three or four Mornings ſucceſſively give a Wine-Glaſsful of this Infuſion to the Perſon afflicted; and if he hath a Thouſand about him, each will contract itſelf in a Coil, and die, and then fall off in the Form of a Boil from the Surface of the Skin.

As to Ciegoes, they will appear to the naked Eye like a ſmall Flea: Theſe are the continual Companions of Filth and Lazineſs, troubleſome chiefly to Negroes, and particularly to ſuch new Negroes as are brought hither from *Guiney*, whoſe Deſpondency, and careleſs Regard for Life, from the Uncertainty, or rather Certainty of what they are to undergo, makes Life, at that Juncture, irkſome to them; though a few Years Experience convinces them, that if there can be a Degree of Happineſs without Liberty, they enjoy a more certain Tenure, and a better Condition of Life, here, than in their native *Africa*.

Ciegoes how diſcovered.

This little Animal, vexatiouſly teazing, unleſs ſoon eradicated, generally faſtens and ſettles on the Toes, or ſome Part of the Feet. It is firſt diſcovered by a moderate Itching in the Part affected, afterwards with a throbbing itching Pain; and if not ſoon carefully picked out with a Pin or Needle, it will, in time, burrow or neſtle one Eighth of an Inch into the Fleſh, and there lay innumerable Quantities of Nits or Eggs, which, when hatched, become equally voracious, and eat the Feet into ſo many little Cells, like Honeycombs.

It is not likewiſe uncommon for theſe little Vermin to get into the Feet of People of the beſt Condition; but as they are ſoon taken out by their Slaves, it ſeldom proves to be of bad Conſequence: Tho' Strangers, not being ſenſible what occaſions the Itching, will ſometimes let them remain, till they become very (8) troubleſome, and perhaps require the Hand of the Surgeon.

Likewiſe the Body-Yaws, and the Running-Yaws, ſo common here, are Diſorders unknown in Northern Climates.

The Body-Yaws.

The Body-Yaws appear in many protuberant fleſhy Knobs, every way as large as a ſmall Thimble, emboſſing the Face, Breaſts, Arms, and other Parts of the Body. Theſe, in time, by the Uſe of Simples, hereafter to be mentioned, dry and fall off.

The Running, or the Wet, Yaws affect chiefly the Joints, eſpecially the Knees or Elbows, from whence continually diſtils a ſanious Humour.

This Diſtemper is thought to be tranſmitted in an hereditary Way from Parents, who have had the Venereal Diſeaſe, to their unhappy Children.

It is obſervable, that no ſmall Care, Skill, and Time, are requiſite to cure it.

O F

(8) *Caſtile* Soap and Lamp-Oil, boiled to a Conſiſtency, and applied hot to the Feet, is one of the beſt Methods of deſtroying the Ciegoes.

O F

WATERS, FOSSILS, *and* MINERALS.

Of WATERS.

IT being unneceſſary to inquire into the conſtituent Parts of Water in general, in a Treatiſe of this kind, I ſhall proceed to conſider only the Nature and Qualities of that which we are furniſhed with in this Iſland, from Springs, Rivers, Wells, and Ponds..

The moſt remarkable Springs of Water are the following: *Cole's Cave* Spring, Mr. *Colliton*'s, Mr. *Eſtwick*'s, Mr. *Brathwait*'s, the Reverend Mr. *Foſter*'s, Mr. *Whitaker*'s, *Pory* Spring, and that at *Codrington's College.*

The ſpecific Gravity of Fifteen cubical Inches of theſe reſpective Waters, as well as compared together with that of the moſt remarkable Spaws, is inſerted in the annexed Table.

The ſpecific Gravity of *Barbados* Water, compared with ſome remarkable Spaws.

As *Cole's Cave* Water hath been generally eſteemed (tho' not ſo in Fact the beſt Water in the Iſland, I have been very careful in the Examination of it. The Spring which iſſues ſo plentifully from the Side of the Cave, affords very tranſparent Water, having ſomething of a ſtyptic Quality; and it hath been obſerved, that bathing in it is of great Service in cutaneous Diſorders.

Among the above-mentioned Springs, that in the Eſtate of the Reverend Mr. *Foſter*, called *Belly-ach Hole*, affords excellent Water. This Spring riſing, as moſt of the reſt do, out of the Side of an Hill facing the Eaſt, hath a fine marly Bottom, mixed with Sand: The Water is very light, ſoft, and pure.

There

There is likewise on the same Estate, near the Mansion-house, another Spring, whose Water is heavier, and somewhat purgative.

Pory Spring, in *St. Thomas's* Parish, and another at Mr. *Estwick*'s, in *St. John's*, as well as Mr. *Whitaker*'s Spring, in *St. Michael's* Parish, afford good light Water: The latter, by its Vicinity to the Sea, is of great Service to supply the Navy, as well as other Vessels, in the adjacent Bay.

The chief Rivers in the Island.

The chief Rivers are *St. Joseph*'s River, another running through the Parish of *St. Andrew*'s, a Third taking its Rise from Mr. *Brathwait*'s Spring, and the Fourth near *Codrington's College*.

All the above-mentioned Streams, with some others of lesser Note, on this Side of the Island, fall into the Sea to the Eastward.

Wells and Ponds.

Our Springs and Rivers of any Note being thus taken briefly notice of; let us now proceed to consider our other Supplies, our Wells and Ponds.

The former are dug with great Expence through, generally, a soft marly Rock, intermixt sometimes with Veins of an harder Nature. These Wells are commonly in Depth from Fifteen to Fifty (9) Fathoms, and their respective Water (especially if the lower *Strata*, from whence it gushes, be Clay, Gravel, or Rock) is preferable, as to its Purity, to River-Water, the latter having generally several Impurities blown into it by the Wind, as well as contracted from the different Soils it passes through, whereas the Well-Waters are very limpid and light.

Mr. *Osborne*'s Well-Water proved to excel all other.

Among the Waters of many of these that I have examined, that from Mr. *Osborne*'s, in *St. Peter's* Parish, excels every other, whether of Wells, Springs, Rivers, or Ponds, in this Island. It is lighter by one Grain, in so small a Quantity as Sixteen cubical Inches, than the pure *German* Spaw Water. The Depth of this Well is not above Fifteen Fathoms, thro' a somewhat soft rocky *Stratum*, terminating in an hard firm Rock: An Hole being made in this, the Water gushes out from beneath; and the same being again stopped with a Plug, and the remaining Water drawn out, the Bottom of the Well may be, and often is, clean washed; and then the Water is permitted to gush out afresh.

I am of Opinion, that the Purity of this Water, above all other in the Island, is owing partly to the Situation of the Place: For, as the Well is dug thro' several rocky *Strata*, beginning in so very high Ground, that its Bottom is higher than the adjacent Plain; consequently it can receive no other Supply but Rain-Water, which (Snow-Water excepted) is the most simple of all others. And when this gradually penetrates thro' many thick *Laminæ* of porous Rocks, by this natural Distillation, as it runs, it refines; that which is purest pervading the closest *Strata* by the Minuteness of its Parts, whilst the grosser Particles, being arrested in their Passage, subside, and cleave to the Rocks.

Another

(9) It is not always that they come to Water, tho' they dig a great Depth. An Instance of this I have observed in the Estate of *Tobias Frere*, Esq; in the Parish of *Christ-Church*; where, after digging Thirty-five Fathoms, they found little or no Water.

Another Reason why Spring-Waters, at their Fountain-Heads, as well as Wells, are purer (especially in hot Climates) than the Waters of Rivers or Ponds, is, because the latter, whether in its current Stream, or in Reservoirs, is exposed to the Heat of the Sun, whose exhaling Power is here so great, that it raises up the lightest and simplest, that is, the best Parts of it in Vapours, leaving the Remainder less pure.

What is here called Pond-Water, is Rain-Water settled in artificial or natural Basins, which, in a Country so sparingly furnished with Springs, or Rivers, are of great Service to Man and Beast.

These Reservoirs are generally dug near a Descent, that they may be supplied with Water as often as it rains, which it generally does in greater Quantity, in the Months of *June*, *July*, and *August*; but tho' we call this Time of the Year the Rainy Season, yet the periodical Return of wet Weather in those Months is not near so certain here, as it is in some Parts of *Africa*. The Rains likewise in those hot Months, when they are most wanted to cool the Atmosphere, fall at such times, by reason of the Rarefaction of the Air, in very large Drops, like those sudden Summer Showers in *England*. We find the above-mentioned Conveniencies of Reservoirs made use of in *Palestine*, as early as the Patriarchal Age.

Ponds or Reservoirs in Use among the Antients.

The Inhabitants of *St. Andrew's*, and some Part of *St. Joseph's* Parish, are plentifully supplied with fresh Water, by digging Holes in the Sand, from Ten Inches to Three Feet deep: These are almost instantaneously filled with fresh Water percolated thro' the Sand.

Some Parts of *Barbary* are in the like manner supplied: And *Cæsar*, being besieged in *Alexandria*, when the Enemy, by turning the Sea-Water into his Springs, rendered them useless, saved himself, and his Army, by digging such Pits or Holes in the Sand, from whence he had a constant Supply of fresh Water.

Tho' these and such Supplies of Water proceed partly from the Sea; yet I am of Opinion, that the above-mentioned Supply, in this Island, is greatly, if not chiefly, owing to the Rain-Water that descends from the adjacent Hills, and then settles in the Sand. This is evident from the greater Plenty that is to be found, and that nearer the Surface, after great Rains: Its Purity likewise may, in some measure, be owing to the Place where it is found; for Sand hath a great deal less of Matter capable of Solution in it than Earth.

As any artificial Reservoir of Water is often, in Scripture, called a Well, when the Station of the *Israelites*, in their Way to *Canaan*, was upon the Borders of the *Red Sea*, the Waters there mentioned to be in such Plenty, that the Princes digged a Well with their Staves, must probably (since they digged with these Instruments) be in such a sandy Situation as the above-mentioned; otherwise Staves would not, in Places less soft and porous, be proper to dig Holes sufficient for that Purpose. The digging these, or almost any other Basin, for the same Use in any other kind of Soil, must, in all Probability, be attended with Art, and great Labour, ſince we find, that it cauſed a Contention between *Abraham* and *Abimelech*.

Many, I believe, are the Places in the *Old Teſtament*, where, with great Submiſſion to our learned Commentators, from their Want of ſufficiently conſidering the Situation of *Paleſtine*, the Nature of the Climate, and its Productions, the real Meaning of ſeveral Texts is either very imperfectly or often not at all, underſtood.

Of all the Waters we have treated of, that of Mr. *Osborne*'s Well, as I before obſerved, is the lighteſt and beſt; and it hath been proved of great Service in Nephritic Caſes.

The Water of *Belly-ach Hole*, in the Eſtate of the Reverend Mr. *Foſter*, is next in Goodneſs, being a ſoft limpid Water.

Each of theſe, by their great Lightneſs and Purity, enter the fineſt Veſſels, where they diſſolve, and waſh away, the ſtagnant Humours, take off the Sharpneſs of the Juices, and break the Coagulations and Acidity of the Blood.

Pure Water likewiſe dilutes, prepares, and corrects the crude and ill-concocted Juices, diſſolves their Salts, and blunts their Force: And as the Decay of human Nature is owing to Obſtructions; and Wrinkles, old Age, and even the Diſſolution of the human Frame, are, in general, chiefly owing to the Want of ſufficient Motion in the Fluids; it is highly probable, that Water (how little ſoever eſteemed by the Generality of Mankind) more excellently divides the Blood and Juices, than any other Liquid whatſoever; and therefore is of the greateſt Service to preſerve Health and Life.

Whoever would be curious enough to inquire into its various efficacious Excellencies, either in Scorbutic, or Hypochondriacal Diſeaſes, and even in inflammatory Fevers, may conſult the ingenious Dr. *Hoffman*'s and Dr. *Shaw*'s Experiments upon different kinds of Water.

Mineral Water.

After the ſtricteſt Inquiry, I found in the whole Iſland but Two Springs, that had any Appearance of a Mineral Principle; the one in the Eſtate of Mr. *Richard Richards*, in the Pariſh of *St. Andrew's*; the other at Mr. *Perry*'s Eſtate, in *St. Joſeph's* Pariſh: The former turns of a faint Purple with the Powder of Galls; the latter inſtantly of a deep Purple; and, like the *Pyrmont*, reſumes its firſt Colour, upon dropping into it a few Drops of the Spirit of Vitriol. The Diſcovery of this Water may, and I hope will, be of great Service in all Caſes where Chalybeats are required.

Aquatic Animalcules.

I ſhall conclude my Obſervations upon theſe Heads with a curſory Deſcription of the ſeveral aquatic Animalcules, which I have obſerved, more or leſs, to inhabit every Reſervoir of Water, eſpecially Ponds.

The Sides of theſe are often covered with a greeniſh Incruſtation in Appearance, as if there were many coarſe Grains of Sand cemented to their ſtony or rocky Sides: Theſe, even to the naked Eye, upon a narrow Inſpection, ſeem to be alive; and the Quantity of a Pin's Head being diluted

diluted in Water, and viewed through a Microſcope, there appear about Eight or Nine little Animals, moving, or rather darting from Place to Place, with a very quick irregular Motion: Their whole Bodies are pellucid, and look in the Water like ſo many Drops of Quickſilver, their Legs being Four of a Side, moving with great Rapidity.

What is called a Water-Bug, is about an Inch long, and Three-quarters broad: This is to be found in moſt Ponds in the Iſland; but ſeldom or never in Well-Water. The Origin of this Bug is a Pond-Worm of about an Inch and a Quarter long, the Body jointed, and the Head guarded with a *Forceps*: The whole Worm, as moſt Aquatics are, is of the Shrimp-kind. This, a little before its Transformation, inwraps itſelf in an hollow Lump of Clay in the Bottom of the Pond: Some time afterwards it caſts off both its Shape and *Exuviæ*, and becomes a Bug: In this laſt Stage, by the Armour upon its Back, and the Fins it paddles with in ſwimming, and its frequent Recourſe to the Surface for Air, it ſomewhat reſembles a Tortoiſe in Miniature.

The Water of a great many of theſe Ponds, at different Times of the Year, eſpecially in a dry Seaſon, when the Water is low, and the Weather ſomewhat calm, throws up a greeniſh Scum to the Surface: This is ſo ſtrong a Poiſon, that, if ſwallowed with the Water by Poultry, ſuch as Turkies, or even by Black Cattle, they in half an Hour's time expire.

Having examined this Scum with a Microſcope, on the ſame Day that a Two-years old Bull had been poiſoned by drinking in the ſame Pond, I obſerved, that a Drop of it contained ſeveral very compact *Faſces*, or ſmall Fagot-like Bundles; the Extremities of their conſtituent Parts being, in Appearance, ſo many cryſtallized *Spicula*.

I likewiſe diſcovered, in the ſame Drop of this Scum, which was diluted in fair Water, ſeveral Annulets, each Link of the Chain being of an orbicular Form, barely touching one another.

It is ſaid, that this Scum is at times full of ſmall Animalcules: However, I cannot affirm this to be true from any Experience of my own.

A TABLE

A TABLE *of the ſpecific Weight of Fifteen cubical Inches of the Waters of the principal Springs, Wells,* &c. *in the Iſland of* Barbados, *as well as the Weight of theſe, compared with ſome other Mineral Waters.*

Fifteen cubical Inches of	Weighed—Ounces.	Drams.	Grains.
Mr. *Robert Osborne*'s Well-Water, in *St. Peter's* Pariſh —	8	3	12
Cole's Cave Water —	8	3	14
The Spring at *Codrington's College* —	8	3	15
Belly-ach Hole —	8	3	13½
Pory Spring Water —	8	3	14½
Pond-Water —	8	3	19½
Pond-Water filtrated thro' a Water-ſtone —	8	3	15½
Rain-Water, received from the Eaves, after long Rain —	8	3	14½
Pyrmont-Water —	8	3	19
Bay-Water, by Percolation thro' the Sand —	8	3	14½
A weak chalybeat Water, in the Eſtate of Mr. *Richard Richards* —	8	3	14
The ſame Quantity of *Barbados* Proof Rum weighed Seven Ounces and an half, Two Drachms, Twelve Grains and an half.			

I recommend to the Inhabitants of the Sugar-Colonies this Method of proving their Rum: Which would prevent the many Differences that ariſe between the Planters and the Merchants; for, by this means, if this general Standard were fixed in ſome Office, recourſe might be had to it, to determine any Diſpute of this Nature.

The preſent Method of proving Rum in *Barbados* is liable to many Inconveniencies; for now the ſame Rum that is not, in reality, ſaleable, whilſt in the Store-houſe, may, by being expoſed to the Heat of the Sun, as is often done, be made to appear far better than it really is; and, by this means, the unwary Purchaſer is not only impoſed upon, but the Credit of the Iſland, as to good Rum, is leſſened in foreign Markets.

It is by ſuch wiſe Regulations that the *Jamaica* Rum is every-where preferred to ours: And, indeed, it is far preferable to our Market Rum; tho' not to be compared with the Cane-Juice-Spirit made in *Barbados* by the Planters for their own Uſe, tho' ſeldom or never exported for Sale.

Of

OF

FOSSILS.

HAVING in the First Book proved the Trade-Wind to be coeval with the Deluge, as well as from the remaining *Vestigia* upon the Surface of the Island, that the Course and Current of both ran conjunctly Westward, our Variety of Sea-Shells, and other Marine Spoils, dug up at different, though small Depths, prove likewise that this Place hath undergone a Share of that universal Shock, which Nature felt at the Deluge.

But as our Fossils are found here much nearer the Surface, than those of the same Kind, and specific Gravity, in more variable Northern Climates, I shall endeavour (tho' a thing hitherto unattempted) to account for the different Degrees of Force of the Deluge in the Northern, and those Southern Climates near the Equator: For, tho' the Face of the whole Earth was covered with Water; yet it undeniably appears, from the greater Number of Fossils dug out from among deeper *Strata* in the former, than the latter, that the Disunion or Liquefaction of the Globe must be then greater and deeper in the Northern, than in these Southern mild Climates. Fossils found generally near the Surface between the Tropics.

This Difference I conceive to be occasioned, first, by the sudden Shiftings of the Winds (a thing common in variable Climates), which, in their Turns, must needs diversly determine and impel the Course of the Waters (perhaps during the Forty Days Continuance of the Deluge) to all the cardinal Points. The Reason why the Force of the Deluge was less between the Tropics.

Add to this the more rapid Rise and Fall of the Waters, in the Flux and Reflux of the Tides, during that Inundation; for, as the Position of the Moon, with regard to the Earth, was not then changed, and as we read of no Miracle to suspend her Power, her Influence over the Element of Water muſt *(cæteris paribus)* have the ſame Effect as it hath now; that is, in ſome Northern Countries, the Attraction of the Moon, upon the Surface of the Water, is ſo great, that it riſes at high Tides, eſpecially at the Vernal and Autumnal Equinoxes, to Forty Feet; whereas in this and the adjacent Iſlands, when the Moon deſcribes its largeſt Circle, it never riſes above Four Feet, and about Five Inches: Therefore, as the Force of Four is to Forty, ſo much more ſtrongly muſt ſuch Northern Climates feel theſe Percuſſions.

After repeated Searches into the Sides of broken Clifts, Caves, and deep Wells, I never found any Veins of uncommon Earth, ſuch as *Terra Lemnia*, or *Bolus Armoniacus*, or of any other worth mentioning, except Three Veins of coarſe red Oaker, Two in *St. Lucy's* Pariſh, and One near *Codrington's College*, in *St. John's* Pariſh: This, where better Paint is wanted, ſerves tolerably well to daub over Wind-mill Vanes, or ſuch rough Work.

Green Tar a Bituminous Foſſil. As our moſt remarkable Foſſils are of the Bituminous Kind, I ſhall begin with the green Tar.

This is an oily Bituminous Exudation, iſſuing from ſome Hills in *St. Andrew's* and *St. Joſeph's* Pariſhes, of a dirty Black, inclining to a Green.

The Method of procuring it is, to dig an Hole or Trench in, or very near, the Place where it ouſes out of the Earth: This by degrees fills with Water, having a thick Film, or Cream, of this liquid Bitumen ſwimming upon the Surface; from whence it is ſkimmed off, and preſerved in earthen Jars, or other Veſſels.

The moſt convenient Seaſon for gathering it is in the Months of *January*, *February*, and *March*.

It is of ſo inflammable a Nature, that it ſerves to burn in Lamps.

Its medicinal Qualities. As to its medicinal Qualities, it is chiefly made uſe of with great Succeſs in paralytic and nervous Diſorders, as well as in curing cutaneous Eruptions. It is of ſo penetrating a Nature, that when an Horſe, that hath been doſed with it, begins to be warm upon his Journey, the Rider will ſmell the Tar very ſtrongly,

This, and one of a blacker Colour, in *St. Joſeph's* Pariſh, are all the liquid Sorts found in this Iſland.

Solid Bitumen found in *Barbados*. There is likewiſe another Species of Bitumen, of a ſolid Subſtance, called here *Munjack*.

This is dug out of Beds, or *Strata*, of Earth, at different Depths, in the Sides of Hills in *St. John's* and *St. Andrew's* Pariſhes; and nearly anſwers the Deſcription of that Bitumen, which the Reverend Mr. *Maundrel* found on the mountainous Sides of the *Lacus Aſphaltites*, or the *Dead Sea*, where formerly ſtood the Cities of *Sodom* and *Gomorrah*. This Sort, in a great meaſure, anſwers the Uſe of Coals.

The Land barren where Bitumens are found. Where the liquid Kinds are thrown up out of the Earth, the Surface of the Ground is one continued Quagmire, bearing very little, if any Graſs; and

and where the more ſolid is dug out, if the Veins are upon, or very near the Surface, ſcarce any Vegetable grows upon it.

If by Accident any of theſe Veins take Fire, they continue to burn a long time, tho' in a dull ſlow manner: For, as the Veins are ſurrounded with Earth, this crumbles, and, falling into the Flame, ſtifles it.

There was an Inſtance of this in *St. John's* Pariſh, where a Slave roaſting Potatoes upon the Side of an Hill, a ſmall Vein of Bitumen, lying very near the Surface, took Fire, and continued ſlowly to burn, tho' ſometimes ſcarce perceptibly, for above Five Years, without the leaſt Danger to the Neighbourhood.

If the inquiſitive Monſ. *Le Clerc* had more narrowly inquired into the Nature of Bituminous Foſſils, and the Soils wherein they are always found, he would not, in all Probability, have been ſo tenacious of his favourite, tho' groundleſs Opinion; *viz.* That the Deſtruction of *Sodom* and *Gomorrah* was not ſupernatural; but that the Plain, upon which they ſtood, was full of Bitumen, which, enkindled by Lightning, deſtroyed the Cities and Plains thereof. **M. *Le Clerc's* Opinion confuted.**

Let us now examine this fine-ſpun Conjecture: If this Deſtruction was cauſed in this natural Way, the Suppoſition will require as much a Miracle to bring it about, as if he had ſuffered *Moſes*'s Deſcription to be true: For here muſt be ſuch Veins of Bitumen found, as will kindle in an Inſtant; and the Lightning muſt be, as it were, as extenſive as the whole Plain; whereas the Bitumen, that is now in ſmall Quantities found in the Sides of the adjoining Hills (and, in all Probability, ever was found there, and not in the Valley), is of a Coal-like Subſtance, like that found in *Barbados*; and is far from being capable, in a natural Way, of cauſing ſo quick a Deſtruction: For if theſe Plains and Cities were by this natural Means deſtroyed, the Cauſe muſt be as extenſive as the Effect: And, as theſe Plains were Seventy-four Miles long, and Eighteen broad, they muſt be wholly, or in ſuch a Part, impregnated with this imflammable Matter, as to be capable by its Quality, and ſufficient by its Quantity, to cauſe ſo general a Deſtruction: But that theſe Plains were not wholly, nor in ſo great a Part, ſufficiently ſtored with ſuch combuſtible Ingredients, will evidently appear, if we allow, as ſurely we muſt, that Nature is as conſiſtent in her Productions of this kind, as ſhe is in other Minerals, Foſſils, and Vegetables.

Neither the Cedars of *Lebanon*, nor the Mountain Oaks, are found in any Climate growing in wet Marſhes; nor Reeds nor Ruſhes upon the Tops of dry Mountains.

This is an Obſervation as early as the Time of *Job: Can the Ruſh grow without Mire, and the Flag without Water?*

Virgil likewiſe gives it in the following Lines:

Nec vero terræ ferre omnes omnia poſſunt.
Fluminibus ſalices, craſſiſq; paludibus alni
Naſcuntur, ſteriles ſaxoſis montibus orni.

Littora

Littora myrtetis lætissima: deniq; apertos
Bacchus amat colles; Aquilonem & frigora taxi.

Virg. G. II. 109.

Nor ev'ry Plant on ev'ry Soil will grow:
The Swallow loves the watry Ground, and low;
The Marshes, Alders: Nature seems t'ordain
The rocky Clift for the wild Ash's Reign;
The baleful Yeugh to Northern Blasts assigns;
To Shores the Myrtles; and to Mounts the Vines.

From hence we may learn, that there are Soils particularly adapted to the different Kinds of Vegetables; and consequently conclude, that Fossils likewise, Minerals, *&c.* will not thrive in an improper Soil.

This leads us to consider the Situation of these Plains: And when we find, that the Whole was an Inland Country, without Trade or Navigation, but what the River *Jordan* afforded, we must of course conclude, that the Subsistence of the numerous Inhabitants must depend upon the Fertility of the Land, and not on foreign Supplies: Therefore, how consistent with that Prudence, which every Age of the World possessed, would it have been to settle in so barren a Spot, not capable of producing the common Necessaries of Life! Yet so it must have been, if the whole Plain was impregnated with this Bitumen: Or, if we suppose, that there were only Veins of it interspersed through this spacious Plain, how fierce and rapid soever the enkindled Flames of these Veins might be; yet their destructive Influence and Power would be confined almost intirely within their own proper Chanels; so that the rest of the Country would be in no Danger of so quick and so general a Calamity, as befel it in this supposed natural Way.

We might pursue this Argument farther (if any additional Proofs were wanting); and observe, that the present, and in all Probability, the former Veins of Bitumen were found, not in the Plain (which was very justly compared with *the Garden of the Lord* for Fruitfulness), but in the Hills on the East and West Sides of it: And, as the River *Jordan*, at this time, somewhat above the Lake, is almost as wide as the *Thames* at *Chelsea*, and ran formerly from North to South thro' the Plain; if we even suppose, that whole Torrents of this liquid Fire burst out from the Sides of either of these Hills, their destructive Course would be stopped, when it reached such a Body of Water as that River contained: So that, unless we extravagantly suppose, that the Mountains on the East, and their opposite on the West, which were Eighteen Miles asunder, took Fire very critically at the same time, the Country on one Side or other of the River must, in all Probability, have been safe, by the Interposition of so great a Quantity of Water.

Lastly,

Laſtly, If we ſuppoſe, that the whole Plain was ſufficiently ſtored with Bitumen to cauſe, in a natural Way, ſuch a Deſolation ; then we muſt neceſſarily grant, that it muſt be a very barren Spot, and improper to be inhabited, contrary to its Situation, with regard to the Climate, and contrary to the Deſcription which is given of it: For, *When* Lot *lifted up his Eyes, he beheld all the Plain of* Jordan, *that it was well watered everywhere* ; and therefore made Choice of it.

I hope I have ſaid enough to ſhew, from the Nature of this Kind of Foſſil, and the Barrenneſs of the Place wherein it is always found, as well as from the great Extent and Suddenneſs of the Deſtruction, that it could not proceed, as Monſ. *Le Clerc* imagined, in a natural Way, from the Firing of theſe ſuppoſed Bituminous Veins, but from the miraculous Power of God.

Of MINERALS.

The Uses of Minerals. BESIDES the Use of Minerals in Physic, they are designed for several other Ends, as well as for Ornament; but their chief Value is in Money, which is become the common and most prudent Method of Exchange: Nor are their Advantages in Utensils to be forgotten, whether they are designed for the common Purposes of Life, or made into Instruments necessary for the Improvements of Arts and Sciences.

In the various and different Forms of Minerals, as well as their Qualities, we find such evident Footsteps of Divine Wisdom, as leave us no room to suspect, that they are the Effect of blind Chance, tho' formed in the deep Recesses of the Earth.

Men have universally consented to stamp the greatest Value upon Gold: And tho' they have as universally assigned the hottest Climates, as the most natural to ripen, and to give Colour to, this illustrious Ore; yet, in this warm Island, neither the Soil upon the Surface, nor at any Depth in the Earth, at least in general, as far as Two (1) hundred Feet deep, give any promising Indications of either Gold or Silver.

Gold found in *Barbados*. However, we have some Cause to believe, that we are not intirely destitute of the former; for there was found in Colonel *Abel Alleyne*'s Estate, on the Surface of the Earth, a Piece of Ore, which, upon Trial in *England*, proved very pregnant with Gold; but, tho' diligent Search was made by digging and otherwise, there was no more discovered.

There is but one other Instance, that I can venture to mention with any Certainty; and that is, a small Piece of Gold, now in the Possession of *James Bruce*, Esq; which was extracted from some Ore sent to *England* from hence by Dr. *Bruce*: But as that Gentleman died before the Experiment was made, it is not well known in what Part of the Island he found it.

There

(1) There is not a Parish in the Island, nor scarce a large Estate, but hath a Well dug in it: And a great many of these are very deep; especially one in the Estate of *Francis M'Mahon*, Esq; which is above Two hundred Feet deep.

There is likewise a Species of Mineral found in *St. Andrew's* Parish, in *Scotland.* This is as bright as polished Brass, and generally of cubical Figures: By its Weight and Colour it is apt to give Hopes of being rich in some valuable Metal; but, being with great Care tried in the Crucible, it produced no manner of Metal. I take this to be a sulphureous Marcasite, or the *Pyrites flavus*; tho' it is seldom met with here in globular Figures, as it is most commonly found in *England.*

There are also Stones taken out of the Sea, that are very hard and ponderous, containing, by their dusky ferrugineous Colour, probably much Iron, and by their Smell, when broken, much Sulphur. The same are sometimes to be met with on Land, especially at the Estate of the Honourable *Samuel Rouse*, Esq;

The Soil likewise, in a great many barren and hilly Parts of *Scotland*, is often mixed with small Flakes of Icinglass, as well as Pieces of transparent Talc; the latter especially, in the Estate of *Miles James*, Esq; where are to be seen solid pellucid Pieces, weighing no less than Fourteen or Fifteen Ounces.

And in *St. George's* Parish are often dug up Lumps of a transparent resinous Substance, at first Sight exactly resembling Resin. What chiefly distinguishes it from Resin, is the Fragrancy of its Smell. Upon comparing it with the Gum of the Birch Gum-tree, I found it to be of the same Species.

In sawing Stones for Building, there are often found, in the middle of solid Blocks of Stones, several kinds of Shells; some of them not to be found on our Shores, especially the long Muscle-shell.

I have likewise a middle-sized petrified Echinite, taken from the middle of a Piece of a split Rock.

Remarkable Caves in *Barbados.*

The Description of our Ores and Fossils being finished, it may not be improper, before we leave our subterraneous Inquiries, to take notice of some of our most remarkable Caves; especially as these are many in Number, and some of them not only curious in the various Appearances of their petrified Icicles (if I may so call them), and other Conglaciations, but likewise, together with our deep Wells, very serviceable to prevent Earthquakes, by giving at so many Mouths Vent to those (2) Vapours, which would otherwise be confined in the Bowels of the Earth.

My Inquiries into these Caves were the most laborious and dangerous; they were also by far the least pleasing to myself; and I fear the Description of them (*Cole's Cave* excepted) will prove least instructive or amusing to my Readers: But as common Report, long before my Intention of undertaking this Work, was strong in favour of several very extraordinary Representations in some Caves in my own Parish; especially, that there was, in one, the Effigy of a Woman with a Child in her Arms; and, a little

(2) In the Time of great Droughts, the Air is not only very hot, but sometimes it smells sulphureous: This I particularly observed on the Third Day of *August* 1747. These sulphureous Vapours were dissipated as soon as the Clifts of the parched Earth were saturated with the Rain, which, about that Time, fell in great Plenty.

little beyond it, a lively Repreſentation of a Lion; and as it would ill become me to deliver ſuch traditional Reports as true, without the utmoſt Certainty; I went down into ſeveral of theſe Caves, but not without Difficulty, and Danger of being ſuffocated, being often taken with a violent Vomiting, when I came out into the open Air; with the additional Mortification of finding, that the petrified conglaciated Subſtances, ſo ſtrongly affirmed to bear ſuch Portraitures, might, without the Aſſiſtance of any far-fetched Ideas, be as well ſaid to repreſent a Cock and a Bull, as a Woman and a Lion.

The Inadvertency and credulous Propenſity of the Vulgar, on one hand, to believe every Story, that hath ſomething marvelous in it; and, on the other hand, that infatuating Fondneſs and Pride ſome Men have to be thought more knowing than their Neighbours; together with lucrative Conſiderations; gave the crafty Part of Mankind an early Opportunity of uſhering into the Heathen World the Belief of Harpyes, Centaurs, and Satyrs, as well as, in a more enlightened Age, the leſs pernicious, tho' not leſs ridiculous, Exiſtence of Unicorns, Griffins, and Flying Dragons, with a great many other ſuch ſenſeleſs Chimera's, which ſerve to aſtoniſh, and ſet the weak and ignorant Part of Mankind a gaping.

Cole's Cave. As *Cole's Cave* is by far the largeſt, and moſt worthy our Notice, I ſhall confine myſelf to the Deſcription of that alone: Firſt, obſerving that I never took a Survey of it, but it ſtrongly affected my Imagination, and recalled to my Mind that awful Deſcription, which *Virgil* gives us of the firſt Entrance into the Shades below:

Spelunca alta fuit, vaſtoq; immanis hiatu.

Its Situation is almoſt in the Bottom of a melancholy hideous Gully (3), which is about an Hundred and Sixty-five Feet deep; where, above you, nothing is to be ſeen but the Tops of high Rocks, and impending Cliſts, thro' the gloomy Branches of lofty Trees.

> *Such was the ſolemn Silence, which o'erſpread*
> *The Shrine of* Ammon, *or* Dodona'*s Shades*;
> *When anxious Mortals from the Mouth of* Jove
> *Their Doom explor'd.* - - - -
>
> GLOVER'S Leonid.

The Deſcent, towards its Entrance, is by a ſteep craggy Precipice of great Height, where your Security from a Fall depends much upon the good Hold you take of the Roots of Trees, and Branches of Underwood.

Having rather ſlid than walked down in this manner a conſiderable Way, you are, on a ſudden, within an Incloſure of very high perpendicular Rocks, the Sky-light being admitted by two Holes in the Roof of

(3) A deep Chaſm made between Hills, by repeated Torrents of Rain.

of it. In the Weſt Side of this gloomy Apartment lies the Mouth of the Cave, which is an Hole of a conſiderable Bigneſs.

Upon our firſt Entrance into it, as we deſcended, the Light of the Day began to grow weak, and proved but a faint and uncertain Guide. Twenty Yards farther it appeared no ſtronger nor brighter, than the Glimmering of a Star in an hazy dull Night: A few Steps more inveloped us in a total Darkneſs.

Quale per incertam lunam, ſub luce maligna,
Eſt iter in ſylvis, ubi cœlum condidit umbra
Jupiter, & rebus nox abſtulit atra colorem. VIRG.

Thus wander Travellers in Woods by Night,
By the Moon's doubtful and malignant Light:
When *Jove* in dusky Clouds involves the Skies;
And Colours fade before their darken'd Eyes.

From hence, with the Aſſiſtance of a Candle and Lantern, I began my ſubterraneous Tour; and ſoon after my Entrance I turned upon the Left Hand to take a View of that Branch of it called the *Dry Cave.*

The Top or Roof of this is crouded with innumerable petrefied (4) Icicles hanging downwards. The Sides of the Cave were likewiſe in ſeveral Places thick ſet therewith, eſpecially where there was a Cavity: There they grew from the upper to the under Shelves of the Rock, like ſo many Baluſtrades, or Baluſters of a Staircaſe, but more in Number, and irregular, ſometimes in two, ſometimes in threefold Diviſions.

Upon breaking ſeveral of theſe Cones, which are of different Lengths and Magnitude, I found thoſe hanging, from the Roof downwards, intirely perforated, and a ſmall Quantity of the moſt tranſparent Water continually dropping through them.

The inner Circumference of theſe Holes was a pellucid ſtony Body, from Top to Bottom; and tho' it was far from having the Luſtre of a *Briſtol* Stone, yet it appeared ſhining, tranſparent, and hard; and its Parts were projected in form of Rays from the Centre to the Circumference.

By comparing theſe Icicles with many that I have ſeen in *England*, *Wales*, and *North America*, as well as thoſe deſcribed by the Reverend Mr. *Maundrel*, found cleaving to the inner Sides of what are called *King* Solomon's *Aqueducts*, near *Tyre*, I find, that theſe Petrefactions, or con-

(4) At a Cave, in the Eſtate of *Reynold Alleyne*, Eſq; called *Mount Standfaſt*, I obſerved Icicles different from thoſe in *Cole's Cave*; the latter being very large, and of a brown Colour on the Outſide, the former exactly reſembling, in Shape, Colour, and Bigneſs, the hollow Part of a middle-ſized Gooſe-quill. Thro' theſe Tubes is continually diſtilled a ſmall Quantity of the moſt tranſparent Water. Here likewiſe I ſaw a blunt Cone of petrefied Water, which continues daily to grow; and the ſeveral *Striæ* made by the falling Drops are viſible. This reſembles, in Miniature, thoſe large petrefied Rocks of the ſame Shape, obſerved by the ingenious Dr. *Shaw* in *Barbary*, which the *Arabs* believe to be ſo many Tents of their Anceſtors turned into Stone.

glaciated Subſtances, are, in general, of the ſame Make and Nature in every Part of the World.

The next thing remarkable in theſe ſubterraneous Apartments is the Wet Cave. This, near its firſt Separation from the Dry, already deſcribed, is very ſpacious and lofty; but its Bottom very much furrowed, and torn up, by the repeated Torrents of Rain, which in wet Seaſons run through it.

Soon after we entered this, we were agreeably ſurpriſed with the Murmurs of a diſtant Stream, which a little farther we found to be a conſiderable Spring of the moſt tranſparent Water, iſſuing from a large projected Rock, or rather the impending Side of the Cave.

Having drunk of this, we might here again juſtly apply another Paſſage in *Virgil:*

Intus aquæ dulces, vivoq; ſedilia ſaxo,
Nympharumq; domus. - - - - - -

A Grot is form'd beneath, with moſſy Seats,
To reſt the *Nereids*, and exclude the Heats.
Down thro' the Crannies of the living Walls
The cryſtal Streams deſcend in murm'ring Falls. DRYD.

The Roof of the Cave, near this Place, is deeply and remarkably pitted with ſeveral Holes, repreſenting ſhallow Cones of Diameters, from Nine to Twenty Inches, whoſe greateſt Depth was not above Twenty-four Inches.

I imagine that theſe Holes owe their Origin to large cavous Icicles, which formerly hung down from them, but were broken off by ſome Convulſions of the Earth.

The Spring here made a ſmall Baſin, or Bathing-place; and the Air is likewiſe there made pure and clear by the Coldneſs of the Water.

From hence forwards, the Cave gradually leſſened in Height and Breadth; and the Icicles hanging from the Top, and irregular Sides, were more in Number, but leſs in Magnitude.

Here I began to want Air; and at laſt the Paſſage became ſo narrow and low, that I was obliged to ſtoop much; and the Icicles were ſo ſmall, that the longeſt of them did not exceed my little Finger in Length, and in Diameter.

This Place, diſtant near a Quarter of a Mile from the Entrance, was my *Ne plus ultra*, being ſo much fatigued, and wanting Air ſo much, that I durſt not, without Preſumption, proceed farther.

I cannot help fanſying, that if this Cave had been ſituated in antient *Greece*, or *Italy*, imbrowned with Shades of Cypreſs Groves, and melancholy Yew, it would, in all Probability, have been the Rendezvous of all the buſy and inquiſitive World: Here oracular *Phœbus* had fixed

fixed his sacred *Tripos*: Here the Fate of Kingdoms, and of Empires, had been fondly consulted, and no less artfully determined: Here the Greedy and Ambitious, deluded by ambiguous Promises of Success, would have grasped fantastic Crowns and Sceptres at the Expence of real Treasures, Anxiety of Mind, and even of human Blood: And from hence perhaps the crafty Statesman would have derived divine Authority to human Laws and Institutions: In fine, *Cumæ*, and neglected *Delphi*, had never been immortalized in Verse or Prose.

As Fraud and Imposture can never bear the Testimony and Face of Day, the Heathen Priests made Choice of such gloomy and dismal Recesses for the Execution of their diabolical Collusions, as might naturally affect the Imagination with an enthusiastic Horror and Amazement; and, at the same time, conceal from Observation those Instruments of Delusion, wherewith they were to impose on the Weakness of their infatuated Querists: And what Place could more effectually promote the Design of these Sorcerers, than such a dreadful Den of Darkness, as I have been now describing? Where, in the Words of *Virgil*,

Horror ubiq; animos, simul ipsa silentia terrent.

All things were full of Horror and Affright,
And dreadful ev'n the Silence of the Night.

The End of the SECOND BOOK.

THE

THE

NATURAL HISTORY

OF THE

Island of *BARBADOS*.

BOOK III.

Of LAND ANIMALS.

ANIMALS are sensitive organic Bodies, endued with spontaneous Motion.

By Animals, in the following Book, I would be understood to mean, without descending to minuter unnecessary Divisions, such only as are generally termed Quadrupeds, Volatiles, and Insects.

In each of these may be traced the Workmanship of a Divine Architecture, each formed in Number, Weight, and Measure; without Defect, without Superfluity, exactly fitted and enabled to answer the various Purposes of their Creation, to execute the Will of their Creator, to minister to the Delight and Service of Man, and to contribute to the Beauty and Harmony of the universal System.

How surprising an Instance of Almighty Power, and how wonderful a Piece of Mechanism, is to be seen in some of the minutest Animals! For Instance, the Potato-Louse, an Insect bred upon Vegetables in this Island, which is so small, that it is scarce discernible by the naked Eye!

Yet this is every way as perfect as an Ox, a Whale, or an Elephant. What leſs than infinite Wiſdom and Power, could diſpoſe a little Portion of Matter, almoſt too ſmall to be viewed by the naked Eye, into that infinite Variety of Parts that are neceſſary to form an organic Body!

Let us conſider how inexpreſſibly fine, ſlender, and delicate muſt the ſeveral Parts be, that are neceſſary to form the Organs, to proportion the Structure, to direct the Machinery, and preſerve and ſupply the vital and animal Action, in one of theſe very ſmall Animals: Yet every Part that is neceſſary to animal Life is as truly found in one of them, as in *Behemoth* and *Leviathan.* I very much doubt, whether any Wiſdom, but that which framed them, can fully comprehend the Structure, the Symmetry, the Beauties, of ſuch almoſt imperceptible Generations: And I think it muſt needs exceed any finite Underſtanding to conceive, much leſs explain, how ſuch an infinite Variety of Parts, and Exerciſe of Powers, could be contained or exerted within ſo narrow a Space.

Firſt, The Heart, the Fountain of Life; then the Muſcles, neceſſary to produce Motion; the Glands, for the Secretion of Juices; the Ventricle and Inteſtines, for digeſting their Nouriſhment; and numberleſs other Parts which are neceſſary to form an organic Body.---This Knowlege is too wonderful and excellent for any human Underſtanding, when we conſider, that each of thoſe Members are themſelves alſo "organical Bo-" dies; that they conſiſt of Fibres, Membranes, Coats, Veins, Arteries, " Nerves, and numberleſs Springs, Tubes, and Pullies, too fine for Fancy " itſelf to conceive."

Can we likewiſe form the leaſt Gueſs how infinitely ſubtile and fine muſt the Parts of thoſe Fluids be, that circulate thro' theſe Tubes, "as the " Blood, the *Lympha*, and Animal Spirits, which in the largeſt Animals " are ſo exquiſitely fine, that no Imagination can explain or conceive?"

Can any Knowlege, any Power, leſs than infinite, produce or explain ſuch wonderful Effects and Appearances as theſe?

We may conclude therefore, that infinite Wiſdom is as truly and wonderfully diſplayed in the ſmalleſt as in the greateſt Works of the Creation; and nothing leſs than the ſame Wiſdom that formed the univerſal Syſtem, could poſſibly produce the ſmalleſt and moſt contemptible Being in Nature.

In treating of our Animals, I ſhall begin with the domeſtic and laborious Kind.

Theſe are much the ſame Breed with thoſe of the ſame Species in *England*, tho' not quite ſo large.

Among the Diſtempers which infect theſe, there is one of a very contagious and peſtilential (1) Kind; for a Beaſt ſhall ſeemingly, by his feeding

(1) If the Planters would make a ſmall continual ſmothering Fire, with Pitch and Tar, and Vegetables of the Terebinthine Kind, to the Eaſtward of the Pens or Places where their Cattle are tied, and fed upon, I am of Opinion, that it might be of great Service to prevent or leſſen this Diſtemper. The moſt common, and I believe the moſt ſucceſsful Drench, given to theſe diſtempered Cattle, is the following:

Take the Quantity of an Egg of Caſtile *Soap, a Pint of Rum, half a Pint of Lime-Juice, a Pint of Arrow-Root-Juice, and a Pint and an half of Vervain-Juice, mixed and incorporated together, and given in a Drench.*

This hath often been of great Service.

N. B. The Soap muſt be diſſolved.

ing heartily, and in Appearance, be otherwise well; yet in a few Hours time, without any Symptom of a previous Disorder, drop down and die.

These, when dead, are by the most judicious Planters immediately buried; and often there is a Watchman appointed, to prevent the new-bought Negroes, and others of the poorer sort, from digging up the Carcases, and feeding upon them; for when this happens, it generally costs them their Lives; especially if they eat the Liver, or any Part of the Entrails: In this Case the Distemper breaks out in the Shape of Plague-boils, near the Arm-pits or Temples.

I have known one very extraordinary Instance of its Virulency: A Negro Woman carrying upon her Head, in a Wicker-basket, a Piece of this Flesh, that had been newly cut off from a dead distempered Carcase, a few bloody sanious Drops fell through the Basket upon her Left Breast. In a few Hours she was swelled all over, and was not able to move a Limb; and in about Two Days there appeared mortifying Ulcers on every Part where the Drops fell: And, tho' speedy Methods were used by Fomentations, and by several other means, to prevent its further Progress; yet neither these Cautions, nor taking off the infected Part, could put a Stop to it; at last, the whole Breast, and adjacent infected Part, were taken off close to the Bones. In this deplorable Condition the Surgeon gave her over. Her Mistress was one of those notable Women, who love to act out of their Sphere (which only an absolute Necessity can justify); who, by some accidental Success in the Use of Simples, called That Skill and Knowlege, which, with greater Propriety, might be attributed rather to good Luck: However, in this Case, where there were seemingly no Hopes of saving the Patient, she very fortunately applied to her Breast a Cataplasm of the Four following Herbs, *Smooth Elder-leaves*, *Soldier-bush*, *Dialthea*, and *Christmas-bush*, boiled and brought into a Consistency with Bees-wax and Hog's-lard; and, in a short time, intirely cured her. The First of these Plants is very detergent, and the Three last very good Vulneraries.

The Sheep that are natural to this Climate, and are chiefly bred here, are hairy like Goats. To be covered with Wool, would be as prejudicial to them in these hot Climates, as it is useful in Winter Countries for Shelter and Warmth: Yet, as Cloathing is necessary (especially in the wet Seasons) to the Inhabitants of the warmest Climates, this intire Want of Wool upon all Sheep naturally bred here, is abundantly supplied by the Cotton-tree, which yearly, and in great Quantity, produces the finest Wool in the World.

Among the Number of Animals, either peculiar to, or brought to this Island, we are happy, that there is not one that is mortally venomous; whilst many of our neighbouring *French* Islands are miserably infested with Vipers, and other Snakes.

There are here but Three Reptiles that can be properly called venomous; the Black Spider, the Forty-leg, and the *Surinam* Scorpion.

The

The SCORPION *of the* LIZARD *Kind.*

A Full-grown Scorpion is about Ten Inches long; the Skin ſoft, but ſcaly, and of a dull Copper-colour. The very young ones are ſurpriſingly preſerved from Danger; for, when this threatens, the Parent Scorpion opens her Mouth, and ſwallows her Young, which ſhe voids out again when the Danger is over.

This is not peculiar to this Animal alone; for the Shark at Sea preſerves its Young in the ſame manner.

The SPOTTED LIZARD; *commonly called, The* POISONOUS LIZARD.

THIS exceeds not Ten Inches in Length.

Its Head is ſomewhat flatter than that of the green Lizard, and the Eyes more prominent.

The Back is very thickly ſpeckled with Ruſſet and White.

Its Feet differ remarkably from other Lizards, being very finely croſs-indented like a File, but infinitely more curiouſly: By this means, as well as by a viſcous Matter iſſuing from theſe ſpongy Indentures, they can creep along a ſmooth perpendicular, or even a projected Cieling.

As the Appearance of theſe is more diſagreeable and ugly than the common green Lizard, they are, I believe, upon that account alone, called *The Poiſonous Lizard*; for, upon the ſtricteſt Inquiry, I could never perceive, that they were really poiſonous.

The SURINAM SCORPION.

A Full-grown Scorpion is Three Inches long from the Head to the Extremity of the Tail.

From the Fore-part of the Neck, cloſe to the Head, riſe Two Claws of about Three Quarters of an Inch long, divided into Three Joints, each Claw near its Extremity ending in a long ſlender whitiſh *Forceps*.

It hath Two Eyes, black, ſmall, and ſhining; likewiſe Four Pair of Legs, the hindermoſt Pair being the longeſt: Each Joint of the Legs, as well as the Back, is marked with ſeveral faint whitiſh Liſts, the intermediate Space being of a ruſſet Colour, mixt with blackiſh Spots.

The Tail is divided into Six Joints: At the Extremity of the laſt appear two ſmall horny, ſomewhat crooked, Stings or Darts, in form of a wide-extended *Forceps*, the upper being above double the Length of the lowermoſt: Nor are they ſeemingly jointed together at the Root; for the longeſt comes from the upper Side of the laſt Joint of the Tail, the other from the lowermoſt Part, the intermediate Space being fleſhy.

When

When a *Surinam* Scorpion walks or runs, she generally curls up her Tail in a Ring; and, when provoked, she extends it to its full Length; and then with a very quick Motion darts its Sting or *Forceps* into her Adversary.

Persons who have the Misfortune to be stung by them, endure very acute Pain for several Hours; and the Flesh, near the Wound, turns generally livid: However, I have known no Instance here of its proving mortal.

Upon examining this horny Sting with a good Microscope, I could not perceive either Hole or Fissure, by which the Poison might be conveyed to the Wound: Yet, upon pressing very hard upon the Sting of an enraged Scorpion with the Blade of a Penknife, I plainly perceived a thin bluish Liquid to issue from the Extremity of the Sting, which, upon a greater Pressure, was followed by a whitish Liquid, of a thicker Consistency: The former I take to be principally the Poison that it injects into the Wound it makes, and the latter to be only the Juices of the Body.

It is the general Opinion here, as well as elsewhere, that if these Scorpions are surrounded, tho' at a Distance, with Fire, or any other way greatly molested, they will sting themselves in the Head, and immediately expire. This is attested by a great many Persons of undoubted Veracity: Yet, as this Opinion contradicts that Principle of Self-preservation, which appears to be so natural, and so general, strongly influencing the whole Brute Creation to preserve their Being, it may be justly questioned, whether the many Experiments made, in order to establish this Opinion, were attended to with that Accuracy which the Subject required.

As for my own Part, I am apt to believe, that the Heat of the surrounding Fire hath, at least sometimes, a Share in the Destruction of these Creatures upon such Experiments.

These Scorpions are chiefly to be seen among old Boards, old Books, or other loose Papers.

They never bear any Young ones but once: The She carries her Young upon her Back; and, as these grow in Strength, the Parent decays and dies.

These Scorpions are but small in this Part of the World, when compared with those in the *Spanish West-Indies*.

An Instance of the Bigness of the latter is seen in a very large one, now in the Repository of the Royal Society; which is about Eight Inches long, and every way proportionable.

The GREEN LIZARD.

THIS is of the oviparous Kind, laying small whitish Eggs, which they deposit in loose dusty Earth, where they are impregnated by the Heat of the Sun.

A full-grown Lizard, in *Barbados*, is about Ten Inches long; tho' in the Island of *Antigua* they are often above Fourteen, and in Bulk Three times as large as ours.

The Head is ſomewhat flattiſh, and hollow between the Eyes: Theſe in all Kinds of Lizards are very prominent.

The Back is of a changeable ſhining Green, imboſſed with ſmall pointed Riſings, like Shagreen: It changes towards the Belly into a pale Silver White.

As Lizards are very harmleſs Creatures, ſeveral Authors are of Opinion, that they are, in particular, great Lovers of Mankind. And, to prove this, they aſſure us, that when Men in hot Countries ſleep under the Shade of Trees, the Lizards will creep upon ſeveral Parts of their Bodies: Nay, ſo fanciful are ſome of theſe Writers, that they gravely affirm, that Lizards will leap upon a ſleeping Perſon to awake him, if he ſees a Snake approaching him; tho', upon a narrow Scrutiny, we ſhall find, that this ſeeming Familiarity and Philanthropy are the Effect of no other Principle, than that of Self-preſervation; for, as theſe Lizards chiefly live upon Flies, their Familiarity in creeping, or Daringneſs in leaping, upon a ſleeping Perſon, is only in order to catch Flies; which they do by creeping near, and then leaping upon, their Prey.

It is likewiſe ſuppoſed, that they are great Lovers of Muſic.

This Opinion is ſupported by ſeveral probable Circumſtances; for they will draw near to the Sound of any muſical Inſtrument.

I have known one Inſtance myſelf of a Lizard drawing nearer and nearer by Degrees, and at laſt leaping upon the Spinet which a Lady was playing upon.

However, I would not, from this one, perhaps caſual, Inſtance, be ſuppoſed to draw any certain regular Inference in favour of a general Certainty of the above-mentioned Opinion.

Monkeys.

THESE are not very numerous in this Iſland: They chiefly reſide in inacceſſible Gullies; eſpecially where there are many Fruit-trees.

The greateſt Miſchief they do to the neighbouring Planters is digging out of the Earth their Yams and Potatoes, and ſometimes breaking and carrying off a great many ripe Sugar-canes.

As a Law of this Iſland provides a *Præmium* for deſtroying theſe, as well as Racoons, they yearly rather decreaſe than multiply.

Rats.

THESE are ſo very numerous, and ſo very deſtructive to Sugar-canes, that the yearly Loſs to the Inhabitants of the Pariſhes of *St. Joſeph's* and *St. Andrew's* alone, is computed to be no leſs than Two or Three thouſand Pounds.

That

That they are in greater Numbers in thoſe Pariſhes than elſewhere, may be attributed to their hilly Situation, interſperſed with high inacceſſible Rocks, in whoſe Cavities they ſhelter, and there breed.

The JESSAMIN *Inſect.*

TO this very curious Animal I am at a Loſs for a Name, having neither ſeen it myſelf, nor ever heard it mentioned by any one, all the time I was in that Iſland. I am obliged for the following Account of it, ſince I came to *England*, to a (2) Gentleman, whoſe Veracity may be relied on, and who took one of them from a Jeſſamin-tree, that was againſt Colonel *Maynard*'s Houſe on his Eſtate in *St. Peter's* Pariſh.

The Body of it is about Three Inches long, and ſo nearly reſembles the Bark of a Jeſſamin Twig, as not eaſily to be diſtinguiſhed from it, inſomuch that it may be fairly denominated from it. The Make of it is much like that of the Inſect commonly called *John Cook's Horſe*, excepting that this Jeſſamin Inſect is furniſhed with a moſt curious Tongue of the Thickneſs of a Horſe-hair, near Two Inches long, ſtanding (as far as it hath been yet obſerved) always projected out of its Mouth: For the Preſervation of which moſt extraordinary Member, as it is, on the one hand, no doubt, greatly inſtrumental in procuring Food for the Owner, and on the other, from its delicate Texture, liable to frequent Accidents; Nature has moſt kindly, and at the ſame time moſt wonderfully, formed a very curious Caſe or Scabbard for it out of its two Fore Feet, which were each of them hollowed exactly for the Purpoſe; and which that Gentleman ſays he ſaw it make uſe of ſeveral times, by lifting them up, extending them, and clapping them together to cover the Tongue. After obſerving it for ſeveral Hours, in order to preſerve it perfect, it was ſtuck to a Board with a Pin: The Pain (as it is ſuppoſed) occaſioned its laying a very ſmall beautiful ſpeckled Egg.

JOHN COOK'S HORSE, *or* HAG'S HORSE.

THO' this hath all its Limbs in Perfection; yet it is ſo ſhapeleſs an Animal, that, without a narrow Inſpection, it can hardly be diſtinguiſhed at firſt from a dry half-rotten Piece of Straw of about Three Inches long.

Its Legs, which are Four in Number, are very near as fine as thoſe of a large Spider. It ſeems to be every way very inoffenſive; and it is generally to be found upon Shrubs and Buſhes.

A great many Negroes have a Notion, that, if they kill one of theſe, they will be very unlucky in breaking all Earthen Wares they handle: Of this they are ſo ſtrongly perſuaded, that I have ſeen a Negro Wench ſuffer a Whipping, rather than, when commanded to do it, kill one of them.

The whole Body and Legs are ſpeckled alternately with a ruſſet Brown, and a dull White; but not diſcernible at any great Diſtance.

From

(2) The Reverend Mr. Dowding, formerly Rector of *St. Peter's* Pariſh, in *Barbados*.

From the Head rise Two hairy Feelers of about an Inch long: From the Neck likewise, close to the Head, come the Two foremost Pair of Legs, which are about an Inch and a half long, and jointed in the Middle: At about an Inch farther Distance from these come the Two hindermost Legs.

The whole Body of this Animal, at about half an Inch Distance from the hindermost Legs, ends in a somewhat forked Tail.

The Cave-Bat.

As this hath nothing in common with Birds, but its Wings, and these differing from theirs by the Want of Plumage; and as it likewise lays neither Eggs, nor hath a Bill, which are the chief Characteristics of Birds; I have ventured to remove it from that Class, where many Writers have given it a Place.

The great *Bochart* observes, that its very Name in *Hebrew* is עטלף *Atalleeph*, i. e. *a Bird of Darkness*.

This Bat hath its Name from its Place of Residence (3).

It is often as big as a young Pigeon. Its Body is covered with a Snuff-coloured soft Hair: Its Ears are more upright, and larger, in proportion, than those of a Rat; and its whole Head, especially its Mouth and Nose, shorter and thicker. From the Extremity of one Wing to the other extended, measures Eighteen Inches: Its Feet are guarded with Six sharp Talons, each turning inwards like Fish-hooks.

Some of the common People are of Opinion, that as Bats are always awake at Night, therefore their Flesh dried to Powder, and given to Dogs, will make them likewise watchful at Night: And this Powder is very often, by credulous People, mixed with their Food for that Purpose.

The Mouse-Bat.

This is of the same Species with those in *England*. Their Wings, at the middle Joints, are provided with sharp Hooks, by which they cling to the Roof of Caves, or under the Eaves of Houses.

(3) From hence that Phrase in the Prophecy of *Isaiah*, Chap. ii. 20. *In that Day*, i. e. *in the Distress of the* Israelites, *shall a Man cast his Idols of Silver and Gold to the Bats and Moles*, i. e. *hide them in the Holes and Caverns of the Earth.*

Of

OF

BIRDS.

THESE are, in general, divided into Terreſtrial and Aquatic. An Inſpection into the Structure, Nature, and Qualities of every Species will convince us, that every Individual is ſtamped with Marks of infinite Wiſdom: And Experience teaches us, that the ſame Power, which at firſt created them, hath, and, I may venture to ſay, ever will preſerve, not only the ſtrongeſt, but the ſeemingly weakeſt, and moſt helpleſs, in a regular Succeſſion, till Time ſhall be no more.

Such is the exact Symmetry, and juſt Proportion, obſervable in the Body of Animals, that every Part is ſubſervient to the Whole. The carnivorous and rapacious Kind, ſuch as Eagles, Hawks, and Vulturs, have ſtrong crooked Bills, and ſharp Talons, to ſecure and tear their Prey: But as theſe Birds are deſtructive to the more uſeful domeſtic Kind, we find that they are by far leſs numerous than the latter.

The melodious Notes, and the beautiful Plumage, of ſome, are not leſs pleaſing, the one to the Ear, and the other to the Eye, than the Fleſh of others is delicate to the Taſte.

Aquatic Birds are Web-footed; and it is obſervable, that thoſe that are obliged to ſeek their Food at a great Diſtance from Land, have Wings remarkably large and ſtrong, in proportion to their Bodies, to enable them to bear the Fatigue of a long Flight.

Thoſe which feed in the miry Skirts of muddy Ponds or Rivers, are provided with long Bills, Necks, and Legs, each being neceſſary either to ſearch for, or ſecure, their Prey in ſuch a Situation.

The Curious and Learned may find theſe Inquiries and Reflections carried on to a laudable Height by the ingenious Mr. *Willoughby*, the learned Dr. *Derham*, and others.

Neither this, nor any of our neighbouring Iſlands, is ſtored with any great Variety of Birds; and the few that we have are not remarkable for their Notes, nor (the Humming-Bird excepted) for the Beauty of their Feathers; and our tame-bred Fowls, except the *Guiney* Fowls, *Muſcovy* Ducks, and rumpleſs Fowls, are much the ſame as thoſe in *England*.

T GUINEY

GUINEY *Fowl.*

THIS is naturally a Wild-fowl, both in Taſte, Appearance, and Qualities. It is at preſent ſo well known, that it would be needleſs to deſcribe it.

All that I ſhall therefore obſerve is, that it is thought by ſome to be the ſame Bird ſo much prized by the *Romans*, and by them called *Gallina Mauritana.*

The Dark-coloured GAULDING.

THE Bill of this Bird is about Three Inches long, blackiſh, and ſharp-pointed. From the Extremity of this to the Feet, extended, are Twenty-two Inches; and the Wings, extended, meaſure Two Feet.

The Cock hath a blue-coloured Tuft upon the Head. The Neck is of a very browniſh Red, ſomewhat near a Snuff-colour, with a pale whitiſh Liſt downwards under the Throat. The Eyes are incircled with a yellow *Iris.* The Feathers covering the Back are long and bluiſh.

The whole Bird is of the Crane-kind, and generally found feeding upon Worms about the Edges of Ponds, as well as upon Mice, Lizards, and Scorpions.

The GREY GAULDING.

THIS Bird is ſeldom ſeen in this Iſland, tho' very frequently at *Antigua.*

It differs from that already deſcribed, chiefly in its Bigneſs and Colour; this being a great deal larger, and its Colour of a greyiſh White.

It feeds generally upon ſmall Crabs and Shrimps in the Salt Marſhes; but we have very few, if any, ſuch Places in this Iſland. We are, I ſuppoſe, for this Reaſon, ſeldom viſited by theſe Birds.

The LESSER TURTLE-DOVE.

THE Plumage of this Bird is always of a Lead-colour, except under the Breaſt and Belly, where they are inclinable to a pale dark Red, eſpecially the Cock. The Hen is generally of a lighter Colour.

This Bird, from the Tip of the Tail to the End of the Bill, is Six Inches and an half in Length, and Ten Inches to the Extremities of the Wings, extended.

They are generally very fat, and near as big as a Lark; and are juſtly eſteemed the moſt delicious of any Birds in this Iſland, as well as perhaps inferior to few, if any, in other Parts of the World.

They feed chiefly upon *Belly-ach* Berries.

The LARGE TURTLE-DOVE.

THIS, by the regular Proportion of every Limb, completing the Harmony of the Whole, justly deserves to be reckoned among the most beautiful Birds. The Plumage of the Neck, Back, and Wings, is of a dark Lead-colour, spotted with Black; the Breast and Belly somewhat lighter.

These differ in nothing material from those of the same Species in *Palestine*, used among the *Jews* for Sacrifice (4), except in their Colour; for the Wings of the latter, as the Royal Psalmist describes them, are like Silver, which, with very little Allowance, answers to the Colour of what we now call *Arabian* or *Barbary Doves*, so common in all *Palestine*.

The Length of this Bird from the Bill to the Tail is Ten Inches, and from Wing to Wing, extended, Fifteen Inches.

COOTS, *or* MOOR-HENS.

THESE Birds are to be chiefly seen about the Skirts of rushy Ponds, feeding upon Pond-bugs, and such Worms; and, when there is a Scarcity of these, they alight upon, or most commonly climb up, Plantain, Banana, or Guava Trees, and feed upon the ripe Fruit, as well as sometimes upon Pulse and Potatoes.

They are distinguished into Three Kinds; the White, the Red, and the Blue-pated. The Bill of each is strait and strong, and of a greenish yellow Colour near the Point; the other Part of a florid Red; the Whole about an Inch long.

The Crown of the Head, but especially near the setting on of the Bill, instead of Feathers, is covered with a white, blue, or red tough Skin, of a circular Form, about the Bigness of an *English* Silver Groat.

The Plumage under the Neck and Breast is of a very deep Blue, inclineable to a Purple. The Feathers on the Back are of a greenish Lead-colour. From the Extremity of one Wing to another, it measures Eighteen Inches, and the Length from the Bill to the Feet, Fifteen. The under Feathers of the Tail are Snow-white. Its Legs are yellow. It hath Three Toes before, and one behind, guarded with strong Claws.

The TWO-PENY CHICK.

THIS Bird hath a strong yellowish sharp-pointed Bill, near an Inch long: Its Length from this to the Feet, extended, is Nine Inches; and from the Extremity of the Wings, Eleven.

The Head is marked from the Bill to the hinder Part of it, with a black List. The Back and small Feathers upon the Wings with pale Russet,

(4) We find that the Heathens had a Notion, that the most perfect ought to be used in Sacrifice. This is evident from *Homer*; for, when *Achilles* speaks about the Sacrifice to *Apollo*, he says,

——— ἀρνῶν κνίσης αἰγῶν τε τελείων.

ſet, or rather dark-coloured Olive, intermixed with black and white Liſts. The Breaſt is of a pale Dove-colour, growing whitiſh, and more ſpeckled towards the Belly.

Its Legs are long, in proportion to its Bulk; and of a greeniſh yellow Colour: It hath Three Toes before, and One behind, each bordered with a ſcalloped lateral Membrane to aſſiſt in Swimming.

This Bird dives with ſuch Quickneſs, that it is very difficult to be ſhot.

It is of the Bigneſs, and much of the Colour, of the *American* Quail.

The THRUSH.

WE have Two Species of Thruſhes in this Iſland; the one much reſembling in her Note the *Engliſh* Thruſh.

As ſoon as the Day appears, ſhe mounts up like a Lark into the Air, almoſt out of Sight.

The other is a ſolitary Bird, and is known by the Name of the *Quaking Thruſh.*

The COTTON-TREE-BIRD.

THIS is a yellowiſh Bird, about the Bigneſs of a ſmall Sparrow; and is chiefly to be ſeen among Cotton-trees.

The BLACK-BIRD.

THE Plumage of this is of a gloſſy Black: Its Bill is likewiſe black and ſtrong; of a ſomewhat triangular Shape, and near an Inch long.

The Eyes are ſurrounded with a white *Iris*, as well as provided with a white Membrane, which, at Will, covers the whole outward *Superficies* of the Eye, both to keep it from Harm, and perhaps to keep it moiſt in ſo warm a Climate.

Its Length, from the Extremity of the Bill to that of the Tail, is Seven Inches; and from the Extremities of the Wings, extended, Eleven.

Theſe Birds are very numerous in this Iſland (tho' there are none at *Antigua)*, and in ſome other Leeward Iſlands.

They are ſerviceable in deſtroying Crickets, and other Vermin: Yet this ſcarce compenſates for the Ravage they make in conſuming our *Indian* and *Guiney* Corn; the former chiefly when young; the latter, ſoon after it is planted, as well as when ripe.

The GOLDFINCH.

THIS beautiful Bird is ſomewhat bigger than a large Sparrow; and is ſeldom, if ever, ſeen, except in the Months of *December* or *January*; and then generally in the moſt woody and eaſterly Part of the Iſland.

The

The only one that I ſaw was in the Plantation of *Richard Eſtwick*, Eſq;

The Head is of a fine Orange-colour, and the reſt of the Body of a dark Colour, except upon the Pinions, where the Feathers are of a deep Red.

The PIVET.

ALtho' this Bird feeds upon Fruit, its Bill is of the ſame Make with thoſe of the carnivorous Kind; and it intirely ſubſiſts by Berries, or ſuch-like Fruit of Trees.

It is no ſmall Inſtance of the Wiſdom of Providence, that there is not a Month in the Year, but ſome Trees, or Shrubs, bear ripe Fruit of one kind or other; ſo that theſe have literally their Food provided for them *in due Seaſon.*

A Pivet is about the Bigneſs and Colour of our largeſt Kind of Sparrows, and hath a wild chirping Note.

They are chiefly to be ſeen where there is the moſt Variety of ripe Fruit.

The WREN.

THIS, excepting its Note and Bill, differs very little from the Thruſh, as to its Plumage and Bigneſs: Its Bill is ſomewhat more ſharp-pointed and longer, than that of the Thruſh.

It is moſt commonly to be ſeen in the Wood near *Hackleton's Clift*, and feeds chiefly upon Oranges, and ſuch ripe Fruit, as well as upon Lizards.

The SPANISH LACKER.

THIS differs very little, if any thing, from a Pivet, but by its Note. It feeds chiefly upon Poiſon-tree Berries, and ſuch wild Fruit.

This Bird is moſt commonly to be ſeen near *Hackleton's Clift.*

The PARAKITE.

THIS is of the frugivorous Kind, and about the Bigneſs of a Thruſh, having a longer and more crooked Bill.

It feeds upon almoſt all manner of Berries, Popaws, and ripe Plantain, reſiding chiefly in inacceſſible Gullies.

The Bird borrows its Name from its Reſemblance in Make, but not in Plumage, to the ſmall green Parakite.

The SWALLOW.

AS the Make of this Bird every way anſwers the Deſcription of thoſe of the ſame Species in *England*, it would be needleſs to deſcribe it. The Cauſe of the Diſappearance of the whole Species, during the Winter

Months, hath been variouſly repreſented by moſt Authors of Natural Hiſtory.

Some have, with great ſeeming Probability, attributed their State of Inſenſibility, or at leaſt their Abſence at that time, to the wiſe Diſpenſation of Providence, in making theſe Birds incapable of any of the Functions of Life, at a Seaſon of the Year when they could not be ſupplied with their daily and proper Food, which conſiſts chiefly of Pond-flies, Butterflies, and Bees.

This ingenious Hypotheſis hath a great Appearance of Truth, and is indeed the beſt that can be given in a Northern Climate, where it is evident that their Food can only be found in Summer Months: But this their recluſe Life, from *October* to *March*, is no leſs evident, than it is almoſt general in this Iſland; in which Months their Prey is no leſs plentiful, and the Weather but very little colder, than in our Summer: Yet they retire to their Holes in the Rocks, and do not appear during theſe Months.

The SUGAR-BIRD.

THIS derives its Name from its frequenting and picking up the looſe Sugar about the Sugar-curing Houſes, and elſewhere.

The SPARROW.

OUR largeſt Sparrow differs very little from thoſe in *England*, being of a pale reddiſh Lead-colour, Four Inches long from the Bill to the Feet, extended; and Seven Inches over the Wings, from the Extremity of one to the Extremity of the other.

Its Bill is very ſtrong and ſharp-pointed; its Legs blackiſh; and its Feet have Three Claws before, and One behind.

The LESSER SPARROW.

THIS likewiſe, called the *Tinker-Sparrow*, differs very little from the former, except in Bigneſs; this now deſcribed being leſs.

The HUMMING-BIRD.

THIS Bird derives its Name from the humming Noiſe it makes as it flies.

Pliny juſtly obſerves, *Natura nuſquam magis quam in minimis tota eſt:* This is ſufficiently evinced in the Make and Qualities of this, which is the ſmalleſt of Birds; for what it wants in Strength and Bigneſs, is ſufficiently made up in its Swiftneſs in Flying, and its Dexterity in making uſe of its ſharp Bill; by which means it is capable of overcoming the largeſt and ſtrongeſt Bird that flies in theſe Parts of the World.

This

This Species of Birds are diſtinguiſhed into the large and ſmall Sort.

One of the latter, which I have now before me, weighs but Forty-eight Grains. From the Tip of one Wing to the oppoſite, extended, is Five Inches; and its Length, from the Tail to the Extremity of its Bill, near Four: Its Feet are compoſed of Three Claws before, and One behind. The Feathers under the Belly, the Tail, and the long Quill-feathers, are of a ſooty Black: Thoſe which cover the Back and Neck have a fine Mixture of Green: The Bill is about half an Inch long, and black: The Tongue, which is about an Inch long, is forked. This, darted into the Bloſſoms, ſucks up the Honey-dew from moſt Flowers: There are a few Feathers, which jut out higher than the reſt, a little above the Bill, of an almoſt inimitable ſhining Green: A little higher up ſtands erect another Feather of a flaming Purple: Theſe look very beautiful, eſpecially when reflected to the Eye by the Rays of the Sun.

The other Sort have a longer Bill, and are every way conſiderably larger than this now deſcribed.

Theſe Birds make their Neſts under the Eaves of Houſes thatched with Straw, or in the Boughs of ſhady Trees: Theſe Neſts are very artfully made, compoſed of Straw, Hair, and Cotton, which laſt covers the Inſide.

I have ſeen one of the Neſts, which was very extraordinary; for it was worked quite round the under Part of the Twig, to ſecure the Neſt, which was above: Had it been tied round with a String of any ſort, it would not have been ſo ſtrong. What was very particular in it was, that the Neſt was not built amongſt ſmall Twigs or Leaves, but upon the upper Part of a ſingle Branch, perfectly free from all Leaves or Twigs.

The Female lays Two ſmall Eggs, ſomewhat bigger than the largeſt Pea, and longer.

I have taken ſeveral young ones when fledged, and endeavoured to raiſe them; but never could effect it; for no Art can prepare a Liquid ſo nouriſhing as that which the Parent Bird extracts from the Flowers.

The SMALL YELLOW-BIRD.

THIS is a very ſmall Bird, whoſe Plumage hath a beautiful Mixture of Yellow and Red, eſpecially about the Head.

BIRDS *of* PASSAGE.

WE cannot ſufficiently admire the Wiſdom of Providence in enduing theſe with a Sagacity proper to know their ſtated Seaſons to migrate from perhaps a colder Climate to warmer, or from a Scarcity of Food in one Place to a Plenty in another, but more eſpecially at the approaching Seaſon for breeding.

What afforded me this laſt Hint, and induced me to believe, that their Migration is chiefly about that Time, was, that ſome Years ago there came

up

up a prodigious Number of these Birds late on *Saturday* Night, and alighted on a small Spot of Pasture-Land: A neighbouring Constable prevented their being disturbed on *Sunday*, during which time they were observed to be almost incessantly treading.

Their Flight is always from the South-west towards the East: But what seems most surprising is, that they direct their Course, unvaried, and unbyassed either by contrary Winds or Storms, over wide-extended Oceans, reaching yearly to their respective Stations, with an Exactness scarce credible to an heedless Observer; tho', after the most diligent Search into the anatomical Structure of their Bodies, there can be found no Analogy between their several Organs, and those designed in the human Species, for those exalted Ends of Thinking, and providently Reasoning.

Therefore the sublimest Genius must, with Humility, own, that *we know but in Part*; and that *by searching we cannot* (till God is pleased farther to unveil his now secret, tho' always wise Ends) find out the Causes and Effects of several *Phænomena*, which future Ages may possibly discover.

Most of these Birds of Passage never fail to appear here between the Nineteenth and the Twenty-seventh of *August*, especially if it be then wet Weather; but if about that time it is very dry, the greatest Part of them are seen to fly very high, and to keep their direct Course towards the East; and, as *Milton* expresses it,

- - - - - In Figure wedge their Way,
Intelligent of Seasons; and set forth
Their airy Caravan, high over Seas
Flying, and over Lands; with mutual Wing
Easing their Flight: So steers the prudent Crane
Her annual Voyage, borne on Winds: The Air
Floats as they pass, fann'd with unnumber'd Plumes.

The WILD WOOD-PIGEON.

THIS is about the Bigness of an House-Pigeon.

The Head is of a blackish Colour; and, from the under Bill to the Breast, of a light Mouse-colour; from thence to the Belly and the under Part of the Tail, of an Ash-colour; the upper Side of the Neck, Back, and Wings, of a dark Ash-colour, growing lighter towards the Extremities of the Wings.

These come hither, tho' in no great Number, about the latter End of *July* or *August*, always alighting upon Trees, and feeding upon the Berries of them.

The

The WILD-DUCK.

THE very few that migrate to this Island, of this Kind, are of the same Make and Species with those in *England*.

The TEAL.

AS this differs little, or not at all, from those in *England*, it will not be necessary to give a particular Description of it.

It will be sufficient to observe, that in *England* they are seen every Winter, tho' hither they come but once a Year, and very few in Number; and some Years there are none at all.

The CROOKED-BILL CURLIEU.

THE Bills of every Species of Curlieus are somewhat crooked; but of this more than any other: From thence therefore it derived its Name.

Its Bill is about Five Inches long; the Head small, in proportion to the Body; the Neck long and slender; as are likewise the Legs.

Its Feet have each Three Toes before, and One behind.

The Feathers of the Head, Neck, Back, and Wings, are thickly speckled with blackish, russet, and white Spots.

The Length from the Tip of the Bill to the End of the Tail is Eleven Inches; and from the Extremity of each Wing, extended, Fifteen Inches.

It weighs, when fat, about Three Quarters of a Pound, and sometimes more.

This, among the rest of the Birds of Passage, visits this Island in the latter End of *July* and *August*; especially if we have any Southerly Winds, and rainy Weather, at that time.

The SHIVERING CURLIEU.

THESE, as well as most other Curlieus, often come over in great Flocks.

They are next in Bigness to the Crooked-bill Curlieus, having their Feathers on the Back, Breast, and Tail, of a brownish Red, slightly streaked with Black.

The STONE CURLIEU.

THIS is somewhat smaller than the last described, and marked with larger and blacker Spots upon the Back, having likewise some of the Pen-feathers in their Wings, as well as some of the Tail-feathers white.

These feed most commonly on the Sea-shore; and therefore have a Taste somewhat fishy.

The BLACK-BREAST PLOVER.

THO' all the Species of Plovers, that make their periodical Visits to this Island, are more or less speckled; yet as this is more remarkably so, and as they are by far more numerous than any other Species, I shall begin with its Description.

The Length of this from the Tip of the Bill, which is about an Inch long, to the End of the Tail, is Nine Inches; and the Breadth over the Wings, extended, Eighteen.

The Head is somewhat large, in proportion to its Body.

The whole Bird is speckled all over with white and black Spots.

The RING-NECKED PLOVER.

THIS is much less than the Sandy-breast; and derives its Name from a white Ring incircling its Neck.

The Colour of the Feathers upon the Back is not so much speckled as the large Plover already described.

This, as well as the other Plovers, has no hinder Claw; so that they cannot alight upon Trees.

The SANDY-BREAST PLOVER.

THIS differs chiefly from that already described, by being somewhat smaller, and the Colour of its Breast not speckled, as in the former, but of a pale White.

The Head of this is somewhat less than the other, and more proportionable to its Body.

They likewise come somewhat later to the Island than the speckled Sort.

This is about Eight Inches long from the Bill to the End of the Tail, and Sixteen over the Wings, extended.

Of the MOPUSES.

THERE are three Sorts of these, distinguished into the Large and Small, and the Hiding Mopus.

The latter, which is the biggest, derives its Name from its immediately concealing itself, as soon as it alights, in swampy wet Places, amongst Rushes or Sedge-grass; nor will they attempt to rise, till you come very near them. Their Bill is Two Inches long; from this to the Extremity of the Tail is Ten Inches; and over the Wings, extended, Fourteen. The Feathers upon the Back are of a reddish Brown, marked lengthways with black Lists.

The Two other Kinds of Mopuses are lesser than this, differing not in Colour, or in scarce any other Circumstance, except that these do not conceal themselves, but alight often in open Grounds, and in Flocks.

The

The Long-Legs.

THESE are diſtinguiſhed into the large and the ſmall Kind, both differing from the Plover, in having a hind Claw.

The largeſt, which is called the *Cackling Plover*, weighs often near Seven Ounces.

They derive their Name from the Length of their Legs.

Nature hath given theſe, as well as all other Birds that ſeek their Food in muddy wet Skirts of Ponds, long Legs, Necks, and Bills, theſe being abſolutely neceſſary in ſuch a Situation; their Legs to prevent their ſinking in the Mud; and the Length of their Bills enabling them to penetrate deep into the Mire, in Search of Worms and Fiſh, which are their Food.

The Great Snipe.

THERE are ſeveral Species of Snipes, that come in the wet Seaſons to this Iſland.

Theſe are diſtinguiſhed by many uncertain and different Names, as well as Marks; tho' the real Difference between ſo many is but ſmall.

I ſhall therefore reduce them into the large and ſmall Sort.

The Little Snipe.

THIS is likewiſe called a Nit.

The Name was perhaps given them from their very diminutive Size, when out of Feathers; for they then ſeldom weigh above two Ounces each.

A few of theſe are to be ſeen here in every Month of the Year, without any Regard to the periodical Seaſon.

They are of a greyiſh Colour, ſpeckled with Black upon the Back and Wings, and the Breaſt and Belly of a whitiſh Colour.

Their Length from the Tip of the Bill to the Feet, extended, is Six Inches; and, over the Wings, Nine.

Theſe feed generally about the Skirts of Ponds.

OF
FLYING INSECTS.

THO' unthinking and incurious People may look upon these to be the most useless Part of the Animal Creation; yet, upon a narrow Inspection into the Use of every Species, we find a providential Design, in not only their Make and Qualities, but likewise in the Time and Season of their Appearance, which never happens till so late in the Spring, that he, who *filleth every thing living with Plenteousness*, hath provided them with Leaves of Trees, their proper Food, whilst in their reptile-State.

They likewise at that Season (except what Providence reserves to continue and propagate the Species) become exceedingly useful, being then the destined necessary Food of young Birds.

But, when these crawling Worms come to their full Growth, they spin; and, after a wonderful manner, inwrap themselves in a Tomb of their own making.

Here they rest for a while; and, in a short time after, they break thro' this temporary Prison, and become Butterflies.

In this State, decked in all the gaudy Pride of Colours, they wanton in the Air; and, by their careless irregular Flights, seem to have scarce any other Business in Life, but sportive Diversions. This perhaps may be intended by Providence to make Amends for the short Duration of their Existence.

Flies, likewise, especially those of the smaller Kind, are a very useful Part of the Creation; especially as they are in the Summer Season the most common Food of Fresh-water Fish.

The LARGE BLACK BEE.

THIS is the largest that we have here of the Bee-kind. The Head is large and flattish.

The lower Part of it, somewhat about the Mouth, is provided with a strong *Forceps*: With this it makes deep Holes in several Kinds of dry soft

soft Timber, which serve as a Shelter from the Weather, as well as Nests to deposit their Honey, and rear their Young in.

In the upper Part of the Mouth there is a Dart, or rather *Proboscis*, pointing perpendicularly downwards: This is about a Quarter of an Inch long, black and shining; and, as the above-mentioned Nippers or *Forcipes* were designed to bore Holes in Trees, so this, by being divided, or capable of opening as a *Forceps*, may perhaps be designed to take and kill its Prey; for I always found their opposite Sides to be hollow, capable of retaining the Juices of their Prey, with this additional Conveniency, that, from the Centre between these Two Claspers, darts out a bearded reddish Tongue.

Its Eyes, which are Two, are small, shining, and hemispherical, situated near the upper Part of the Head, for the Conveniency of seeing before, as well as behind.

Below these stand Two Horns, or Feelers, of about an Inch long.

The Back is black and shining, joined by a strong Ligament to the *Abdomen*, which is made up of Six *Annuli*, or Sections.

The Rapidity of its Flight depends upon Four glossy Wings.

The Body, likewise, when upon the Ground, is supported by Eight shaggy, or rather hairy Legs, each ending in a Claw or Fork. The Two hindermost and opposite Legs being the longest, these about the Thighs are generally loaded with a granulated yellow Substance, like Bees-bread.

The Extremity of the *Abdomen* is guarded with a black Bee-like Sting.

This Bee makes an humming Noise as it flies.

The MASTICH-FLY.

THIS is about an Inch and a Quarter long, and of a glossy purple Colour, inclinable to a Green.

It hath Two Feelers, each Five Inches long. It derives its Name from the Tree it feeds upon.

The COCK-ROCH.

THE Body of this Fly is near an Inch long, and somewhat flattish, and of a very dark Snuff-colour.

It lays many small brownish Eggs.

It flies about in the Evenings, especially in wet Weather, and generally into the Houses, where they breed, and are very troublesome, being voracious of most kinds of dressed Victuals, as well as destructive to Cloaths, Books, or any kind of Paper, by gnawing it to Pieces; besides they are very offensive, by the stinking Smell they leave behind upon every thing they touch, especially Meat: So that the Words of *Virgil* may be justly applied to these:

Diripiuntque dapes, contactuque omnia fœdant
Immundo. - - - - - - - -

The beſt Method of deſtroying them is to grind to Powder the Back-bone of a Gar-fiſh, and then mix it with ſomething that they like beſt to feed upon: If they eat of this Mixture, they generally die.

It is ſaid, that the *Abdomen*, boiled in Oil, and poured into the Ear, cures the Patient of any aching or throbbing Pain.

The FIRE-FLY, *or* JACK LANTERN.

THESE Flies, which are of the Glow-worm Kind, are in great Number in *North-America* upon the Continent; but in the warmer Climates ſo very rare, that I have never ſeen any in this Iſland; tho' I am credibly informed, that they are ſometimes to be found here.

The KNOCKER.

THIS Fly is ſomewhat larger and blacker than a Cock-Roch; and derives its Name from the tinkling Noiſe it makes.

The Head and Back are hard and ſhining; the former divided from the latter by a broad cloſe Joint.

As it bends its Head backward, the upper Joint falls as a regular Spring into the Socket of the lower; and, when it bows its Head forward, it opens with a ſharp tinkling Note, as the Spring of the outward Caſe of a Watch, when preſſed.

The HARD-BACK.

THIS Fly is about half an Inch long, having a very ſmall Head, provided with Two ſmall Feelers.

Its Eyes are round, black, and ſhining; the Neck thick, and ſpeckled with White and Black: Its membranaceous Wings are defended with Sheaths, or Shell wings.

Without ſuch a Covering, the exquiſitely fine Texture of the Wings would continually be liable to be torn by the leaſt Touch of an harder Subſtance, ſuch as the Blades of Canes, Corn, or the Branches of Trees.

Its Body is ſupported by Six Legs: Each of theſe, near the Extremity, is divided into Three cloſe Joints, white, and flat underneath: Theſe are very finely croſs-indented, like the Surface of a File.

Beſides, each Foot ends in a ſharp, ſomewhat crooked Point; by which means it ſtrongly graſps its Prey.

FIDDLERS.

THIS Fly, in Shape, Colour, and Number of Legs, much reſembles a Cock-Roch, except that it is ſmaller, and longer, in proportion to its Bulk.

It

It is called a Fiddler from the ſhrill Noiſe it makes, when held and ſqueezed between the Fingers.

It generally flies about in the Evening.

The CUCKOLD-FLY.

THIS is of the Beetle-kind, of about half an Inch long, and of a dark-red Colour.

The Back and *Abdomen* are covered with a Pair of hard Wings, or rather Covers for its membranaceous Wings, which are large, and of an elegant fine Texture. The Head, Neck, and *Abdomen*, are cloſely joined together.

It is ſupported by Four Legs on each Side; and flies chiefly in the Duſk of the Evening.

The CARRION *or* FLESH-FLY.

THIS is the ſame with the common Carrion-fly in *England*; at leaſt, it differs in no material Circumſtance.

The MASON-FLY.

IT is called a *Maſon*, from the great Quantity of Mire and Morter which it carries into Houſes and elſewhere, wherewith to build its Neſts, which it generally does on the Cielings, or the Roofs.

In theſe Neſts, which are above an Inch long, they lay Two or Three Eggs; and then carry in a great Number of young Spiders; and afterwards cloſe up the Entrance into the Neſts with Clay.

From each Egg, thus depoſited, there is in time hatched a ſmall white Maggot, which feeds upon the Spiders, till it comes to its *Aurelia*-ſtate: Then they ſpin and wrap themſelves up in a tough Bag, or Web, till they turn into young Maſons.

The BAT-FLY.

THE Original of this is a Tobacco or a Fig-tree Worm; for this, after its *Aurelia*-ſtate, turns into this Fly, which is about an Inch and a Quarter long.

The Head is guarded with a Pair of Horns, or Feelers, finely indented, about Three Quarters of an Inch in Length.

Its Two Eyes are large, ſpherical, and black.

The Back, as well as the *Abdomen*, which is divided into Six *Annuli*, is of a reddiſh-yellow Colour, covered with a ſoft whitiſh Down.

It

It hath Two Pair of membranaceous Wings, the upper by far larger than the under.

From its Mouth it darts a twirling *Proboscis*, with which it sucks the Honey-dew from most Flowers, but especially from Nightshade-flowers.

Its Legs are Four, Two on each Side.

If kept confined in a Box, or otherwise, it turns from a Fly-bat into a long black Worm, such as is found in the Earth in Potato-ground.

The BUONAVISTA CHINK.

THIS is a small green flattish Fly, which generally feeds upon the green Pods of *Buonavista*; and smells, when killed, like a Bug. It is very destructive to several Kinds of Pulse.

The POND-FLY.

THIS is an Inch and an half long, and of different Colours.

The Head, near the Mouth, is of a light Sea-green.

The Mouth is provided with Two Pair of Nippers, or *Forcipes*, with which it destroys its Prey.

The Eyes, being Two, are very large, prominent, and shining; and of an oval Form.

The Head is fixed to the Body by a very slender Ligament.

Its Breast is likewise very prominent, as well as the opposite Part of the Back high and convex; the upper Part of which supports Four narrow membranaceous Wings.

The whole Body is strait, about an Inch long, and ends in a forked Tail, which is alternately spotted with Green and Black.

It hath Six Legs, the hindermost by far the longest; the lower Part of each being thickly covered with stiff Hairs, the Extremity of each Foot being forked.

The Wings above-mentioned are narrow, and a full Inch long; the Texture resembling a Sea-feather, but infinitely finer.

The POPE-FLY.

THIS Insect is better known to the Inhabitants, by the great Destruction it causes in almost every sort of Grain, than by its Shape; for it spares neither Pulse nor Grain of any kind, if they have been for any considerable time gathered in.

However, in general, they resemble a Wevil in their Make.

The LOCUST, *called here*, the Ash-coloured GRASHOPPER.

THIS is about Two Inches long.

Its Eyes are black and oval.

The Head is provided with Two hairy Feelers, and is covered, as far as the Back, with an hard unpliable Cowl, streaked with Russet and Black.

Its

Its Wings are large, in proportion to the Body; and of a greyiſh-ruſſet Colour.

The Body is ſupported by Three Pair of Legs; the hinder, with which he ſprings or jumps, are about Two Inches and an half long.

The Thighs, which are of a ruſſet Colour, finely poliſhed, and ſtained with black Liſts, extending lengthways, are of an hexagonal Form; and the Legs are elegantly ſtudded with a great many white ſmall Prickles, ſet on, like the Teeth of a Saw, alternately.

This Kind of Graſhopper is generally to be ſeen among the Blades of Sugar-canes; and, when the *Guiney* Corn is ripe, they feed much upon it.

The moſt ignorant and ſuperſtitious of the Inhabitants are very apprehenſive of ſome approaching Illneſs to the Family, whenever they fly into their Houſes in the Evening, or in the Night.

There were bred, or came to the Iſland, ſuch vaſt Swarms of theſe in the Year 1734-5. that they deſtroyed almoſt every green tender Plant. So great was the Deſtruction that they cauſed, eſpecially among the Potato-Vines, upon whoſe Roots the poor People chiefly ſubſiſt, and ſuch the Scarcity of Food, occaſioned thereby, eſpecially in *St. Philip's* Pariſh, that there was a Collection made for the Sufferers thro' the reſt of the Iſland. Theſe differ not in their Make from the Locuſts that are to be ſeen in the Repoſitories of the Curious.

The GREEN GRASHOPPER.

THIS is ſomewhat leſs than the Aſh-coloured Graſhopper already deſcribed; but differs in no other material Circumſtance.

The BLACK-SPOTTED BUTTERFLY.

THIS is larger than any hereafter deſcribed.

Its Wings have a few pale-red and whitiſh Spots intermixed with the Black.

The GREENISH-YELLOW BUTTERFLY.

THIS is about Three-quarters of an Inch long.

The Back is covered with a ſoft greeniſh Down.

The *Abdomen* is divided into ſeveral *Annuli*, or Sections, tho' ſcarce perceivable.

Its *Antennæ* are about half an Inch long, and its Legs Six in Number.

It hath Four very thin membranaceous Wings, covered with a fine yellow Mealineſs.

This mealy Duſt, when viewed thro' a Microſcope, appears to be ſo many regular Quills, feathered with the utmoſt Exactneſs and Proportion.

The Body of this, as well as of the following ones, is decked with a Profusion of Beauty: And all, in the Words of the great *Milton*,

> - - - - - - - *wave their limber Fans*
> *For Wings, and smallest Lineaments, exact*
> *In all the Liv'ries deck'd of Summer's Pride,*
> *With Spots of Gold and Purple, Azure and Green.*

Should any one impertinently ask, What Use these are of in the Creation? it may be answered, in the Words of the ingenious Mr. *Ray*, that they are designed *ad ornatum universi, & ut hominibus spectaculo sint; ad rura illustranda, velut tot bracteæ, inservientes. Quis enim eximiam earum pulcritudinem & varietatem contemplans mira voluptate non afficiatur? Quis tot colorum & schematum elegantias naturæ ipsius ingenio excogitatas, & artificii penicillo depictas, curiosis oculis intuens, divinæ artis vestigia eis impressa non agnoscat & miretur?*

The White Butterfly.

THIS exactly resembles the last described, in every Particular, except its Colour.

These are chiefly to be seen flying about Ponds of stagnated Waters in the most beaten Roads.

The Dark-red black-spotted Butterfly.

THIS is about an Inch long, from the Head to the Tail.

Its *Antennæ* are Three-quarters of an Inch long; and its Two Eyes black, round, and shining.

The Wings are of a dirty Red, irregularly impanelled with black Lists; and the Margin or Border of each Wing much darker than the rest; and here-and-there adorned with many white Spots, as well as the Head, Back, and Breast.

The *Abdomen* is of a dark ferrugineous Colour, and composed of Seven *Annuli*.

The Clinker, *or* Gully-bell.

THIS is of the Cricket-kind, and derives its Name from the tinkling Noise it makes at Night, which much resembles the shrill Note of any musical Instrument.

This Noise, which is repeated Three or Four times without ceasing, may, in a still Evening, be heard above an Hundred Yards off.

But as these Creatures live generally in inaccessible Gullies, they are very seldom, if ever, caught; and therefore their Shape is known but to very few, much less the Cause of so surprising and regular a Note.

The

The LONG-SPOTTED FLY *or* LADY-BIRD.

THIS Fly exceeds not half an Inch in Length.

Its Back and Head are finely ſpotted with Red : The intermediate Spaces are of a ruſſet Colour.

There is likewiſe a ſmall red Lady-bird generally to be ſeen upon Ochra-leaves.

The GOLDEN LADY-BIRD.

THIS very beautiful Inſect is very ſeldom ſeen.

It is about the Bigneſs of the laſt deſcribed.

Its Head, Wings, and Feet are, as it were, covered with burniſhed Gold.

The CANE-FLY.

THIS is a ſmall whitiſh Fly, with Two membranaceous Wings.

It is chiefly to be ſeen among thick-planted ripe Canes.

The CORN-FLY.

THIS is a ſmaller Fly than the laſt, and is to be ſeen hovering over Heaps of *Guiney* Corn, when in the Granaries.

Its Colour is much the ſame as that of the laſt-mentioned.

The MUSCHETTO-FLY.

THIS derives its Name from *Muſca*, a general Name in the *Latin* Language for a Fly.

Perhaps the Name of *Merry-wing*, applied to another Fly, from the Noiſe it makes with its Wings, would be more properly applied to this, if the Acceptation of the Word had not rendered it almoſt univerſal, at leaſt among *Engliſh* Writers.

WEVILS.

THIS is ſo commonly found in decayed Corn and Flour in every Part of the World, that it ſcarce deſerves a particular Deſcription.

It is a black ſcaly Fly, of near a Quarter of an Inch long.

The MERRY-WING.

THIS is a very minute Fly, very troubleſome, eſpecially in calm Mornings and Evenings, to thoſe Inhabitants who live on a ſandy Soil near the Sea.

They

They seem to be exactly of the same Species with the Gnat in *England*.

They swarm upon the Face and Hands, as well as upon every other Part of the Skin exposed to the Air; and almost in an Instant dart a small *Proboscis* from their Mouths, and penetrate so deep as to draw Blood, leaving generally an angry little Tumor behind.

As the Wind rises, they are no more to be seen, except behind Doors or Window-shutters, till the Return of the calm Mornings or Evenings, not being able to stand against the Wind that blows in the Day-time.

Of REPTILES *and* INSECTS.

THE former are denominated from their creeping or crawling upon their Bellies, either with Feet, as the Caterpillers, or without Feet, as the Snakes and Earth-worms.

The latter are likewise generally divided into Aquatic and Terrestrial.

These, tho' often vile in Appearance, yet all of them, in their several Ranks and Stations, proclaim aloud the Wisdom and Power of their Creator.

Their Variety, their Dispositions, their Sagacity, their Policy, their Industry, the wonderful Proportion of their Organs, the Delicacy of their Structure, and a Thousand other Curiosities, observable in every Species, are Matter of wonderful Delight and Pleasure to a curious and inquisitive Genius: But, were we able to examine them in a nearer View; could we be capable of knowing all the direct Purposes of infinite Wisdom in their Creation, in the Relation they bear, and the harmonious Proportion they stand in, to the universal System; it would afford us endless Matter of Astonishment and Surprize, as well as of religious Reverence and Adoration to their omnipotent Creator.

Tho' small and contemptible they appear to us, they are really formed with the most exquisite Symmetry, the most delicate Proportion. An attentive Eye, assisted with the Use of Microscopes, discovers in them astonishing Marks of Wisdom, arming, cloathing, and accommodating them with all the Instruments and Faculties necessary to their Condition.

The SNAKE.

WE have but one Species of the Snake-kind in this Island, of which I have not seen above Seven in Seven Years.

The largest that I saw was not above Three Feet long.

They are not at all hurtful, except to young Pigeons and Poultry, or small Birds, Mice, *&c.*

I cannot here omit a remarkable Instance of Superstition of a Negro, with regard to one of those Snakes.

A Man

A Man who had killed one of theſe, happened to be afflicted with the Rheumatiſm ſoon after, eſpecially in his Arm: He then immediately concluded, that it was a Puniſhment upon him for killing the Snake: He perſuaded himſelf, that the different Degrees of Intenſeneſs of the Pain he felt in his Arms, were exactly in proportion to the Pain he gave the Snake. It hath been his Cuſtom ever ſince, by way of Atonement, to feed all the Snakes that come near his Hut, with ſome Offals of his Repaſt, and Water; eſpecially the latter, which he daily puts in the moſt likely Place for them to find.

The FORTY-LEGS.

THIS takes its Name from the Number of its Feet, being every way much of the ſame Make with what we call in *England Multipes*, or *Centipes*; but thoſe here are far larger, being ſometimes Four Inches long.

The Back is of a dark Copper, ſcaly and jointed, the ſeveral *Annuli* being equal in Number to the oppoſite Feet.

The Head is guarded with a Pair of hairy Feelers, and a ſtrong *Forceps:* This they can open or extend above a Quarter of an Inch wide, and again cloſe it very nimbly together, either to hold its Prey, or annoy its Enemy.

The Forty-Legs often lay above an Hundred ſmall white Eggs, which are depoſited in an Hole in the Ground, generally under a looſe Stone: Round theſe the Female coils herſelf, until they are hatched: If at that time ſhe is diſturbed, ſhe immediately ſwallows her Young.

The old ones, among other things, feed upon young Cock-Roches and Spiders.

The Bite of a Forty-Leg is very painful for at leaſt an Hour, or ſometimes a great deal longer.

There are ſome of theſe Vermin that are ſlightly tinged with a bluiſh Colour: Theſe are ſmaller than the above-mentioned, and more poiſonous.

The Forty-Legs in *Surinam* are a great deal larger than what are bred in *Barbados*. Of this we may ſee an Inſtance in the Repoſitory of the Royal Society.

The INDIAN-ROOT CATERPILLER.

THIS Worm feeds upon the Leaves of that Plant which we call here the *Indian Root*.

It is generally about Two Inches long.

Its Head is guarded with Two black ſoft Horns or Feelers, of about a Quarter of an Inch long.

The different *Annuli*, or Joints of the Back, are ſtreaked with yellowiſh and whitiſh Liſts. The Tail-part hath likewiſe a black Pair of Feelers, or Horns.

This from its *Aurelia*-ſtate turns into a reddiſh Butterfly.

The FIELD CRICKET.

THIS is about Three-quarters of an Inch long, and of a dark-brown Colour; the Head and Body cloſely joined together.

Its Two Eyes are oval and prominent.

From under theſe riſe Two half-inch long hairy Feelers.

The *Abdomen* is divided into many *Annuli*; and from its Extremity iſſues a Spear of half an Inch long, and on each Side of it Two other ſhorter Darts.

Its Two upper brown ſhining Wings, or rather Coverings, guard Two other membranaceous ones.

Its Legs are Six in Number, Three on a Side; the hinder Pair being ſtrong, and of a great Length, in proportion to the reſt: The ſmaller Parts of this Pair are thickly and regularly ſtudded with ſmall ſharp Teeth, like thoſe of a Saw.

The Breaſt and Thighs are of a dark-yellowiſh Colour.

What is moſt remarkable in this kind of Cricket is, that it not only produces its own Likeneſs from Eggs; but that it gives Birth alſo to another Animal, quite different in its Nature from the Cricket itſelf: For the ſmall wriggling Worms, which are about Seven or Eight Inches long, and often to be ſeen in ſtanding Water, and by the Vulgar ſuppoſed to be animated Horſehair, and therefore called *Horſehair Snakes*; theſe are evidently propagated by Crickets; ſince they are found in them, and have been often obſerved to be voided by them (when fit to ſhift for themſelves) thro' the *Anus*: Theſe afterwards are cheriſhed by the Heat of the Sun, and grow to the Length above-mentioned, and I ſuppoſe lay Eggs before they die.

Theſe may be, and, in all Probability, are, picked up by the Crickets, whoſe Bodies prove a proper *Nidus* for them: By this very ſurpriſing foſtering Care is this Species preſerved.

Theſe Crickets are generally to be met with under looſe Stones, and Clods of Earth; and are much coveted by Poultry of every Kind.

The ASH-COLOURED, *or* SICKLY CRICKET.

THIS is near One-third larger, eſpecially in the Length of its Legs, than either the Field, or the Houſe-Cricket; and, if purſued, ſecures itſelf more by running, than by taking ſtarting Leaps, which is always the Refuge of the Two other Sorts; tho' this, if in Danger, ſaves itſelf at laſt by the ſame Method.

This Kind makes a diſagreeable ſcreaking Noiſe; and ſome of the Inhabitants are ſo weak as to believe, that, when their unwelcome Sound is heard in their Houſes, it is an Omen of Death to ſome of the Family.

The HOUSE-CRICKET.

THIS differs very little, if any thing, in Shape, from the Field-Cricket already deſcribed, except that the Noiſe it makes is loud and purring.

When they make this Noiſe, I have obſerved their Poſition to be ſtanding very upright upon their Legs, with their Wings extended archwiſe, the middle Part ſwelling from their Bodies, and their Extremities touching them near the lower Part of the *Abdomen*.

The FLY-CATCHER.

THIS is of the Spider-kind, differing from it chiefly by a Pair of *Forcipes*, which riſe from each Side of the Neck.

It hath Four Legs on each Side.

They jump or dart forward often full Ten Inches, to ſeize the Flies, which are their Prey.

The HOUSE-SPIDER.

THIS is a dark-brown hairy Spider, having a large Head joined to the *Abdomen* by a ſlender Ligament.

The Body is ſupported by Eight Legs; the Two foremoſt extending above Two Inches aſunder.

Near the Mouth come out Two ſhort, tho' ſtrong, Feelers.

It carries its Young in a white ſmall Bag under its Belly.

Tho' the Sight of theſe Spiders is not agreeable; yet are they ſerviceable in Houſes, by deſtroying Cock-Roches, which are both offenſive by their Smell, and deſtructive to Victuals, Cloaths, and Books.

The FIELD-SPIDER.

THIS very much reſembles thoſe of the ſame Species in *England*.

The FLAT-BACK'D SPIDER.

THIS, in the Number of its Legs, and the Make of its *Forceps*, reſembles the *Italian* Tarantula; differing chiefly in the Number of its Eyes, which in the latter are Eight, in the former Two.

The Bite of this is very near as painful as the Stinging of the *Surinam* Scorpion, cauſing an immediate Swelling, with acute throbbing Pain, which continues for ſeveral Hours without Intermiſſion.

The GOLDEN SPIDER.

THIS is called the *Golden Spider*, from the bright Gold-colour Liſts, with which its Legs, and ſome Part of the Body, are marked.

It

It spins its Web in the open Air, generally, among the Branches of the large *American* Torch-thistle.

The Slender-legg'd Spider.

THIS spins a Web, and very much resembles that so often found among Hay or Stubble-ground in *England*, in Harvest-time.

The Potato-Louse.

THIS is a small reddish Insect, scarce perceptible to the naked Eye: Yet, when viewed thro' a Microscope, it is evident, that its external and internal Parts, whether we consider the Disposition of its Limbs, or the curious inward Fabric of its Heart, Lungs, Veins, and Arteries, are as regular, and as perfect, as the largest Elephant, or the huge *Leviathan*.

Had *Pliny* known the Use of Microscopes, with what greater Justice, Surprize, and Admiration, would he have said, *In magnis siquidem corporibus facilis officina sequaci materia fuit: in his tam parvis, atq; tam nullis, quæ ratio, quanta vis, quam inexplicabilis perfectio* (5) !

The Indian Corn-Worm.

THIS is a short thick Worm, preying upon the Pith in the Inside of the Stalk, and by this means rendering the Ear at the Top very imperfect, having scarce any Grain on it.

The Earth-Worm.

THIS differs, at least to the naked Eye, in nothing from the Earth-Worms in *England*.

The Ground-Ass, *or the* Lion-Pismire.

THIS is of the Colour of an Hog-Louse; in Length about a Quarter of an Inch.

The Back is convex, not ill resembling the Back of a Tortoise in Miniature; but somewhat longer in proportion.

The Head is small and long, when compared to the Bulk of the Body.

This is supported by Six Legs, the Pair next the Shoulder being the longest.

What is most remarkable in this Animal, is its Motion, which is always retrograde; and this not by walking, but by quick Starts, springing back.

These generally live in very loose Dust or Sand, under Logs of Wood of such Coverings: In these Places they artfully make a circular Hole of about

(5) *Pliny*, Lib. XI. Cap. II.

about an Inch Diameter, of a Funnel-ſhape; in the Centre of which this Creature lies unſeen, watchfully waiting for his Prey.

As ſoon as an Ant, or ſometimes a Fly, walks upon or near the circular Brink of this Hole, the Lion-Piſmire either ſees his Prey, or more probably hath Notice of it, by ſome Grains of Duſt falling to the Centre: He then, by a ſudden Start, at the Bottom, undermines the duſty Sides of the whole Fabric: Then the Prey upon the Brink falls precipitately, with the falling Duſt of this ſhelving Funnel, into the Centre: Whilſt the Prey is thus immerſed in Duſt, he ceaſes his Work, and devours it; and ſoon after repairs to his Cell, and lies concealed like a Spider in the Centre of it, till the next Prey calls him forth to repeat his unhoſpitable Talents.

This Animal is well deſcribed in *Spectacle de la Nature.*

The WOOD-ANTS.

THE Wood-Ants are the moſt pernicious of all others, being ſo very deſtructive to Timber of moſt ſorts, that, if not prevented, they will in a few Years time deſtroy the whole Roof of an Houſe, eſpecially if it be ſoft Timber.

They have likewiſe cauſed great Loſſes to Shopkeepers, by boring Holes thro' whole Bales of Linen, as well as Woolen Cloths.

They are very expeditious in building their Neſts, which are long hollow Tubes, the Outſide being an Incruſtation of a gritty clayey Matter.

The Method of deſtroying them is, to make a ſmall Hole near the upper End of the Neſts, and pour into it a little Arſenic, which generally kills thoſe that are preſent; and the reſt, that follow, eat up the Carcaſes of the ſlain, and almoſt inſtantly ſwell, burſt, and die.

The GREAT-HEADED ANT.

THE Head of this (which is joined to the Body by a ſmall Ligature) is very large, in proportion to the Bulk of the Body.

It no-way differs from the laſt deſcribed, but by the Bigneſs of its Head, from whence it derived its Name.

The SMALL RED ANT.

THIS is a very ſmall Ant: Yet the Part of the Skin it bites continues painful for near Four Hours afterwards.

If theſe are likewiſe killed, and rubbed upon the Skin, they raiſe a Bliſter.

The Bodies of theſe Ants are thickly covered with ſharp fine-pointed Briſtles, imperceptible to the naked Eye.

The STINGING ANT.

THIS appears to be the ſame with what is to be ſeen in *England* in the Summer Seaſon in moſt Paſture-Lands.

The HORSE-ANT.

THIS is the largeſt-ſized Ant, and is often to be met with both within and without Doors; and hath nothing peculiar in its Make or Qualities.

The SUGAR-ANT.

THIS is a ſmall whitiſh Ant, very fond of Sugar, or any ſweet and oily Liquids; and conſequently very troubleſome to good Houſewives, it being difficult to keep them from every kind of Victuals.

The WINGED ANT.

THO' there are ſcarce any Neſts of Ants of any Kind but have ſome winged Ants among them, which are eſteemed the She-Ants; yet, when we ſpeak of theſe as diſtinct from the ſeveral other already deſcribed, we may obſerve, that they generally live under-ground, and are ſeldom ſeen to come out of their Holes but in rainy Weather.

The LAND-SNAIL.

THESE are generally of an Aſh-colour, or black: The latter is rather a Diſtinction of Age than Species.

Theſe are chiefly to be ſeen among green Herbage, eſpecially Potato-Vines.

The SMALL SPIRAL SNAIL.

THESE are not above half an Inch long, very ſlender, and ſharp-pointed.

The Colour of the Shell is of a ruſſet Grey.

They are often found cleaving to the Bark of large Cotton-trees.

The DUNG-SNAIL.

THESE are very ſmall, and reſemble, in Appearance, a flat cruſty Wart, or ſome ſuch Excreſcence.

The ſmall Part is very ſoft, when compared to any other of the Snail-kind.

The Inſide likewiſe is of a tough, reddiſh, Jelly-like Conſiſtence.

They ſtick to, feed upon, and thus deſtroy, ſeveral Kinds of ſucculent Vines, eſpecially the Granadillo-Vines.

HAVING now taken a View of the Brute-Animal Part of the Creation, let us turn our Thoughts to the great Author of all Beings; and gratefully acknowlege his Bounty, in making Man the Lord over the Whole: Let us admire that Wiſdom, which is always preſent to every one of them, ſupports and governs them.

And

And left, upon Comparison with their supposed Meanness, we should think too highly of our own Knowlege; lest our scanty Portion of Wisdom and Power make us vain; let us make a Comparison between ourselves and our Creator.

After this Survey we shall find very little Occasion of Boasting; we shall find, that our greatest Knowlege is but splendid Ignorance; and that we see no farther than the Surface and the Outside of Things, as directed by the general Laws of Motion. All beyond this is mere Guesswork, Conjecture, and Uncertainty.

Let the wisest Man go out of himself, and survey the immense Extent of Nature, the Variety of its Works, the Regularity of its Motions, and the Harmony of Providence; and let him seriously pronounce, how little is his real Knowlege, how great his Ignorance!

Let him take a Prospect of the vast Dimensions of those astonishing Heaps of Matter, that lie within the Reach of his Senses: Let him consider the stupendous Motion that agitates the vast Mass of Matter, and whirls about the numberless Bodies, that take their Courses thro' the unmeasurable Space; and carry his Thoughts into that Immensity, where Imagination itself can find no Limits: Let him consider that infinite Duration, which is before and after him; and, finding his own Life included in it, let him observe the little Scantling of it, that falls to his Share.

It is just to acknowlege (what hath been undesignedly omitted), that some of the above Reflections, as well as Part of those in Page 88. are owing to the learned Mr. *Hildrop*'s Works.

The End of the THIRD BOOK.

THE

THE

NATURAL HISTORY

OF THE

Iſland of *BARBADOS.*

BOOK IV.

Of VEGETABLES.

IN treating of theſe uſeful and beautiful Parts of the Creation, I ſhall take particular Notice of ſuch as are curious in their Make, or uſeful in their Phyſical, or other remarkable Qualities: And when I am obliged to differ materially, or circumſtantially, from other Writers, it is with no other View but that of diſcovering the Truth; and after the joint laborious Inquiries of the paſt, as well as the preſent Age, into ſo extenſive a Subject, we may ſtill ſay with *Seneca*, *Multa venientis ævi populus ignota nobis ſciet. Multa ſeculis tunc, cum memoria noſtri exoleverit, reſervantur*; *i. e.* "Many are thoſe "Secrets which are hid in Obſcurity from the preſent Age, and are reſerved "to bleſs Poſterity with their Diſcoveries."

In our Inquiries into this Part of the Creation, we ſhall be entertained with an agreeable Mixture of Knowlege, Profit, and Pleaſure. We ſhall

 find

find every Individual replete with unerring Deſign, Elegancy, and Wiſdom, each contributing to the Harmony of the Whole.

The Eye is raviſhed with beholding the majeſtic Height, and agreeable Verdure, of ſome; whilſt the Flowers of others delight us with their Profuſion of Sweets, and varied Beauties: "And, (6) could our Eyes attain "to the Sight of the admirable Texture of the Parts, on which the ſpe- "cific Difference in Plants depends, what an amazing and beautiful Scene "of inimitable Embroidery ſhould we behold! What Variety of maſterly "Strokes of Machinery! What evident Marks of conſummate Wiſdom "ſhould we be entertained with! The little that we do know of their "Make and Structure is ſurpriſing!"

What Power leſs than Divine could provide them with ſuch Number and Variety of Veſſels, endued with dilating and contracting Qualities, capable, by Expanſion, to give room to the excrementitious Juices to fly off, and by Contraction to force back their remaining unconcocted Parts to be purified by a farther Circulation and Secretion!

How greatly are we at a Loſs, even to conjecture, much more explain, how an Aloe-Plant, and a Sugar-Cane, tho' planted in the ſame Tub of Earth, and abſorbing the ſame common Juices; yet each ſhall concoct theſe Juices, ſo as to render them ſuitable to their reſpective Natures; the one producing a very ſweet, the other the moſt bitter of Juices!

If from the amazing Structure, and inimitable Beauty, of Plants, we explore their other Uſes, we ſhall find their great Author's Bounty, not, like the Patriarch's, confined to one Bleſſing, but as boundleſs as the World: Theſe are not only pleaſant to the Eye, but good for Food. The Delicacy and Poignancy of their Fruits, the Fragrancy of their Gums, Oils, and Balſams, their healing Qualities, and the grateful Taſte of their Juices in Liquors---all proclaim, that theſe are thy Works, O God! and convince us, that in Wiſdom thou haſt made them all.

As this Treatiſe, with regard to Vegetables, is confined within a narrow Sphere, it cannot be ſuppoſed to compriſe Rules for a regular Syſtem of Botany; nor doth it therefore require me to deſcend to every minute Circumſtance neceſſary, and expected only to be expatiated upon by thoſe who undertake to write Botanic Syſtems.

The Method I have taken, in claſſing the Plants by their Fruits, is, I hope, the ſhorteſt and plaineſt, and conſequently the eaſieſt to be underſtood by common Readers.

It would perhaps ſeem an invidious Task every-where to particularize many of thoſe leſſer Circumſtances, wherein I am obliged to differ from other Authors. Mr. *Miller*, of *Chelſea*, as far as he hath ſeen our *Weſt-India* Plants, hath deſcribed them with great Judgment and Accuracy.

As to my *Weſt-India* Readers; when theſe ſhall compare my Deſcriptions with the Originals, they muſt do me and themſelves the Juſtice to chooſe Specimens from perfect Plants, growing in ſuch a Situation as they naturally love: For the ſame Species of Plants, which would flouriſh, and grow

(6) *Vide* Dr. *Hale's* Veget. Stat.

grow luxuriant, in a particular Soil, or in the Shade, might perhaps be scarce kept alive, if too much exposed to the Sun, or unnaturally situated: Nor are Specimens from Hot-houses in cold Climates to be depended upon; for the Inconstancy of artificial Heat, causing sudden Transitions from the required natural Degree of Heat to unnatural Cold, frequently occasions great Alterations in them. This, among many other Instances, may be clearly seen in the irregular Growth of the *American* Torch in most Green-houses.

And, indeed, after all our Accuracy, it must be owned, that as the Growth of Plants depends so much, even in their natural Climates, upon the different and uncertain Seasons of the Year, it would be more rash than judicious to pretend absolutely to ascertain the determinate Size of the Leaves, or the exact Growth of any Species of Plants.

Nor is the Colour of the same Plant always the same. Thus the young Leaves of the Sea-side Grape are intirely red; but, when old, altogether green, except the middle and transverse Ribs.

It will likewise be of Use to observe, that as, in the Infancy of settling this Island, Caprice and Chance had a greater Share, than Botanic Skill and Judgment, in imposing Names upon our Vegetables; several of our Plants have many and different Names in different Parts of the Island.

With this Caution to my Readers, I shall, I hope, not only myself escape the Charge of Omissions, but they will likewise avoid that Confusion, which would otherwise, in the Course of this Work, be almost unavoidable.

As to our several Trees, Shrubs, and Plants, that abound with Oil, Gums, or Balsams; the curious Inquirer must likewise take Notice, that these, generally speaking, are not to be expected but from Vegetables that are full-grown, and at particular Seasons of the Year.

After all my expensive Care and Trouble, and a strict Attachment to Truth in all my Observations; yet I make no doubt but this Treatise hath many unavoidable Imperfections. For, to speak with the Poet,

Whoever thinks a faultless Piece to see,
Thinks what ne'er was, nor is, nor e'er shall be.

POPE's Essay on Crit.

Of Nuciferous TREES, SHRUBS, *and* PLANTS.

THO' theſe are not generally eſteemed among the moſt delicious Fruits; yet the Kernels of many of them are not only agreeable to the Taſte, but valuable alſo for the great Quantity of Oil they afford; nor are they deſigned for the Uſe of Man only, but likewiſe to be the moſt common, and almoſt the peculiar Food of a certain Species of (1) Animals.

Thoſe who think meanly of this, or any Part of the Creation, are generally Men of narrow Conceptions, and ſcanty Knowlege, which barely skims over the Surface of Things

Hence ariſes that partial Judgment, which would confine the Goodneſs of God, as the (2) *Syrians* of old would his Power, within narrow and local Bounds.

Theſe ſhort-ſighted Obſervers conſider not, that the Chain, of which they ſee but a Part, reaches to the Ends of the Earth.

When we extend our Views, and give our Inquiries a greater Scope, the more will our Ideas be inlarged, when we find the common Father of All, every-where preſent, pouring forth his Bleſſings with a bountiful Profuſion; then every diſtinct Part, which we had before inadvertently looked upon as mean and trivial, will gradually riſe in our Eſteem, and appear in its native Beauty.

For Inſtance; Let an Inhabitant of a Northern Climate, where theſe Fruits we are treating of, are not very inviting to the Palate, trace the different Species of this kind into other Regions, growing under the benign Influence of a warmer Sun; let him examine the various Species which the lofty Palms alone produce, and he will find them no leſs numerous than uſeful, till he comes at laſt to the moſt remarkable and valuable of all others, the Coco-nut.

When he ſees and conſiders the admirable Texture of the Whole of this ſurpriſing Fruit, eſpecially the thick husky Covering with which it is caſed, to prevent the piercing Rays of the Sun from abſorbing the limpid Water it contains, ſo grateful to Man, and ſo neceſſary to give Riſe and Nouriſhment to its delicious Kernel: This, and its Growth, will appear ſtill more wonderful, as well as more replete with Wiſdom and Goodneſs; eſpecially when he conſiders the Situation of thoſe Countries, and the Soil wherein they generally grow; and how great will his Admiration be, when he finds, that Providence hath ordained theſe to thrive and flouriſh in the moſt barren and ſandy Deſarts of *Ethiopia*, *Africa*, and *Arabia*, where Nature denies almoſt every other Herbage!

Let us likewiſe (at leaſt in Imagination) follow our Fellow-Creatures the Inhabitants of theſe Climates; let us fanſy we view the fainting Traveller paſſing thro' theſe inhoſpitable Deſarts, where neither a grateful Verdure chears the Eye, nor the Murmurs of even a diſtant Stream the Ear; view him hungry and thirſty, doubting, murmuring, and almoſt ſaying aloud with the *Iſraelites*, *Can God find Food in the Wilderneſs?* when, on a ſudden, ſome intervening Hillock is paſt, which forbad the pleaſing Proſpect; a

(1) Squirrels. (2) 1 *Kings* xx. 23.

thick Grove of these Trees appears. The Beauty of their grateful Verdure, their waving Foliage, and the cool Shade which they promise, infuse fresh Vigour; a few Moments more he is reclined under its welcome Shade, and there refreshed with the most delightful Viands these Nuts afford.

Thus, as in the respective Stages of human Life, Providence hath provided Pleasures and Enjoyments suitable to every distinct Period of it; so in every Climate the same unerring Hand hath adapted to each, by Weight and Measure, its proper peculiar Advantages, generally far overbalancing its Inconveniencies.

If *Egypt* is denied the refreshing Rains; the *Nile* supplies that seeming Want: If *Palæstine* is not blessed with such a providential Alotment; yet its Hills and Valleys copiously drink of the Dew of Heaven: The Hills, by their high Summits, arrest the passing Clouds, and cause them to descend on the Valleys in the former and the latter Rain.

Where dreary Desarts deny Support to the generous Horse, there the laborious Camel, patient of great Labour, and long Thirst, is adapted by Nature to bear the Wants and Fatigues of Journeys thro' such Sun-burnt Regions.

In short, were we capable of taking a general Survey of the Face of Nature, we should find, that Providence hath enriched every Climate with Blessings peculiar to itself, and adapted to the Necessities of its Inhabitants; and for this Reason we may conclude, that the Abundance of Coco Nut-trees found in this Island, and such Places, was graciously calculated and intended for the Comfort and Refreshment of those whom Providence hath there placed.

The DATE-TREE.

THIS Tree grows to a considerable Height, often Fifty Feet.

Its Leaves much resemble those of the Cabbage-tree; but their *Pinnæ* are harder, and aculeated.

The Branches, as well as the Spines or middle Ribs of the Leaves, are guarded with Three-inch-long triangular Prickles, always chanelled on one Side.

The straitest and youngest Branches which grow near the Summit of the Tree, are much used here by the *Jews*, upon their Feast of the Tabernacles: These they usually gild, and adorn them with various Flowers, and then carry them in Procession to their Synagogue. Whether this is the same kind of Palm that was used by the *Israelites*, we know not, or whether it is not here succedaneously used as bearing the nearest Resemblance to it. The Branches are set on *squammatim*, leaving, when they fall, very deep Impressions upon the Trunk.

The Dates grow in large spiral Clusters, about the Bigness and Shape of a middling Olive; but never come to the Perfection here of being eatable; for they have a very austere and acerb Taste; yet the Swine feed upon them greedily: This crude rough Acidity of the Fruit shews, that they have a great deal of essential Salts.

THE

THE

COCO NUT-TREE; *Lat.* PALMA NUCIFERA.

THO' every Part of this Tree, either in its Make or Uſe, is ſtamped with ſo many evident Signatures of Divine Wiſdom, as to make it juſtly the Object of our Admiration; yet, in deſcribing it, we ought not to add to it (as moſt Writers have done) pretended Qualities, and Excellencies foreign to its Nature.

For as the Light of the Sun cannot be heightened by that of a Candle, ſo neither can the perfect Works of God require the Aid of our weak, much leſs falſe Embelliſhments, to ſet them forth in a ſtronger and more advantageous Light, leſt he ſhould ſay unto us, Who requireth this Thing at your Hand?

As the ingenious Mr. *Ray* was one of the firſt Perſons of Character who was impoſed upon, moſt other Authors ſince his Time, depending upon his Deſcription, have repreſented this Tree as actually capable of, and really producing, Bread, Water, Wine, Vinegar, Brandy, Oil, Honey, Cups, Spoons, Beſoms, Maſts, Nails, Needles, and Covering for Houſes.

But, ſince many of theſe Qualities are merely chimerical, I ſhall endeavour to deſcribe this Tree according to its outward Appearance, and real Properties.

Its Roots are many, very much reſembling, in Colour, Make, Texture, and Extent, thoſe of the Cabbage-tree, being very ſmall, and many in Number.

The Trunk of this likewiſe ſomewhat reſembles the former, having near the Top, ſurrounding the Heart of the Tree, many Flakes of that eatable Cabbage-like Subſtance.

But the Body of this Tree hath no Claim to a juſt Proportion in Growth, being often near as thick at Thirty Feet high, as it is within Three Feet from the Ground; and it generally leans one Way or another, occaſioned, in ſome meaſure, by the great Weight of Nuts it ſuſtains whilſt young, which determines the bending of the Tree, which Side ſoever they project; and, if likewiſe they grow upon the Brink of a Pond, the Trunk of ſuch Trees always bend towards the Water.

Some of them grow to Sixty Feet high.

The Bark is more deeply pitted by the *Veſtigia* of the fallen-off Branches, than that of the Cabbage-tree.

There is likewiſe this very remarkable Difference: That this keeps its Aſh-colour Bark to the very Place where the Branches begin; the other, as hereafter deſcribed, always diſcovers the Space of near Five Feet, of a beautiful deep Green, between the Aſh-coloured Bark, and the Part where the Branches begin: The lowermoſt Branches likewiſe in a Cabbage-tree expand almoſt horizontally; whereas thoſe of a Coco Nut-tree bend more wavingly and irregular.

The Branches come to their full Growth in about Three Months, and are, in a flouriſhing Tree, about Twenty-eight in Number.

Every Branch hath a great Number of pennated Leaves: Theſe are of different Size, the largeſt always near the Trunk, the reſt growing leſs towards the Extremity of the Branches: The former are often near Three Feet long, and reſemble, in their Make, thoſe of the Cabbage-tree, except that they want thoſe ſtrong Thread-like Filaments, which the Cabbage-leaves afford.

And as the Branches of the Coco Nut-tree are often about Fifteen Feet long, they would be liable to be broken down by high Winds, eſpecially thoſe that are preſſed upon by the Weight of the large Bunches of Coco Nuts. Indulgent Nature hath therefore fortified and wrapped the Foot-ſtalks of every Branch with a ſtrong, cloſe-woven, reticulated, web-like Subſtance ariſing from the Tree. This Subſtance is of a reddiſh Colour.

What is farther remarkable in this is, that it becomes peculiarly adapted to the different Growth of the Branches; for, as theſe increaſe in their Growth, by eaſy gentle Degrees, this Web, tho' of a very ſtrong Conſiſtence, is yet ſo elaſtic and pliable, that, growing with the Branch, it never prevents its Growth by too cloſe a Bandage, nor ſuffers it to be liable to the Injuries of Storms by too lax a Stricture.

This Contrivance of Nature (if I may uſe ſo low an Expreſſion) is not only curious in its web-like Make; but, being often a Foot and an half long, may, where extreme Neceſſity calls for ſuch Shifts, be capable of being ſewed together, and made into a kind of Garment, which would at leaſt keep off the Heat of the Sun: And this is the moſt probable Pretenſion the Writers of Wonders have, to ſay that this Tree affords Cloaths.

As to its being fit for Maſts of Veſſels; this is a moſt prepoſterous Suppoſition, or rather an Impoſition upon Mankind; for, without being well verſed in Sea Affairs, every one is ſenſible, that a Maſt of any Veſſel ought to be, and always is, made tapering from the Bottom to the Top; whereas this Tree is, for the moſt part, almoſt as thick at the ſetting on of the Leaves, as it is near the Ground: And as the Inſide is ſoft and

pithy

pithy, and its whole Strength conſiſting in the outward ſtrong ligneous Part, which is not above Two Inches thick; this prevents the Poſſibility of its being made tapering by the Uſe of Tools.

As to many other of its pretended Qualities, they are as groundleſs as this.

But to return to my Subject: This Tree, if planted in a rich ſandy Soil, will begin to bear Fruit in the Fourth or Fifth Year; ſometimes not ſo ſoon, if the Soil and Situation are not indulgent to it.

Its firſt Appearance of bearing Fruit is a ligneous, pod-like, husky *Spatha*, of about Two Feet long, and about Three Inches broad, narrower towards its Footſtalk than towards its Extremity. The Whole is of a browniſh Green on the Outſide, and white within: It comes to its full Growth in Three Months from its firſt Appearance.

As the Foot-ſtalk is continued thro' this husky *Spatha*, there it wreaths and twines itſelf, as well as its partitional ſurrounding Twigs, into many Bendings. When the *Spatha* comes to its Maturity, it burſts in the Middle, on the upper Side; and the incloſed Twigs appear thick-ſet with pale-white Flowers, inclinable to a yellow Colour.

Each of theſe are compoſed of Three rigid ſharp-pointed Leaves.

From the Middle of theſe riſe Six yellow *Stamnia*, with a ſhort whitiſh Style, cleft at the Top into Three Parts: Theſe Flowers have a weak faint Smell; and, in about Five Days from their firſt Appearance, they fall to the Ground: Then the Nuts, by degrees, are formed: Theſe, when young, are round, and their Shell of a ſoft Texture. The outward husky Covering is of a ſomewhat reddiſh Colour; each Nut round the Foot-ſtalk being cloſely embraced with the *Calyx*, which is compoſed of many ſtiff roundiſh Leaves, which cloſely ſtick to the Nut, like ſo many laminated Scales.

The Cavity in the Nut, whilſt thus young, is narrow and ſmall, growing both larger and rounder with Age.

Each Nut, on the Part next the Stem, hath in it Three Holes cloſely ſtopped; one of theſe being both wider, and more eaſily penetrated, than the reſt: From this, when the Nut is planted, riſes the *Germen*, or young Tree: Thro' theſe Holes likewiſe is the Water copiouſly diſtilled into the Nut from the Roots of the Tree. This is evident from the Saltneſs of the Liquor in theſe Nuts, if the Tree grows near a brackiſh Spring; but, if they are planted in an agreeable Soil, the more ſandy and looſe, the better. The Water in young Nuts, from ſuch Trees, is very limpid, and extremely ſweet.

When the Kernel begins to grow, it incruſts the Inſide of the Nut in a bluiſh jelly-like Subſtance: As this grows harder, the incloſed Liquid becomes ſomewhat acid, but ſtill of a ſweetiſh agreeable Taſte, and far more palatable than any Mixture of theſe Two Ingredients from the beſt of Artiſts.

As

As the Nut ripens, the Kernel becomes ſtill more ſolid; and lines at laſt the whole Inſide of the Nut for above a Quarter of an Inch thick as white as Snow, and of the Flavour of an Almond.

Theſe Nuts, when at their full Growth, which they generally are in Six Months after their firſt Appearance, contain from half a Pint to ſometimes above a Pint of the above-mentioned Liquor.

In about Three Months after the Kernel is in its full Perfection, the husky or outward Coat of the Nut, in a ſhort time, begins to be ſomewhat ſhriveled; and the Nut ſoon after falls from the Tree. This happens in the Tenth Month from its firſt Appearance.

In this Stage the whole Fruit is of a blunt oblong Shape, near, and often above, a Foot in Length, and otherwiſe proportionable: The Outſide is likewiſe bluntly marked in a triangular manner: The husky Part, which covers the Nut, for near an Inch thick, is of a pale Brown on the Surface, and of a reddiſh Colour within. This conſiſts of ſo many ſtrong ſtringy Filaments running lengthwiſe, that it is not without Difficulty ſeparated from the Nut; but when taken off, and cleared of the intermixed pithy Subſtance, it then reſembles coarſe Oakum; and it is ſo called by moſt Authors, and may perhaps, in Caſes of Neceſſity, be uſed as ſuch.

The Nut likewiſe, when cleared of its husky Tegument, appears ſlightly marked on the Outſide with Three Sutures, the whole Nut being of a blackiſh Colour.

If theſe are planted, the Kernel will, ſome time after, firſt bud, and ramiſy, and fill the whole Cavity of the Nut; and then ſhoot out at the above-mentioned Hole in the Top, and ſoon appear above-ground in Two narrow Leaves.

As the Kernels of moſt Nuts are more or leſs pregnant with Oil, it will perhaps be here worth obſerving, that tho' Oils in general, when in the common Way extracted, are not miſcible with Water; yet they may be brought to mingle with it in the Form of Cream.

For Inſtance; Let any Quantity of Coco Nut-kernel be pounded in a Mortar, with a Quantity of Water ſufficient to penetrate into the broken Interſtices of the Kernel: This, being again mixed with a great Quantity of Water, and ſuffered to ſettle for a few Hours, will neceſſary contract a white creamy Subſtance upon the Surface, in which Form the Oil then appears.

Its Parts, by being thus pounded, and abſorbed in the branny Subſtance of the Kernel, and mixt with Water, are rendered too ſmall to cohere, and to form viſible Drops.

On the other hand, if the Oil, drawn from the Kernel by Expreſſion, be ſhaken ever ſo much in Water, it will almoſt inſtantaneouſly collect its ſeparated Parts, and form itſelf into its genuine tenacious oily Drops: Which Oil, when firſt expreſſed, is very mild, and of an inſipid Taſte; but in a few Days, unleſs kept very cool, which is ſcarce practicable in hot Climates, it will become rancid.

The

Pl. 3 Page 107
To Her Royal Highness
The Princess of Wales
This Plate is humbly Inscrib'd
&c.
Mynde Sculp.

The great *Boerhaave* obſerves, that the freſh Oil of Sweet-Almonds, taken in a *Linctus*, proves an excellent Emollient, where the *Fauces* are grown rough with an *Angina* : And yet, if the ſame ſhould be taken by a Perſon in Health, after it had ſtood in the Heat of a Summer's Day, it would corrode and inflame them. Add to this, that the ſweeter theſe Oils are, when freſh, the more acrid they are, when they are old and rancid.

Hence the ſame ingenious Author obſerves, how abominable are Almonds, Walnuts, and Piſtach-nuts, when they once become rancid ; and how eaſily they will cauſe an *Angina*, and raiſe a Fever, by inflameing the Throat, Oeſophagus, Stomach, and Inteſtines.

How inexcuſable therefore muſt thoſe Practitioners in Phyſic here be, who for the above Purpoſes uſe the rancid Oil of Almonds, when they might every Day extract the beſt and ſweeteſt of Oils from the Coco Nut-kernel !

From this ſeeming Digreſſion, concerning the Nature of Oils, we may proceed to obſerve, that, if the Liquor in the Nut be firſt fermented, I make no doubt of its producing, upon Diſtillation, a ſtrong Spirit ; and likewiſe, if a conſiderable Quantity of it were boiled for a long time, it would probably yield a ſmall Quantity of Sugar.

This Tree is delineated in *Plate* III.

The BARREN COCO NUT-TREE.

THIS differs very little, if any thing, in its Shape and Texture, from that already deſcribed : However, this never bears any Nuts, but an husky Subſtance only, reſembling in Shape a common Coco Nut.

Theſe are ſo rare, that I never ſaw any myſelf ; but had them deſcribed by a Perſon of great Veracity.

The CABBAGE-TREE ; *Lat.* PALMA MAXIMA.

THIS Tree is by ſome Authors called *The Palmeto Royal.* And well may it be called Royal from its great Height, majeſtic Appearance, and Beauty of its waving Foliage : Neither the tall Cedars of *Lebanon*, nor any of the Trees of the Foreſt, are equal to it in Height, Beauty, or Proportion ; ſo that it claims among Vegetables that Superiority which *Virgil* gives to *Rome*, among the Cities of *Italy* :

> *Verum hæc tantum alias inter caput extulit urbes,*
> *Quantum lenta ſolent inter viburna cupreſſi.* Ecl. I.

> Imperial *Rome* o'er other Cities tow'rs,
> As lofty Cypreſs humble Shrubs o'erpow'rs.

Its Roots are innumerable, reſembling ſo many round Thongs, of a regular determinate Bigneſs, ſeldom exceeding the Size of the little Finger, but of a great Length, penetrating ſome Yards into the Earth, eſpecially where the Soil is ſandy, or otherwiſe porous: Theſe Roots are of a dark-brown Colour.

The Trunk jets or bulges out a little near the Ground, by which means it hath the becoming Appearance of a ſubſtantial Baſis to ſupport its towering Height. It is generally as ſtrait as an Arrow; and ſcarce can a Pillar of the niceſt Order in Architecture be more regular, eſpecially when it is of about Thirty Years Growth: And as there is a natural involuntary Pleaſure ariſing from the Harmony of juſt geometrical Proportions, ſtrikeing the Eye of the moſt unskilful and ignorant Beholder, it is not ſtrange that theſe Trees are univerſally admired.

Writers of Wonders repreſent ſome of them to be Three hundred Feet in Height: However, the higheſt in this, where they are more numerous than in any of our neighbouring Iſlands, is but an Hundred and Thirty-four Feet.

The Trunk of this, near the Earth, is about Seven Feet in Circumference, the whole Body growing tapering to the Top.

The Subſtance of the Tree, for about Two or Three Inches of the Outſide, but within the Bark, is of a blackiſh Colour, and extremely hard and ſolid: This ſurrounds the inner Subſtance, which is a whitiſh Pith, intermixt with ſome ſmall Veins of a more ligneous Texture.

The Colour of the Bark much reſembles that of an Aſh-tree, and is very faintly clouded, at about the Diſtance of every Four or Five Inches, with the *Veſtigia* of the fallen-off Branches: This Colour of the Bark continues till within about Twenty-five or Thirty Feet of the Extremity of the Tree: There it alters at once from an Aſh-colour to a beautiful deep Sea-green, and continues to be of that Colour to the Top.

About Five Feet from the Beginning of the green Part upwards, the Trunk is ſurrounded with its numerous Branches in a circular Manner; all the lowermoſt ſpreading horizontally with great Regularity; and the Extremities of many of the higher Branches bend wavingly downwards, like ſo many Plumes of Feathers.

Theſe Branches, when full-grown, are Twenty Feet long, more or leſs; and are thickly ſet on the Trunk alternately, riſing gradually ſuperior one to another: Their broad curved Sockets ſo ſurround the Trunk, that the Sight of it, whilſt among theſe, is loſt, which again appears among the very uppermoſt Branches, and is there inveloped in an upright green conic Spire, which beautifully terminates its great Height.

The above-mentioned Branches are ſomewhat round underneath, and ſlightly grooved on the upper Side: They are likewiſe decorated with a very great Number of green pennated Leaves: Some of theſe are near Three Feet long, and an Inch and an half broad, growing narrower towards their Points, as well as gradually decreaſing in Length towards the Extremities of the Branches.

As

As there are many Thousand Leaves upon one Tree, every Branch bearing many Scores upon it, and every Leaf being set at a small and equal Distance from one another, the Beauty of such a regular lofty Group of waving Foliage, susceptible of Motion by the most gentle Gale of Wind, is not to be described.

The middle Rib, in each Leaf, is strong and prominent, supporting it on the under Side, the upper appearing smooth and shining. The pithy Part of the Leaf being scraped off, the inside Texture appears to be so many longitudinal thread-like Filaments.

These, being spun in the same manner as they do Hemp, or Flax, are used in making Cordage of every Kind, as well as Fishing-nets, which are esteemed stronger than those usually made from any other Material of the like Nature.

It is observed, that the lowermost Branch, for the Time being, drops monthly from the Tree, carrying with it an exfoliated circular *Lamen* of the green Part of the Tree, from the Setting on of the Branches to the Ash-colour Part, which is about Five Feet in Length, and, in Breadth, the Circumference of the Tree at that Part. This, and the Branch to which it is always fixed, fall together.

When the Loss of this lower Leaf happens, then the green conic Spire, which issues from among the Centre of the uppermost Branches, and rises superior to all, partially bursts, and thrusts from its Side a young Branch, which continues the uppermost, till another of the lowermost Branches drop off: Then the Spire, the common Parent of all the Branches, sends forth again another Branch, superior in Situation to the last: So that the annual Loss of the Branches below is providentially supplied in this manner by those above.

The green-coloured Part of the Tree, already mentioned, differs from the Ash-coloured Part no less in Substance than Colour: The former, instead of being extremely hard on the Outside, and pithy within, is composed of so many Coats, or separate *Laminæ*, of a tough bark-like Substance of near a Quarter of an Inch thick, and so very closely wrapped together, that they jointly compose and constitute that green Part of the Tree.

As the lowermost, as well as each other higher Branch, when they successively grow to be old, is joined by the broad Socket of its Foot-stalk to this outward Coat, *Lamen*, or Folding, it is observable, that some time before the lowermost Branch is intirely withered, this green circular Coat, which to the Eye appeared some Days before to be a solid Part of the Tree, slits open lengthwise, from the Setting on of the Branches to the Ash-coloured Part beneath, being about Five Feet in Length, and the Circumference of the Tree in Breadth; and, peeling off, it falls with the falling Branch to which it is joined by many strong Cartilages, leaving the next succeeding Coat, to appear for a time as a constituent Part of the Tree, till a succeeding withered Branch carries this off likewise.

Having

Having felled one of these stately Trees, to examine its Make, Texture, *&c.* I observed, that the several Exfoliations of its green Part were equal in Number to the Branches.

The First, Second, Third, and sometimes the Fourth of these *Laminæ* are green on the Outside, and perfectly white within: All the remaining inner Coats, or Foldings, are of a bright Lemon-colour without, and white within.

When these very tough husky Exfoliations are taken off, what is called the Cabbage, lies in many thin, snow-white, brittle Flakes; in Taste something like the Kernel of an Almond, but sweeter: It is so full of Oil, that a curious Observer may see several very small Cells abounding with it. These Flakes are called, from some Resemblance, when boiled, the Cabbage, which then eats somewhat sweet and agreeable: Yet I have always thought it the Height of Extravagancy and Luxury to fell so stately a Tree, which would be an Ornament to the most magnificent Palace in *Europe*, to gratify the Taste of any Epicure, especially as there is but a very small Part of it eatable.

What is called the Cabbage-flower, grows from that Part of the Tree where the Ash-coloured Trunk joins the green Part already described.

Its first Appearance is a green husky *Spatha*, growing to above Twenty Inches long, and about Four broad; the Inside being full of small white stringy Filaments, full of alternate protuberant Knobs, the smallest of these resembling a Fringe of coarse white Thread knotted: These are very numerous, and take their Rise from larger Footstalks; and these Footstalks likewise are all united to different Parts of the large Parent-stalk of all.

As this husky *Spatha* is opened, while thus young, the farinaceous yellow Seed, in Embryo, resembling fine Saw-dust, is very plentifully dispersed among these stringy Filaments, which answer the Use of *Apices* in other more regular Flowers: These Filaments, being cleared of this Dust, are pickled, and esteemed among the best Pickles, either here or in *Europe.*

But if this *Spatha* is not cut down and opened, whilst thus young; if it be suffered to continue on the Tree till it grows ripe and bursts; then the inclosed Part, which, whilst young and tender, is fit for pickling, will, by that time, have acquired an additional Hardness, become soon after ligneous, grow bushy, consisting of very many small Leaves, and in time produce a great Number of small oval thin-shelled Nuts, about the Bigness of unhusked Coffee-berries: These, being planted, produce young Cabbage-trees.

The greatest Number of these Trees uncultivated, are in Mrs. *Alleyne*'s Wood, in that Part of the Island called *Scotland*; at Mr. *Holder*'s, and at *Codrington*'s-*College*; and by far the longest planted Walk is at Mr. *Ball*'s Estate, commonly called *Farmer's Plantation* in *St. Thomas*'s Parish: But the most regular in Growth, Proportion, and Beauty, either in this Island,

Plate IV.
p. 111.
HONI SOIT QUI MAL Y PENSE
ICH DIEN
To His Royal Highness
FREDERICK,
Prince of Wales,
This Plate is most humbly Inscrib'd
&c.
G. D. Ehret. delin. & sculp.

Iſland, or perhaps in the World, is a ſmall Walk at the Eſtate of Mrs. *Warren*, at the *Black-Rock*; and another, tho' much inferior to it, at the Eſtate of *Joſeph Terril*, Eſq; in *St. Lucy's* Pariſh.

It is obſerved, that when theſe ſtately Trees are felled, either intentionally, or by the Violence of the Wind, there breeds in the Pith of the Trunk a kind of Worms, or Grubs, about the Size and Length of the firſt Joints of the Thumbs: Theſe are eaten, and eſteemed great Delicacies, by the *French* of the neighbouring Iſlands.

Theſe Trees grow on the Tops of Hills, as well as in Valleys. The hard outſide Part makes very durable Laths for Houſes.

This beautiful Tree is delineated in *Plate* IV.

The ſmall CABBAGE-TREE.

THIS, even at its greateſt Growth, is far leſs than the Cabbage-tree already deſcribed, having likewiſe the green Part, near the Top, much leſs in proportion to its Bulk, than the former; and the *Spatha* likewiſe is ſomewhat redder.

As the Trunks of theſe Trees are ſo convenient to make long and durable Pipes, or Gutters, to convey Water, and other Liquids, from one Building to another, there are ſcarce any of them left growing in the whole Iſland.

The PALM OIL-TREE; *Lat.* PALMA OLEOSA.

THIS grows about Fifty Feet high.

Its Roots much reſemble thoſe of the Cabbage, or Coco Nut-tree.

The Trunk is leſs tapering than the former, but generally ſtraiter than the latter: It is likewiſe very deeply marked with the *Veſtigia* of the fallen-off Branches, tho' not ſo ſcaly and rough as that of the Date-tree.

Its Branches ſomewhat reſemble thoſe of the Cabbage, or Coco Nut-tree; but they are far more ragged, leſs uniform, ſhorter, and leſs verdant, than either the Cabbage, or Coco Nut-leaves.

The Spine or middle Rib of each Leaf is likewiſe thickly ſtudded with ſharp-pointed Prickles, each Two Inches long.

The firſt Appearance of the Fruit is an husky *Spatha*. When this opens, it expoſes to view a great many ſmall Nuts: Theſe are covered with an husky Tegument of a yellowiſh Colour, containing in its many Interſtices, when ripe, a conſiderable Quantity of fine ſweet Oil, which the Slaves, after the whole Fruit is firſt roaſted in the Embers, greedily ſuck.

When this outward husky Covering is taken off, the Nut appears: This is of a ſomewhat blunt conic Shape, the Inſide being filled with a white Kernel, of the Nature of the Coco Nut-kind, but in Taſte not ſo agreeable.

The Nut, being bored and emptied of its Kernel, is much worn by ſeveral Nations of Negroes, by way of Ornament, about their Necks.

It is evident from the History of *Moses*, that this Tree was of great Value and Esteem among the *Israelites*. This appears from the several Portraitures of it in the Temple, even in the most sacred Place.

Several Cities in *Palæstine* were likewise called the *Cities of the Palm-trees*, especially *Jericho*: And *Deborah*, when she judged *Israel*, dwelt under the Palm-tree, between *Ramah* and *Bethel*.

These, and the Olive-trees, were of great Esteem; because they afforded such a Quantity of Oil, which was of so much Service, both in sacred and common Use.

As to the First, the Nature of my Subject will neither require, or even permit me to search into the Origin of that Custom. As to the latter, we have numberless Instances to prove, that the *Jews* mixed it with Flour, to make, at least, their unleavened Bread; especially from the Answer of the Widow-woman of *Sarepta* to the Prophet *Elijah*, when he desired her, in the time of Famine, to supply him with a Cake of Bread: *And she said, As the Lord thy God liveth, I have not a Cake, but an Handful of Meal in a Barrel, and a little Oil in a Cruse: And, behold, I am gathering two Sticks, that I may go in and dress it for me, and my Son, that we may eat it, and die*, 1 Kings xvii. 12.

Neither was this the only domestic Use of Oil among the antient *Jews*; for they made great Use of it to anoint their Bodies, which is plain from several Expressions of the Royal Psalmist; especially when he says, That he *shall be anointed with fresh Oil*: And in another Place, after enumerating the several temporal Blessings received from God, he mentions that of *Oil* to make his *Face to shine*.

The Inhabitants of *Africa*, where Palm Oil-trees abundantly grow, are living Comments upon these Two Parts of Scripture; for, to this Day, they mix Oil with their boiled Rice, and other Victuals; and anoint their Bodies with it, to supple and relax their stiffened Nerves, as well as to prevent a too plentiful Perspiration.

This is so universal a Custom, that all the Slaves, brought now from any Part of *Africa* to this, or any of our neighbouring Islands, are always, before they are brought to Market, anointed all over with Palm Oil, which, for that Purpose, is brought from *Guiney*: Being thus anointed, their Skins appear sleek and shining.

This Custom, however, was not peculiar, even in the earliest Ages, to the *Jews* alone; for the *Persians*, *Greeks*, and *Romans*, at certain Seasons, especially at their Feasts, anointed themselves with Oil, which was mixed with Spices.

Alexander, when he took the Tent of *Darius*, found there several Caskets of Ointments and Perfumes.

Many Instances from *Homer* make it evident, that it was then in Use: And the warlike *Spartans*, at the Streights of *Thermopylæ*, are represented combing their anointed Hair, in Sight of the numerous Army of *Xerxes*.

And that it was a Custom among the *Romans*, will appear from the following Passages: *Tyrrhena*

Tyrrhena Regum progenies, tibi
Non ante verso lene merum cado
Cum flore, Mæcenas, rosarum, &
Pressa tuis balanus capillis
Jamdudum apud me est. --- Hor. Carm. L. III. Od. 29.

--- *Dum licet, Assyriaque nardo*
Potamus uncti. ---- Hor. Carm. L. II. Od. 11.

These are some, out of the many Instances, such as *perfusus liquidis odoribus---nitidi capilli---* with many others too tedious to be mentioned.

Nor was this Custom intirely disused, nor the Ointment thought of little Value, even so late as the time of our Saviour: For, when a Woman in *Bethany* poured on his Head a Box full of precious Ointment, there were some who had Indignation among themselves, and said, *Why was this Waste of the Ointment made? for it might have been sold for more than Three hundred Pence.*

From what hath been said, with regard to the Custom of Anointing among the *Jews*, I would not be understood to mean, that Palm Oil alone was made use of; for the Oil of Olives is what is chiefly mentioned in Scripture: However, there are some Circumstances, which may induce us to believe, that both these Oils were promiscuously made use of.

This, I think, is very probable, from the several Places of Scripture, where the Growth of the Palm-trees is made a Part of the Blessings of the Land of *Canaan*; and the very natural, easy Method of coming at their Oil makes it more than probable.

These Trees are very scarce in this Island, except at *Drax's-Hall*, the Estate of *Henry Drax*, Esq; and even there they exceed not Twelve in Number.

Among these, there is one young Tree about Twenty-five Feet high, which is remarkably incorporated within a growing Fig-tree; nor was the following manner of its being thus inveloped, less surprising.

About Nine Years ago, a ripe Fig, carried by Birds, or otherwise, was dropped among the upper Branches of the Palm-tree: As these are large and scooping near the Parent Trunk, and being set on in Scales, they retain a considerable Quantity of Dust, and rotten Leaves, as well as Moisture, their curve Shape, and close Juncture to the Trunk, being capable of retaining both. The Fig, deposited in such an earthy moist Place, soon germinated, and took Root; and, in a short time, its new-produced Roots, which extended themselves among the Sockets of the neighbouring Branches, meeting there with the like Nourishment, vigorously grew, and surrounded the Top of the Trunk in several cartilaginous Thongs: From thence, growing downwards, these innumerable fibrous pliable Roots swathed the Tree with many Bandages, which in time reached the Earth, and took fresh Roots: By this means their Growth was soon greatly

greatly advanced, by the Nouriſhment they received from thence. This new Supply gave likewiſe Birth to many Branches of the Fig-tree, riſing from the Top, and growing among the natural Branches of the Palm-tree; the Roots of the other, in the mean time, continuing daily to ſurround the Trunk of the Palm, which, at preſent, being almoſt intirely caſed or covered by the Fig-tree, is deprived of a great Part of its former Nouriſhment; and the few remaining Branches are far from being in a thriving Condition.

- - - - - - So, obſcur'd
By creeping Ivy, or by ſordid Moſs,
Some lordly Palace, or ſtupendous Fane,
Magnificent in Ruin lies. - - - - - Glover's Leonid.

There is another Inſtance, ſimilar to this, on *Horſam's-Hill*, in the Pariſh of *St. Peter*'s; where, in like manner, a Fig, depoſited among the Branches of a Cherry-tree, hath grown, and will, in a few Years, intirely caſe, if not deſtroy it.

The MACKAW-TREE.

THIS is of the Palm-kind, having a great many ſtrong Roots, like thoſe of the Cabbage-tree, or Coco Nut-tree.

The Trunk is very ſtrait, round, and tapering: This is almoſt intirely covered with black Prickles, of about the Length, and very near as fine, as a large Stocking-needle.

The Leaves much reſemble thoſe of a Coco Nut, except that the middle Spine of this is always thickly covered with Prickles, and the Branches always ending in a Square, rather than a ſharp Point, as if their Tops were ſheared or clipped off.

The Inſide of the Body of the Tree affords what we have, in the Coco Nut-tree, called Cabbage.

From among the Branches riſes up an hard falcated *Spatha*, or Sheath, much reſembling that of a *Turkiſh* Scimiter: This is near Three Feet long; and, when it opens, it diſcovers a round cartilaginous ſtrong Twig, which runs the whole Length of the Pod, a great many thin-ſhelled Nuts, and ſomewhat bigger than a Sloe.

The SAND-BOX TREE; *Lat.* HURA.

THIS grows to a large Tree, often to Forty Feet high; yielding a Shade of as many Feet Diameter.

The Trunk, which is of a light-coloured Bark, is thickly beſet with ſhort ſtubbed Prickles; and the Branches, cloathed with ſhining-green ſharp-pointed Leaves, are ſet on alternately: Theſe are about Four Inches long, and Three broad. The Extremity of the Branches bears both Male and Female Flowers.

Plate 5.
Pa. 114
Fructus
To his Grace the Duke of Dorset,
Lord President of his MAJESTY'S
most honourable Privy Council & Kn.t
of y.e most Noble Order of y.e Garter,
This Plate, is humbly Inscrib'd. &c.
G. D. Ehret delin. et Sculp.

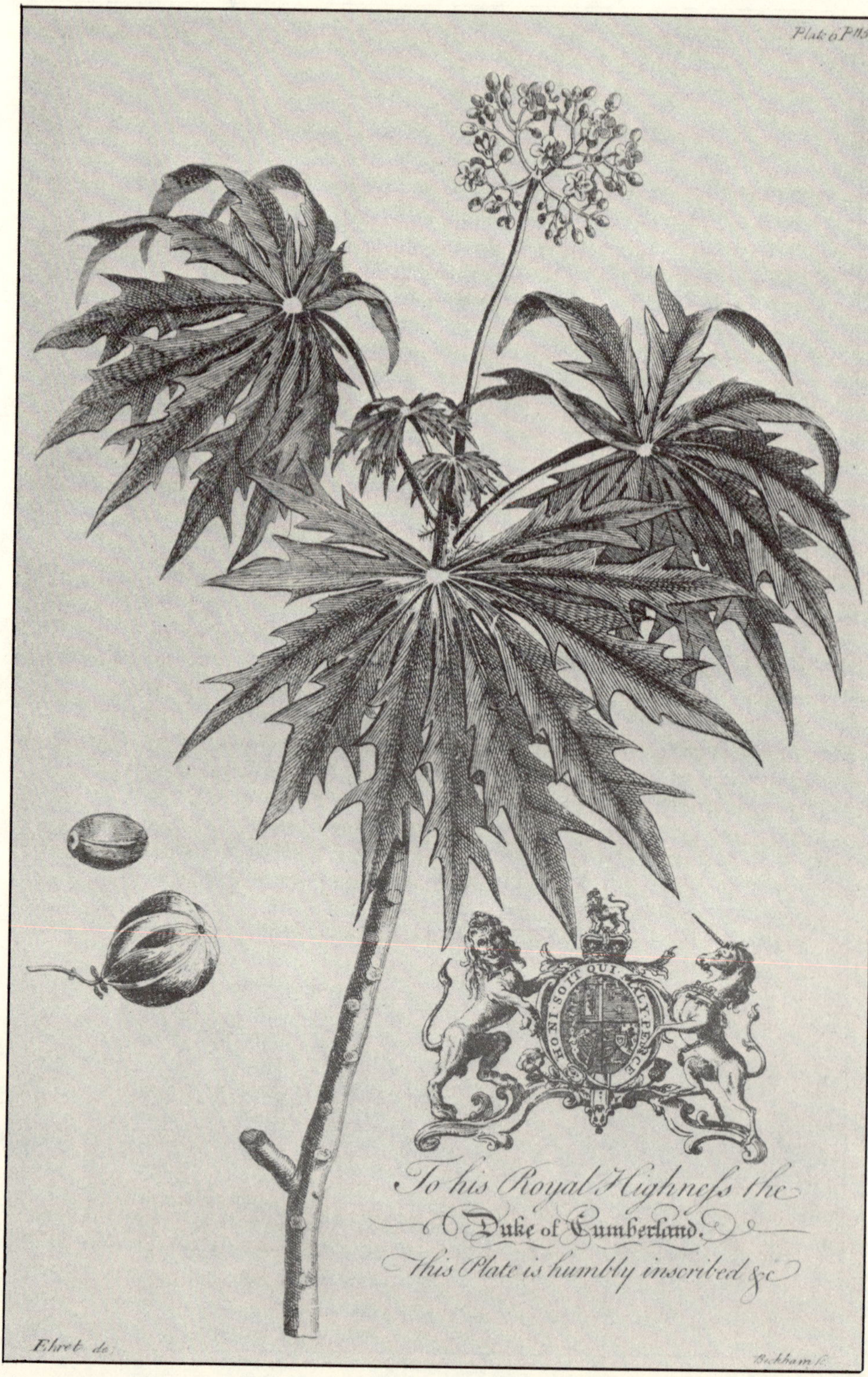
Plate 6 P 115*
HONI SOIT QUI MAL Y PENCE
To his Royal Highness the
Duke of Cumberland.
this Plate is humbly inscribed &c
Ehret de.
Bickham f.

The former are of a blunt conic Shape, having their Surface thickly ſtudded with reddiſh Knobs, each incricled with Two Rows of farinaceous whitiſh Duſt: The Female Flowers are funnel-like Tubes, of a red Colour, whoſe thick *Labia* are much deflected back, and cut into Fifteen or Sixteen blunt Segments.

This is ſucceeded by a very flat round Fruit. deeply ſegmented into, generally, as many regular Diviſions, or ſeparate Cells, each incloſing a flat Kernel.

Theſe, if eaten, operate by way of Purge; tho' it ſometimes proves an Emetic.

This Tree is of very quick Growth, eſpecially if planted in a ſhady Place, and near a Spring of Water.

The Trunk of one of theſe, planted in ſuch a Situation, may be now ſeen at the Eſtate of *Richard Eſtwick*, Eſq; and is not above Fifteen Years old, which girts round Nine Feet within Two Feet of the Earth; and its Top hath ſeveral Branches proportionably large.

There is another at the Eſtate of the Reverend Mr. *Reynold Foſter*, the Diameter of whoſe Shade is above Sixty Feet.

And, tho' I have taken the Dimenſions of theſe Trees myſelf; yet, as their Stature is ſo very extraordinary even here, I thought it proper to mention the Place of their Growth, that the Curious may, if they pleaſe, be further ſatisfied in this Particular: And I am well aſſured, that, if any one takes the ſame Trouble I have done, he will find their Dimenſions conſiderably larger than I have repreſented them.

There is another Tree of this kind, near the Manſion-houſe in the Leeward Eſtate of *George Hannay*, Eſq; in *St. Lucy*'s Pariſh, which is little inferior to the above-mentioned, in its great Growth.

Theſe Trees are called *Sand-boxes*, from the Uſe that is made of their Fruit to that Purpoſe.

This is delineated in *Plate* V.

The PHYSIC-NUT-TREE; *Lat.* RICINOIDES.

THIS is generally a knotty ſhrubby Tree, ſeldom growing above Twelve Feet high.

The Extremities of the Branches are covered with Leaves, ſomewhat round, having their Edges waved, as well as ſlightly indented.

The Flowers are of a green herbaceous Kind, and pentapetalous, ſet on in an Umbel-faſhion round the Extremities of the Branches, but eſpeciall the main Stalks.

Theſe are ſucceeded by as many Nuts, whoſe outward Tegument is green and husky: This, being peeled off, diſcovers the Nut, whoſe Shell is black, and eaſily cracked: This contains an Almond-like Kernel, divided into Two Parts: Between this Separation lie Two milk-white thin membranaceous Leaves, eaſily ſeparable from each other: Theſe have not only a bare Resemblance of perfect Leaves, but have, in particular, every Part, the Stalk, the middle Rib, and transverse ones, as visible as any Leaf whatsoever.

The Kernel is not only eatable, but hath a delicious Almond-taste.

It is the common Opinion, that the purgative or emetic Quality lies in the Two membranaceous Leaves, that separate the Kernel into Two Partitions: But this is a Mistake; for I have, as well as others, tried it both ways; and its physical Effects were, with or without these, always the same.

The Leaves, and tender Buds, emit a milky Juice, which either by Insolation, or being rubbed on the Palm of the Hand, turns to the Consistence of a Salve, which is looked upon to be very good to heal any green Wound.

The Nut likewise, when ripe, pounded and boiled, will yield a considerable Quantity of Oil: A Spoonful of this swallowed, whilst fresh, is looked upon to be a good Purge, to abate Swellings in Dropsical Disorders.

The FRENCH PHYSIC-NUT; *Lat.* RICINOIDES.

THIS Shrub often grows to be Ten Feet high.

The main Stalk divides into very few Branches, and is covered with a greyish white Bark.

The Leaves stand upon Six-inch Footstalks, surrounding the main Stalk, generally near the Top, in an irregular Order: These are very deeply digitated: Each Digit, by the several Subdivisions that grow from its Sides, in Shape, much resembles a Branch of the palmed Hart's-horn.

The Flowers grow in Bunches, Umbel-fashion, upon the Extremities of each large Stalk, very much resembling, at their first Appearance, a Bunch of red Coral: These afterwards open into small five-leaved purple Flowers, and are succeeded by Nuts, which resemble the common Physic-nuts in Shape and Qualities.

This is delineated in *Plate* VI.

The CEDAR.

THIS grows to be a large Tree, with a dark-coloured Bark.

Its Leaves are of a pale Green, about Two Inches long, one broad and sharp-pointed; their Edges being irregularly ferrated.

The Flowers are succeeded by a Fruit as big as a Damascene: These, when ripe, are of a russet Colour, and split into Four Parts, discovering a thin flat Seed.

This differs from the *Carolina* Cedar, chiefly by the far closer Grain of the Wood: It wants also the fragrant Smell of the former.

The

The CALABASH-TREE.

THE Fruit called *Calabaſhes* are of Two Sorts; the one ſmall and round, containing (when the pulpy Subſtance, and numerous white flattiſh Seeds, are ſcooped out) from One to Four Pints; the oval Sort often containing Fifteen Pints.

The Tree bearing the former is taller, and hath a leſſer Leaf, and is alſo of a paler Colour, than the latter.

The Bark of both Trees is rough and ſhagged, if not ſulcous in ſome old Trees; and what ſome Authors call *Capſular* Leaves, are only a Separation in the green husky Part of the *Calyx*; for this Tegument ſeems indeed to have been deſigned by Nature only to protect the Fruit, whilſt tender, and in its infant State; and when that hath expanded itſelf on the Inſide, and is become ſtrong enough to bear the Air, it then burſts its Way thro' the former Covering.

This Tree, when full-grown, is about Eighteen Inches Diameter; and, in Height, from Fifteen to Twenty-five Feet; dividing generally into many ſpiral Branches, ending in a blunt Point.

The Flower is a large yellow Cup, freckled with Red, a Piſtil riſing from the Middle. The *Stamina* take their Riſe from the Side of the Petals.

The Fruit is often ſo large as to contain (when cleared of its Seed, and white ſoft Pulp, with which it is filled) Two or Three Quarts, eſpecially thoſe which grow of a blunt oval Shape; but the round ſort are much ſmaller. . Each make very convenient Drinking-cups, and are ſerviceable to many other Uſes.

The GALL-TREE.

AS there are very few, if any, of theſe Trees, left undeſtroyed in the Iſland, I am obliged to repreſent it from the Deſcription of it given me by a Perſon of equal Knowlege in Plants, and Integrity in deſcribing them.

This Tree ſeldom riſes above Twelve Feet high, covered with Leaves of the ſame Make with, but ſomewhat longer than, thoſe of the baſtard *Lignum Vitæ*.

The Trunk, which is covered with a whitiſh Bark, ſoon divides into very many buſhy Branches: The Extremities of theſe bear many ſmall bluiſh Flowers in Groups, which are ſucceeded by numerous Berries, or Galls hanging on ſhort Footſtalks: Theſe are about the Bigneſs of a Sloe, of an Aſh-colour on the Outſide, and black within.

This, with the Pods of Achaſee, boiled together, makes good Ink.

The NEGRO OIL-BUSH, *Lat.* PALMA CHRISTI.

THIS ſhrubby Plant is diſtinguiſhed into the red and white Sort.

The Oil, extracted from the Berries of the Red, is leſs rank than that of the other; and is ſometimes made uſe of by ſeveral Nations of Negroes in their Soups.

The

The Trunk of this Shrub, which often grows to above Five Feet high, is ſtrong, and cloſely jointed.

The firſt-mentioned hath a reddiſh, and the other a greeniſh Stalk; both generally clouded over with a glaucous Mealineſs.

The Leaves are many, ſurrounding the main Stalks, as well as the Branches (eſpecially towards the Top): Theſe are very large, often above Twenty Inches Diameter, being cut into ſeveral, ſeldom leſs than Eight, ſharp-pointed Diviſions: They are ſupported upon Footſtalks of about Thirteen Inches long, centring in almoſt the Middle of the Leaf.

The Flowers, which ſurround the Summits of the main Stalks, for about Seven Inches in Length, firſt appear in the Form of ſmall green *Capſulæ*, of a blunt conic Shape, which ſoon divide into Five ſharp-pointed green Leaves; and by deflecting towards the Footſtalk, diſcover innumerable ſmall, pale, yellow, ſtamineous Flowers, tipped with *Apices*.

Theſe Flowers, at firſt, are ſucceeded by greeniſh Berries, thick-ſet with harmleſs *Spinulæ*: As they grow riper, theſe drop almoſt intirely off, and the ſoft browniſh Husk of the Nut is diſcovered.

This is of a triangular Shape, divided into Three Parts, touching one another at Right Angles, and compoſing one Body, ſlightly rimmed with a browniſh Fur, the Nut ſtanding upon a pendulous bluiſh Footſtalk.

When the outward Skin is taken off, it diſcovers a ſpotted Kernel.

Theſe are ſo oily, that Ten Pints of them, being bruiſed, and then boiled, afford a Pint of Oil.

This Plant, or rather Shrub, thrives beſt in a ſandy Soil, near the Sea; and as the Heat in ſuch Places is intenſe, from the very ſtrong Reflexion of the Sun from the Sand, I am of Opinion, that Nature intended it ſuch a Situation, to ſerve (as it were) inſtead of an Alembic, to force up to the Extremities of the Branches, where the Berries are, ſuch a great Quantity of Oil.

The BLACK NICKER-TREE.

THIS grows to be a conſiderable large Tree, covered with a Bark of a greyiſh White.

The ſmaller Branches are thickly cloathed with ſeveral Pairs of pennated Leaves: Theſe are Four Inches long, and an Inch and an half broad, ſmooth-edged and ſharp-pointed.

The Flowers are ſucceeded by a round Snuff-coloured Fruit, ſomewhat bigger than a large Cherry, or rather of the Make of the ſmall round white Plumb in *England*.

The outward Subſtance is tough, and very gummy: This, when ripe, ſeparates from the Nut within, which is black and round.

When this dark-red ſhelly Covering is broken, it diſcovers a white bitteriſh Kernel.

If the outward Husk, which is no thicker than a Crown-piece, is put in Water, it will raiſe a Lather much ſooner and ſtronger, than an equal Quantity of any Soap would do.

The

The JACK-IN-A-BOX; *Lat.* HERNANDIA.

THIS is of a very quick Growth; and, when at Maturity, is very feldom fhort of Forty Feet high.

Its main Branches are numerous and ftrong: Thefe are fubdivided into feveral leffer ones, which are alternately cloathed with large Heart-like Leaves, being Ten Inches long, and Six broad.

The Flowers iffue in Bunches from the Extremities of the Branches: Thefe ftand upon Two-inch long Footftalks, and are compofed of three whitifh feemingly capfular Leaves, furrounding Three leffer white Petals: Thefe alfo inclofe feveral fhort *Stamina*, tipped with yellow *Apices*.

The Socket of the Flower difcovers, near the Roots of the *Stamina*, feveral fmall yellowifh farinaceous Grains.

As the Piftil grows in Strength, it becomes the Rudiment of the Fruit; which, when ripe, is of a blackifh Colour, narrow at each End, and Three-quarters of an Inch long.

From each Extremity, it grows in the Middle to a roundifh Body, of about Two Inches and an half in Circumference, deeply furrowed into feveral irregular Chanels.

This blackifh Coat inclofes a Nut, whofe Kernel is very near as large as a fmall Nutmeg, and much refembles it in the Colour of its Veins.

The Whole of this Fruit is furrounded with a thin husky brittle Cafe, or Covering, which never touches the Fruit, nor the Stalk, but where it furrounds it at the Bottom. This hollow *Capfula* is of a tawny yellow Colour, which is capable of containing, at leaft, a Gill of any Liquid, having an Aperture, at its Extremity, of about half an Inch Diameter: Thro' this the Fruit may be eafily feen in the Cavity or Womb of this capfular Husk; which, when ripe, hath a very ftrong fragrant Smell.

The Tree is never known to be without both Fruit and Flowers.

The Wind, blowing into the Cavities of thefe Husks, makes a very fonorous whiftling Noife.

The inclofed Kernels are very oily.

It grows chiefly in fhady Places.

The End of the FOURTH BOOK.

THE NATURAL HISTORY OF THE Island of *BARBADOS*.

BOOK V.

Of TREES, SHRUBS, *and* PLANTS, *of the* POMIFEROUS *Kind*.

BY Pomiferous Fruits, I would be underſtood to mean all of the Apple-kind, ſuch as have a thick fleſhy Subſtance, incloſing many Seeds.

It is obſervable, that where the Climates begin to be too hot for the Growth of ſuch Apples as are produced in ſome Northern, and all temperate Climates; there the gradual Chain is progreſſively carried on, and the delicious *Aſiatic* Fruits begin to take place, ſuch as the Pine-apples, Shaddocks, Forbidden Fruits, and Oranges.

It is alſo worthy our Notice, that moſt of our Summer Fruits, eſpecially in hot Countries, tho' delicate in their Taſte, have a great Mixture of a latent Acid in them, which is neceſſary for the Preſervation of Health in in hot Climates. And as most kinds of such Fruits are best adapted to qualify by their Coolness the Heat of the Body in Summer Seasons; so we find, that they grow, and are in Plenty, in Winter Countries, only at such times as they are most wanted: But as the Climates within the Tropics have, almost without Exception, a continual Summer, so the same Divine Hand hath provided for the Wants of those Places, by bountifully supplying them with a constant Series of various fresh Fruits of this kind.

The MANCHANEEL-TREE; *Lat.* MANCANNILLA.

THIS, among many others, is very imperfectly described by all former Authors: And, tho' the Juice of this Tree is confessedly poisonous; yet the Force and Malignity of its Poison are extravagantly enhanced.

One (1) Historian, in particular, who loved to dwell upon the Marvelous, says, That the Heads of People, who happen to sleep a short time under its Shade, swell; and those who have so slept, grow blind; but if they sleep for a considerable time, they recover their Sight; likewise, that if the Leaves but touch the naked Skin, they raise Pustles, which cause deadly Pains, unless helped with Water and Salt, or fasting Spittle; and that even smelling the Wood is dangerous.

This Description is fabulous in every Article: For sleeping under its Shade hath no manner of bad Effect upon the Head or Eyes; nor is any ill Consequence to be feared from the Leaves touching the naked Body, unless these are bruised, and the white milky Juice they contain is suffered to pervade the Pores; which if it does, it raises Blisters, like those of the Confluent Sort in the Small-pox, causing acute Pains; but the simple Drops of Rain-water, falling from these Leaves upon the Skin, will not, as is falsly asserted, have any bad Effect; which I have often, upon repeated Trials, experienced.

There are some seeming Instances urged to confirm the contrary; especially of a Person washing in *Indian River*, who, upon an unexpected Storm of Rain, and high Wind, sheltered himself naked under the adjacent Shade of this Tree; where he had not been long, before he felt some Part of his Skin, about the Shoulders, smart, and soon afterwards an Eruption of painful corrosive Blisters.

I will lay no Stress upon the great Heat of the Sun, during the time of his Bathing, which often, near those sandy Bays, raises Blisters upon the Part exposed out of the Water; but am satisfied, and well convinced, that these Blisters were occasioned by Drops falling from this Tree; tho' at the same time, I must deny, that they were merely Drops of Rain-water: However, we cannot be at a Loss to account for the Blistering, and its painful Cause. If we recollect, that the Wind was very high in this

(1) *Peter Martyr.*

this Instance, and the Rain very heavy, the former, by violently tossing and beating the tender Buds and Leaves against one another, bruised them; by which means, the milky poisonous Juice issued out in great Plenty, and washed off from the Leaves, with the Drops of Rain, upon the Person sheltering under the Shade.

From these Remarks upon the Mistakes of others, I proceed to give the best Description I am able of this Tree, both with regard to its outward Appearance, and its real Qualities.

The first thing that is to be observed, is, that it is of a very quick Growth, and seldom or never found growing to any Perfection, but in loose sandy Soil, near the Sea, or other Water.

Its Roots are strong and numerous.

The Trunk of a full-grown Tree is, generally speaking, from Two Feet and an half to Three Feet Diameter, branching, most commonly, from Three to Fifteen Feet high from the Ground.

The Heart, which is very small, in proportion to the Bulk of the Tree, is very hard and solid, of a pale Yellow, with a greenish Cast, interspersed with small blackish Veins, the Grain smooth, and the Wood durable; the Bark of a sleek whitish Grey, the Branches many in Number, and full of thick, smooth, shining Leaves.

There are here and there, among the Branches, long pendulous Katkins, which are produced at remote Distances from the Embryo's.

It bears a Fruit much of the same Make as the round sort of Crab-apples, which we have in *England*; and both, when ripe, have the same fragrant Smell.

The Pulp of these Manchaneel-apples exceeds not, in Breadth, one Seventh of an Inch, the Inside being an hard stony Kernel, deeply and irregularly sulcated; in which are included the Crab-like Seeds.

The Juice of the Apple is of the same Colour and Quality, as that of the Leaves.

But however strong the Poison really is at some Seasons, and at certain Growths of these Apples; yet I have known a Woman, big with Child, who longed for them, to have eaten of them, without any apparent bad Effect: However, I cannot say, that the Fragrancy of their Smell, or their tempting Looks, have induced others to try and follow her Example by so dangerous an Experiment.

If some of this crude milky Juice falls upon even an Horse, the Hair, from the Part affected, soon falls off, and the Skin rises up in Blisters, which will require a long time to heal.

One Instance of its Malignancy happened about Two Years ago in *Speights-Town:* A certain Slave, conceiving herself injuriously treated, poured into her Master's Chocolate about a Spoonful of this Juice: Immediately after he had swallowed it, he felt a violent Burning in his Throat and Stomach; and, suspecting he was poisoned, he strove, and with good Success, to vomit; and, having taken after this seasonable Discharge, a regular Emetic, his Stomach was, in a great measure, suddenly cleansed of the Poison, tho' it cost him a long time to perfect the Cure.

I am likewise of Opinion, that the Vehicle, in which it was conveyed, contributed not a little to prevent its otherwise very ill, if not fatal Effects; for, as most Oils are Antidotes against Poison, the Chocolate, being rich of the Nut, might be so very oily, that the poisonous Salts were sheathed or blunted in that Vehicle.

It hath been also observed, that Fish, as the *Barracuda*, and others, which eat these Apples, dropped casually into the Sea, are often found dead in the Wash of the Water; and, if taken, whilst alive, and eaten, often prove poisonous; and even the large white Crab, that burrows in the Sand, is not, if near these Trees, to be made use of for Food.

I shall conclude the Description of this Tree with a remarkable Observation, generally found to be true; which is, that where-ever a Manchaneel-tree grows, there is found a White-wood, or a Fig-tree, near it; the Juice of either of the latter being an infallible Antidote against the Poison of the former.

Salt Water is no less efficacious; and as these always grow up by the Sea-side, this Remedy is near at hand.

Formerly no one dared to cut down these Trees, without first having made a large Fire round them, in order to burn the Bark, and dry up the Spry and Juices that fly from them in cutting: But now naked Negroes venture to cut them down green, only using the Caution of previously rubbing their whole Bodies with Lime-juice, which prevents the Juice from corroding or ulcerating their Skins.

Bruising and mashing the tender Leaves and Boughs, in Fish-ponds, hath often been likewise a roguish Practice of taking and destroying Fish; for the Fish very soon after will grow stupid, float, with their Bellies upwards, on the Top of the Water, and often die.

The SHADDOCK-TREE; *Lat.* AURANTIUM FRUCTU MAXIMO INDIÆ ORIENTALIS.

THE Tree, bearing the large white Shaddock, hath many strong Roots. When full-grown, it is often above Sixteen Feet high, divided near the Top into many strong Branches, cloathed with numerous blunt-pointed Leaves, of about Four Inches long, and near Three in Breadth, of a deep Green above, and pale underneath; resembling, in their Make, that of a sour Orange.

The Flowers are like the Orange, surrounding numerous *Stamina*, tipped with *Apices*.

These are succeeded by the Fruit, somewhat in the Shape of a Pear; but far larger, and somewhat rounder.

The outside Skin is yellowish and smooth.

This Fruit was brought firſt from the *Eaſt-Indies*, by Captain *Shaddock*; from whom it derives its Name.

The Leſſer SHADDOCK-TREE.

THIS is about the Bigneſs of a China Orange-tree; the Leaves differing in this only, that they are not above Two Inches and an half long, and about an Inch and an half broad.

The Flowers are monopetalous, deeply ſegmented into Five ſeeming white Petals, ſurrounding a great many *Stamina*, tipped with *Apices*.

The Fruit is diſtinguiſhed into the red and white Sort, each larger than either an Orange, or a Forbidden Fruit; and is juſtly eſteemed to have a fine delicious Taſte.

The WILD SHADDOCK; *commonly called, The Large* LIME-TREE.

THIS grows ſomewhat larger than an Orange-tree; tho' the Leaves of this, as well as of all other Trees of this Claſs, differ very little in their Shape; and the Difference in their Flowers is hardly perceptible.

This is by far the largeſt Fruit in the Iſland, growing often to be Twenty Inches in Circumference.

The ſeveral rough Indentings and Knobs on its Outſide, as well as the internal Make and Taſte of its Juice, determine it to be rather of the Lemon, than either of the Shaddock or Lime-kind.

I have known near half a Pint of ſour Juice ſqueezed out of one of theſe Fruits.

This ſerves to ſupply the Want of Limes or Lemons in making Punch.

The SOUR ORANGE-TREE; *Lat.* AURANTIUM.

THIS grows in a rich Soil, to the Height of about Seventeen Feet. Its Roots are ſtrong, deeply penetrating into the Earth.

Its Branches are numerous, eſpecially towards the Extremity of the Tree; and are cloathed with ſharp-pointed ſmooth Leaves, of a deep Green above, and ſomewhat pale underneath.

Theſe are about Three Inches long, and near Two broad, ſupported by an Heart-like Footſtalk, an Inch in Length.

The Bark of the Trunk is generally of a dirty-grey Colour, and of the ſmalleſt Branches, towards the Top, of a deep Green.

Its numerous white Flowers are monapetalous, divided into Five Sections, ſo deep, that they reſemble pentapetalous Flowers.

Theſe are ſucceeded by the Fruit, which is too well known to require a Deſcription.

If

If theſe Fruits are not gathered when ripe, thoſe which drop not off, will ſeemingly wither upon the Trees; yet theſe will revive, flouriſh, and be again, as it were, ripe, at the uſual Seaſon in the following Year.

I have ſeen an Inſtance of this kind in the Eſtate of the Honourable Colonel *John Maycock*, in *St. Lucy's* Pariſh.

The larger Branches of this Tree are, in common with all Orange, Lemon, Shaddock, Forbidden Fruit-trees, and Lime-trees, ſtudded with many ſharp ſtrong Prickles.

The CHINA ORANGE-TREE.

THIS Tree, either in regard to its Roots or Size, Diſpoſition or Make of its Leaves, differs very little, if any thing, from the ſour Orange-tree already deſcribed, unleſs that its Leaves have an Heart-like Appendix joined to them, ſerving as Footſtalks: Theſe are ſomewhat narrower, or ſharper pointed, than in the former.

The Flowers are of the ſame kind as thoſe of the ſour Orange already deſcribed, except that theſe are alſo leſs, and the Petals not ſo thick or fleſhy.

The GUINEY ORANGE.

THIS Fruit, I think, ought rather to be called the *Sour Forbidden Fruit*; for in every Circumſtance (except in the Sourneſs of its Juice) it reſembles that Fruit; but it is much leſs than the eatable Forbidden Fruit.

The GOLDEN ORANGE-TREE.

THIS Tree differs not in either its Trunk, Leaves, or Flowers, from that already deſcribed.

It bears a large fine Orange, of a deep Yellow within; from whence it derives the Name of a *Golden Orange.*

This Fruit is neither of the *Seville* or *China* kind, tho' it partakes of both, having the Sweetneſs of the *China* mixt with the agreeable Bitterneſs and Flavour of the *Seville* Orange.

The PUMPLE-NOSE-TREE.

THIS Tree reſembles the Shaddock in every material Circumſtance.

The Fruit nearly repreſents, in its Make, that of the leſſer Shaddock; and is in Bulk ſomewhat leſs than a Shaddock, and larger than the largeſt Orange.

Theſe Fruits are eatable, their inward juicy *Veſiculæ* being much the ſame as thoſe of Shaddocks: However, they are often too ſour to be eſteemed a very deſirable or delicious Repaſt.

The

Plate 7. p. 127
To the right Reverend
Father in God,
George L.d Bishop of Exeter,
This Plate
is humbly inscribed.
Ehret. de
Austen sc.

The BURGAMOT-TREE.

THIS Tree, in its Bulk, Make of its Leaves, or Colour of its Bark, differs very little, if any thing, from the laſt deſcribed.

It bears a Fruit ſomewhat larger than an Orange, whoſe Inſide affords very little Juice; and what it yields is not palatable, being too ſour to be eatable.

What is moſt valuable in this is its Oil, which is extracted, by ſlicing the outward Skin, and ſqueezing both the Oil and Spry into a Glaſs: The Oil will immediately ſeparate from the Juice, and ſwim upon the Surface.

This, being carefully drained off, and preſerved, is what alone ought to be called the genuine *Burgamot Oil*.

The *Portugueſe* have a Method of molding the Rind of this Fruit, to make Snuff-boxes: Theſe Boxes retain, for a very long time, the grateful Fragrancy of the Oil and Spry.

FORBIDDEN-FRUIT-TREE.

THE Trunk, Leaves, and Flowers of this Tree, very much reſemble thoſe of the Orange-tree.

The Fruit, when ripe, is ſomething longer and larger than the largeſt Orange; and exceeds, in the Delicacy of its Taſte, the Fruit of every Tree in this or any of our neighbouring Iſlands.

It hath ſomewhat the Taſte of a Shaddock; but far exceeds that, as well as the beſt Orange, in its delicious Taſte and Flavour.

This is delineated in *Plate* VII.

The CITRON-TREE.

THE Encomiums that are often, by *European* Writers, beſtowed upon this Tree, and generally uſhered in with a pompous Deſcription of the Beauty of its Flowers, Fruit, and delightful Shade, are ſomething like the poetical Repreſentation of a Shepherd's Life, embelliſhed with purling Streams, and ſhady Groves; but not a Word of their coarſe Fare, nor of their enduring the uncertain Viciſſitudes of Heat and Cold!

What bears the Name of a Citron-tree is, in Reality, ſo far from affording a friendly Shade, that it is but a groveling prickly uncouth ſhrubby Tree, having neither Beauty in Appearance, nor Deliciouſneſs in the Fruit, unleſs what the Rind affords by Diſtillation.

The Leaves, which are ſet alternately upon the Branches, are about Four Inches long, and ſharp-pointed.

The Flower reſembles that of an Orange; and the ſucceeding Fruit is large and yellow, when ripe; the Outſide being irregularly indented, mixt with numerous Protuberances: The Inſide reſembles that of a Lemon.

There lies between the outward Coat, and the juicy Part, a thick ſpongy whitiſh pithy Subſtance.

The St. Helena Lemon-tree.

THIS is rather a prickly branching Shrub, than a Standard. It branches very near the Ground; and, as the Branches are many in Number, they interfere with each other; and their Tops, especially when loaded with Fruit, bend wavingly downwards, almost to the Ground.

The Leaves, which are set alternately, are longer and paler than those of the Orange-trees, and want likewise their Heart-like Footstalks.

Its Blossoms resemble those of the just-mentioned Tree; and the Fruit, when ripe, is of a pale-yellow Colour, and oblong in Shape, ending with a Nipple-like Protuberance.

These are but very moderately sour, and very grateful to the Stomach: Therefore it is much made use of in all Sawces, which require any Acid.

It grows best in shady Places.

Gulielmus Piso, in his *Natural History of the* West-Indies, has been profuse in his Encomiums upon the Virtue of Lemons; which appears from the following Words:

Peritissimi nonnulli Indiarum orientalium & occidentalium medicastri, præter vulgares illas toto orbi celebratissimas limonum & citreorum vires, quas quoq; perpetuo commendabiles habent, plus præsidii a duabus his limonum partibus (semine, scilicet, & medulla) ponunt contra malignos morbos, & pestilentes febres, atq; ipsa deniq; venena, quam in lapide bezoardico, & fastidiosa theriacæ compositione. Piso, p. 314.

Ansam mihi suppeditari autumabam aliquid ad artis nostræ incrementum ediscendi; idque aliquoties tentans, non infeliciter processit: in febribus, scilicet, intermittentibus, circa initium paroxysmi, haustu calido succi limonum, cum sacchari & aquæ fontanæ tantillo, semel atq; iterum exhibito, exacerbationes vel in totum cessâsse, vel saltem indies præter expectationem, simul cum siti, deferbuisse, & intra paucos dies plane siluisse, deprehendi, subsequente utiq; manifesto affectu urinæ, scilicet, & imprimis sudorum larga promotione. Ibid. p. 315. Vide p. 313.

The Spanish Lemon-tree; *Lat.* Limon.

THIS groveling shrubby Tree seldom grows to what we may properly call a Standard; for it generally divides near the Earth into many weak Branches, which rise about Ten Feet high; and then, especially if loaden with Fruit, they bend downward, often so low as to touch the Ground.

The Leaves are about Four Inches long, and about Two in Breadth, of a faint yellowish-green Colour, and set on the Branches alternately.

The Flowers are the same with the foregoing; and are succeeded by an oblong yellow Fruit.

The LIME-TREE.

THIS Tree, generally ſpeaking, ſoon divides into many Branches, and theſe into ſtill leſſer ones, thickly cloathed with Prickles, as well as deep-green ſmooth ſharp-pointed Leaves, which are near Two Inches long, and above an Inch broad.

It bears a very great Number of ſmall white monopetalous Flowers, divided into Five ſeeming Petals.

Theſe are ſucceeded by the Limes, which are a round Fruit, about as big as a Crab-apple.

Theſe are yellow, when ripe; and afford a great Quantity, conſidering their Bulk, of a thin and very ſour Juice.

This, or the Juice, or Lemon, is the chief Ingredient of the acid kind in both Sherbet and Punch.

The Inſide affords a great many ſmall white Seeds of the Apple-kind.

The Trees are propagated from theſe Seeds, planted about an Inch deep in the Earth.

The Limes, when young, are preſerved, and make a Part of our *Barbados* Sweet-meats.

The SUGAR-APPLE-TREE; *Lat.* GUANABANUS.

THIS is of a middle-ſized Growth, cloathed with weak pale thin Leaves, of about Four Inches long, one in Breadth, and ſmooth-edged.

Theſe are ſet on the Branches alternately.

The Flowers are compoſed of Three thick narrow herbaceous Leaves, white within, and green on the Outſide.

From the Centre of theſe grows the Fruit; which at firſt appears in a green Button; but as it grows, enlarges, and becomes of a blunt conic Form; the Outſide divided into ſeveral bluiſh irregular ſquamous Protuberances.

Theſe afford a ſoft, ſtringy, ſweet, pulpy Subſtance, incloſing a blackiſh oblong Stone or Kernel.

The Fruit is looked upon to be wholſome.

The MONKEY APPLE-TREE.

THIS grows to about Twenty Feet high.

Its Branches are thickly covered with Leaves, much reſembling thoſe of a Sourſop-tree.

It bears a large round Apple, the Inſide reſembling the Sourſop-fruit.

The Tree, Leaves, and Fruit, emit a very offenſive Smell in rainy Weather.

The Fruit hath its Name from its being eaten by Monkeys.

The

The AVIGATO PEAR-TREE; *Lat.* PERSEA.

THIS grows to be a large wide-extended Tree, whose Branches are cloathed with many large sharp-pointed Leaves.

The Flowers, which are hexapetalous, appear at the Extremities of the Branches, each succeeded by a large Fruit of a Pear-shape.

These are of Three Kinds; the Red, the Purple, and the Green sort: The last is esteemed the best.

The pulpy Part of this Fruit is soft, when ripe.

It is looked upon to be very nourishing.

The most common Method of eating it is, to mix this Pulp with a little Sugar and Lime-juice.

The Stone or Kernel is very large, and of a pale-russet Colour.

The POMGRANATE-TREE; *Lat.* PUNICA.

THIS is a shrubby Tree, seldom arising above Sixteen Feet high.

Its Branches are very many and slender.

Its Flower is of the brightest Red, composed of a blunt red conic husky *Calyx*, divided, at the Margin, into Six sharp-pointed Sections, inclosing as many Rose-like scarlet Leaves.

These likewise surround a great many red short *Stamina*, tipped with yellow *Apices*.

The succeeding Fruit is round, and as big as a large Apple, decorated at the Top with a *Corona*, as a Pear.

The outside Rind is strong and husky, and generally, when ripe, of a maiden-blush Colour.

The Inside is divided into several partitional Cells, inclosing a great Number of granulated small *Acini*, tasting somewhat like, but more delicious than, white Currans in *England*, when not over-ripe.

The outward husky Peel of this Fruit, when dried and pulverized, is good for Fluxes, and much used for that Purpose.

The Fruit of some of these Trees is so sharp, that they are not eatable.

They are generally distinguished into the white and red Sort.

There are of each sort some particular Trees bearing sweet, others bearing sour Fruits.

These Trees flourish in every Part of the Land of *Palæstine*; and were in so great Esteem among the *Jews*, that the Chapiters of the Pillars, in *Solomon*'s Temple, were adorned with the Representation of these, as well as of Palm-trees.

The small POMGRANATE-TREE; *Lat.* PUNICA AMERICANA.

THIS is a small woody perennial Shrub, which seldom rises above Four Feet high; and is chiefly planted for Border-hedges in Gardens.

It

It is thickly covered with ſmall longiſh blunt-pointed green Leaves.

The top Branches ſuſtain a great many red blunt conic husky *Calyxes*.

Each of theſe is divided, at its Margin, into Six regular ſharp-pointed Sections.

This *Calyx* ſurrounds Six Roſe-like ſcarlet Petals.

Theſe likewiſe incloſe a great many red *Stamina*, tipped with yellow *Apices*.

When the Flower drops, the *Calyx* becomes the Rudiment of the Fruit, which grows ſomewhat larger than a large round Crab-apple; the Top decorated with a *Corona*, like a Pear.

The Inſide is divided, by thin membranaceous *Laminæ*, into ſeveral larger partitional Cells.

Theſe are filled with granulated *Acini*, cloſely joined together; and are generally too ſour to be eaten.

The SOURSOP-TREE; *Lat.* GUANABANUS fructu e viridi luteſcente molliter aculeato. *Plum.*

THIS is a middle-ſized Tree, having many Branches, and theſe cloathed with numerous ſharp-pointed Leaves.

The Fruit is of an Heart-like Shape, but ſomewhat longer, and generally bending towards the Point; the Outſide of a glaucous Green, ſtudded here and there with ſoft-pointed Prickles.

The Inſide is a ſoft pulpy Subſtance, which is eaten; and looked upon to be a good Cooler in Fevers.

The ANCHOVEE APPLE-TREE.

THIS grows to the Bigneſs of a large Apple-tree.

Its Branches are many, cloathed with ſmooth green ſharp-pointed Leaves, which are Six Inches long, and almoſt Three in Breadth.

Theſe are ſet on the Branches alternately, whoſe Summits ſuſtain a ſhort conic husky yellowiſh Pod, opening into Three Partitions; the Inſide diſcovering Seven other yellowiſh imperfect husky thick Leaves.

In the Centre of theſe is the Fruit in Miniature, which is then about the Bigneſs of a Button, incruſted all over with a pale-yellow granulated Subſtance.

When the Fruit grows ſufficiently ſtrong, to bear the ſeveral Viciſſitudes of the external Air, the Leaves drop off, and the Fruit continues to grow to the Bigneſs of a very large Ruſſet-apple; nor doth its outward Coat ill reſemble that Apple in Colour.

Its Shape is rather of a very blunt Cone, than round.

The Inſide is of a fine Yellow; and the Pulp, which ſurrounds a great many *Laminæ* of oblong flattiſh Seeds, is of an agreeable ſweet Taſte.

The GUAVA-TREE; *Lat.* GUAJAVA.

THE Fruit called the *Guava* is diſtinguiſhed into Two ſorts; the White and the Red; and theſe, with regard to their Shape, into the round, and the Pear-faſhioned, or perfumed Guavas.

The latter have a thicker Rind, and are looked upon to be of a more delicate Taſte, than any other: However, I think the Difference is very ſmall.

This Tree, if carefully cultivated, and in a good Soil, will grow to about Eighteen Feet in Height.

Its Bark is very ſmooth, and of a reddiſh Grey.

The Leaves are about Three Inches long, ſharp-pointed, and high-ribbed.

The Flowers are white, and pentapetalous, guarded with Four capſular Leaves, green without, and white within.

The Petals ſurround a great many ſhort *Stamina*, tipped with pale-yellow *Apices*.

Theſe incloſe the *Stylus*, which is the Rudiment of the Fruit; which, excepting the perfumed Guava, is round, and about the Bigneſs of a large Tennis-ball; the Rind, or Skin, generally of a Ruſſet, ſtained with Red; the Top adorned with a *Corona*, as a Pear.

The Skin, or Rind, is lined with an Apple-like Subſtance, as thick as a Crown-piece.

The Inſide of this is full of an agreeable Pulp, mixed with innumerable ſmall white *Acini* or Seeds.

The Rind-part of the Guava, when ſtewed, is eaten with Milk, and juſtly looked upon to be preferable to any other ſtew'd Fruit.

From the ſame Part is made Marmalade; and from the whole Fruit, the fineſt Jelly perhaps in the World.

Theſe Trees grow in moſt Parts of the Iſland.

The Fruit is ripe about *October*.

The SAPPADILLA-TREE; *Lat.* CAINITO.

THE Roots of this Tree are conſiderably large.

The Bark is very much furrowed, and of a greyiſh-white Colour.

The Branches are very many, and ſpreading.

Theſe are thickly covered with very ſmooth ſharp-pointed ſhining-green Leaves, of about Three Inches long, and an Inch and an half broad

It bears many ſmall monopetalous Flowers, of a pale White, deeply laciniated, at their Tops, into many Diviſions, reſembling diſtinct Petals ſurrounding a green Piſtil.

From the Sides of theſe ſeeming Petals riſe many ſhort *Stamina*, tipped with yellow *Apices*.

Theſe Petals are guarded with Two Sets of capſular Leaves; one of a brown, and the other of a pale White.

The

Pl. 8 Page 138
NE CEDE MA...
To his Excellency
the Rt. Hon.ble Earl of Albemarle,
this Plate is humbly inscribed
&c.

The Piſtil becomes the Rudiment of the Fruit; which, when full-ripe, hath a ruſſet rough Coat, with a *Corona* upon the Top of it.

This, in Shape, Bigneſs, and Taſte, doth not ill reſemble a Burgamot Pear, except that it is ſweeter and ſofter; the Pulp of it incloſing ſeveral longiſh black Seeds.

The Leaves, as well as their Footſtalk, and tender Buds, emit, when broken, a white clammy Milk.

The MAMMEE-TREE.

THIS grows to be a large ſhady Tree.

The Bark of the Trunk, as well as of the main Branches, is ſomewhat ſulcated.

The Leaves, which are very thickly ſet on in an alternate Order, are round-pointed, and ſmooth, of a very deep Green, about Five Inches long, and near Three broad.

The Flowers are white, and pentapetalous, ſomewhat larger than Orange-flowers.

Theſe are ſucceeded by a round ruſſet Fruit, full as big as the largeſt Ruſſet-apple.

The outward Coat is rather an husky, than a skinny Tegument.

This peels off as the Rind of a Tree in the Spring does, and diſcovers the eatable Part of the Fruit, which is of a pretty ſolid Subſtance, of a fine yellow Colour.

This is about half an Inch thick, ſurrounding Two or Three large Stones, ſo joined together, that the Whole makes a circular Figure.

The eatable Part is well taſted, and hath an agreeable Flavour.

The FRENCH WILLOW.

THIS is a ſhrubby Tree, ſeldom growing above Sixteen Feet high.

Its Branches are very numerous, and very thickly cloathed with Leaves of Six Inches long, and a Quarter of an Inch broad.

Theſe, as well as the tender Stalks, when broken, emit a conſiderable Quantity of thick milky Juice.

The Extremities of theſe Branches ſupport ſeveral yellow Flowers, compoſed of Five Leaves.

Theſe are ſo cloſely joined together, that the Flower exactly reſembles the Bell-faſhion kind; and are ſucceeded by an Apple of about the Bigneſs of a large Crab-apple.

The Pulp of this ſurrounds a flattiſh Kernel, which is white and ſoft; and the Top of it marked with Two unindented Seams croſſing one another at Right Angles.

This is delineated in *Plate* VIII.

The

The Garlick Pear-tree.

THIS is a middle-sized Tree.

The Colour of the Bark, especially on the upper Branches, is of a Dark-grey, interspersed with whitish Specks.

The Leaves are of a fine clear Green, of about Four Inches long, and sharp-pointed.

These are set upon long Footstalks, in a tripartite Order.

The Flower is composed of Five small Spoonlike Petals, growing very slender, and small at the Bottom.

From the Inside of these rise about Fifteen purple *Stamina*, tipped with yellowish-green *Apices*.

From the Centre rises the *Stylus*, which bears upon its Top the Rudiment of the Fruit.

As this grows in Bigness, the *Stylus* grows in proportion; and at last becomes ligneous, able to support the Fruit, which in time grows to be as big as a large *Guava*; the Inside being full of an agreeable Pulp, interspersed with small granulated Seeds.

The tender Buds, from the young Branches, being bruised, and applied as a Cataplasm to any Part of the Body, will in time raise a regular Blister.

It would perhaps be beneficial, if some of the Learned in Physic would inquire, whether it would intirely answer the End of the *Cantharides*; especially as it may not affect the urinary Vessels, and consequently not cause a Strangury.

Whilst this Tree is in Flower, it is much frequented by Humming-Birds, which come to suck the Honey-dew from the Blossoms; and Butterflies lay their Eggs in great Abundance upon the Leaves of this Tree.

The Dunk-tree, *or* Mangustine.

THIS is a middle-sized Tree.

The Branches are numerous; and, after growing to about Fifteen Feet high, they bend wavingly downward, with a considerable Sweep, till they nearly touch the Ground, leaving a circular fine *Area* between that, and the Body of the Tree.

The Bark is of a reddish Grey.

The Flowers, which are many and small, are white; each consisting of one stellated Leaf, whose *Discus* is surrounded with Four short white *Stamina*.

The succeeding Fruit is, in Shape and Colour, like a small Crab-apple, except that both Ends are somewhat more depressed.

Its Taste is very agreeable; and therefore esteemed by most People.

The Tree, when in Blossom, emits a very fetid offensive Smell.

The

The CASHW-TREE; *Lat.* ANACARDIUM.

THIS is a low wide-ſpreading Tree.

Its Branches are generally crooked and ſtraggling, cloathed with oval Leaves, whoſe middle Ribs are ſtrong and prominent; the tranſverſe ones running almoſt directly croſs the Leaf.

It bears ſmall white five-leaved Flowers.

Theſe are ſucceeded by the Fruit, which is diſtinguiſhed into Three Sorts; the Large White, the Large Red, and the Red-and-white.

Theſe are not unlike, in Shape, to an Apple, called in *England*, the *Pearmain*.

The Inſide is very ſtringy, full of rough, aſtringent, yet pleaſant Juice.

Upon the Top of each ſtands a Nut, incloſing a Kernel. This Fruit is the *Anacardium* of the Antients.

Its Shape is like an Hare's Kidney, affording a great Quantity of cauſtic Oil; and the inſide white Kernel is roaſted and eaten.

The COCHENEEL SHRUB.

THIS very much reſembles the Pimploe Shrub already deſcribed; and differs chiefly from it by its greater Height, which is often Twelve Feet.

Its Leaves likewiſe are very prickly.

It bears upon conic Footſtalks, ſuch as the prickly Pear Footſtalks, ſeveral roſaceous Flowers, of a deep Scarlet, ſurrounding a Tuft of purple *Stamina*, tipped with whitiſh *Apices*.

In the Middle ſtands the Piſtil, divided at the Top into Six Starlike Diviſions.

The Flowers are ſucceeded by a Fig-like Fruit; which, when ripe, is full of a deep-purple Pulp, as well as Juice.

It is to be wiſhed, that the fine Colour it ſtains Linen with, could be fixed.

The AMERICAN TORCH; *Lat.* CEREUS MAXIMUS.

THE Roots of this are fibrous and many.

The reticulated Body of the main Stem, as well as the outward *Lamina*, covering the Whole, exactly reſembles the Texture and Colour of Pimploes, already deſcribed. The Body of this, near the Ground, riſes into ſeveral upright leſſer Stalks, from whoſe Top riſe others, till by ſuch a Multiplication the Whole grows often to Eighteen or Nineteen Feet high.

Each of theſe are nearly of a Bigneſs, being about Eight Inches in Circumference; and, from Joint to Joint, about Three Feet in Length.

Theſe are regularly chanelled from Top to Bottom into about Ten deep-gouged Furrows.

The Edges or Eminences of each Furrow are thickly beſet, at the Diſtance of half an Inch, with Star-like Tufts of whitiſh ſlender Prickles, from an Inch to an Inch and an half long.

Theſe, while young, are nurſed and guarded at the Bottom with a Cotton-like woolly Subſtance, which flies off when theſe are ſufficiently ſtrong to endure the Viciſſitudes of the Weather.

From near the Top of ſome of the largeſt Stalks riſes a longiſh ſcaly green Pod-like Subſtance, here-and-there marked with white Spots; from which grow many yellowiſh ſharp ſlender Prickles.

The Flower, which is at the End of this Pod-like Subſtance, is compoſed of a great many round-pointed Petals, of a yellowiſh-red Colour.

The Piſtil and *Stamina* are tipped with yellow *Apices*.

When theſe Flowers drop, the above-mentioned green *Calyx* becomes the Rudiment of the Fruit; which, when ripe, is full of ſmall *Acini*, affording a purpliſh Juice.

The TURK'S HEAD; *Lat.* MELOCACTUS AMERICANUS.

THIS, I ſuppoſe, derives its Name from its diſtant Reſemblance to a high-raiſed Turbant.

Its outward Texture is a very green ſkinny Tegument; its Shape of a very blunt Cone, whoſe Sides are deeply ſtriated; each intermediate Riſeing thickly ſtudded with long ſharp-pointed Prickles.

Near the Summit appear very many red roſaceous ſmall Flowers.

Theſe are ſucceeded by ſmall oval red Berries, full of an agreeable Pulp, interſperſed with ſmall *Acini*.

The Root of theſe Flowers and Fruit is guarded with a ſoft Cotton-like Subſtance.

The MUSKMELON VINE; *Lat.* MELO.

THIS Fruit is diſtinguiſhed, by the Colour of its Pulp, into the white and red Sort; each deriving its Name from the Fragrancy of its Smell.

As this Fruit is ſo well known, both here, and in *England*, a particular Deſcription of it would be ſuperfluous.

I ſhall therefore only obſerve, that the Vine is rough, almoſt to a Pricklineſs, bearing ſmall yellow monopetalous Flowers.

Each of theſe divides into Five Sections, ſo deep, that they reſemble pentapetalous Flowers.

The WATER-MELONS.

THESE, as well as the former, are diſtinguiſhed, by the Colour of their Pulp, into the White and Red; each deriving its Name of a Water-Melon from the great Quantity of that Liquid they contain.

By

By their cooling and diuretic Quality, they are ſo ſerviceable in hot Climates, that the poorer Sort in *Perſia* and *Turky*, eſpecially in the *Levant*, for the Summer Months, live almoſt ſolely upon the Musk and Water-Melon, Cucumbers, and Milk.

The Vine producing the Water-Melon is long and trailing.

PUMKINS; *Lat.* PEPO.

THE different Species of theſe are generally diſtinguiſhed by the Names of the White, the Blue, the Marbled, and the Garden Pumkin.

The latter differs from all the reſt, by having no Seed; but is propagated by Slips.

Theſe are all eaſily produced, and of a very quick Growth.

They make a great Part of the Food of the poorer Sort, in the Summer-time, as well in *Aſia* and *Africa*, as in *America*.

The Vines, bearing theſe different Sorts, are each rough and hairy, both Stalks and Leaves; and the Flowers are yellow, monopetalous, and very large, divided at the Top into Five deep Sections.

The Piſtil is ſurrounded with yellow farinaceous Duſt, which, by Three open Slits at the Bottom, drops into the Ovary.

The ſucceeding Fruit is generally, whilſt young, of a Mixture between a deep Blue, and a pale White.

They are boiled and eaten with Fleſh-meat.

I have ſeen a Species of theſe Fruit at *Paris*, which exceeds any in the *Weſt-Indies* in Bigneſs.

It is much uſed by the poorer Sort in Soups.

SQUASSHES; *Lat.* MELOPEPO.

THE Squaſh-vine is long and trailing, the main Stalk multangular and hairy.

It ſupports itſelf by its numerous Claſpers and Tendrils.

The Leaves are large, and very rough.

Their Edges are irregularly ſerrated, as well as the Leaves in general, ſomewhat ſcalloped.

From the Bloſſom of the Leaves riſe ſeveral Pedicles, ſupporting the Fruit in Miniature; whoſe Top is decorated with a large reddiſh-yellow Bloſſom, which continues on the Fruit, till it is eatable; which it generally is, when as big as a Walnut.

When boiled, they are by moſt People eſteemed to be very delicate eating.

Theſe are of Two ſorts; the Long and the Round.

The

The large GOURD VINE; *Lat.* CUCURBITA.

THIS Vine is long and trailing, of a multangular Shape, and very hairy.

The Leaves, which are large, and almoſt circular, are covered with ſoft *Villi*, or Hairs.

Theſe Leaves ſmell ſtrongly of Musk.

The Flowers, which ſtand upon Footſtalks Six Inches in Length, are monopetalous, divided into Five deep Sections, and of a dirty-white Colour.

Theſe are incloſed in a ſmall hairy *Calyx*, divided alſo into Five Sections; and are ſo tender, that they are cloſed up as ſoon as the Sun ſhines upon them.

When theſe Flowers drop, they are ſucceeded by the Fruit, which grows ſomething Pear-like in Shape.

The outſide Tegument, as it ripens, grows hard, ſomething reſembling a Nut-ſhell, but ſofter. The Inſide is of a ſoft white Subſtance, intermixed with flattiſh Seeds, like thoſe of Melons.

Some of the largeſt Species of theſe Gourds are capable, when cleared of their Pith, to contain Twenty-two Gallons: However, ſuch are very uncommon.

The largeſt of theſe that I ſaw, was at Mr. *Richard Jackman*'s, in *St. Peter's* Pariſh.

The ſmall GOURD; *Lat.* COLOCYNTHIS.

THIS Vine is round, ſoft, and hairy, provided with numerous Claſpers. With theſe they cling to the neighbouring Buſhes.

The Leaves are large, and of almoſt a circular round Shape, ſtanding upon Footſtalks, Four Inches in Length.

Theſe are ſet on alternately; the Flowers, which are white, and compoſed of Five white Petals, ſurrounding ſeveral *Stamina*.

Theſe are ſucceeded by the Gourd, which is yellow, when ripe.

The ſhelly or husky Outſide incloſes a white bitter Pulp, interſperſed with whitiſh flat Seeds.

An Hole being made in one of theſe ripe Gourds, if a Glaſs of Rum be poured in, and ſuffered to remain there for Twenty-four Hours, and then drunk, it proves a ſucceſsful Purge; but is ſo bitter, and leaves ſuch a *Nauſea* behind, that it is ſeldom uſed.

The SWEET GOURD.

THIS Gourd differs from the laſt deſcribed, by its very great Length and Narrowneſs; being often above Two Feet long, and about Six Inches in Circumference.

It differs likewiſe from all others, by its Pulp being rather ſweet than bitter.

When

When these are ripe, the inside soft waterish Pulp and Seed are scooped out: Then there remains, next to the husky or shelly outward Part, a Pulp of a more solid Substance, than what is scooped out.

This, being likewise taken out in Slices, and somewhat dried of its natural Juice, is put into several Syrups; and at last, candied or frosted over, becomes an agreeable Sweetmeat.

BATCHELOR'S PEARS; *Lat.* SOLANUM fructu pyriformi inverso.

THESE are of Two sorts; the one almost spherical; the other hath the Resemblance of a Nipple upon the Top.

The Plant that bears these, grows to about Three Feet in Height, supported by very few Roots.

It hath many Side-Branches, which are very prickly.

These are cloathed with deep segmented rough Leaves, which are about Six Inches long, and Three broad.

Their middle Ribs, as well as the transverse, are guarded with a great many sharp-pointed Prickles.

The Flower, which is composed of one single Leaf of a whitish Blue, expands itself horizontally.

This is supported by Five small sharp-pointed capsular Leaves.

From the Centre rise Five yellow deep-chanelled *Stamina*.

These are succeeded by the Fruit above-mentioned; which, when ripe, is of a golden Colour, and about the Bigness of a Tennis-ball; the Inside being full of small Seed.

This Plant grows chiefly in very rich Ground, especially upon Dunghils.

BREAD AND CHEESE; *or*, SUCKING-BOTTLE.

THIS is a ligneous Wyth, with dark Iron-coloured Leaves, each like those of an Orange, having a longish Heart-like Stalk.

These are about Three Inches long, sharp-pointed, and here-and-there snipped on the Edges.

The Stalk of the Vine, as well as the middle Ribs of the Leaves, is of a purple Colour.

The Flowers are succeeded by yellow conic capsular Pods, somewhat in Shape like a Bottle, each dividing into Three Partitions, having one Seed, which, at one End, is covered with a whitish rough Pith, which is sometimes eaten.

The Root of this Vine, mixed and boiled with Lime-juice, and the Rust of Iron, by way of Plaister, cures the Body-Yaws.

CUSTARD-APPLE; *Lat.* GUANABANUS fructu aureo, & molliter aculeato. *Plum.*

THIS Tree is about the Bigneſs of the laſt-deſcribed.

It bears a Fruit nearly round, and of a yellow Colour, when ripe.

The Inſide is full of a ſoft white pulpy Subſtance, from whence it derives its Name.

The End of the FIFTH BOOK.

THE

THE

NATURAL HISTORY

OF THE

Iſland of *BARBADOS.*

BOOK VI.

Of TREES, SHRUBS, *and* PLANTS, *of the* BACCIFEROUS *Kind.*

BY Berries we are to underſtand thoſe kinds of Fruits, whoſe outward Texture is generally of a thin flexible Make, and whoſe Inſides are full of Pulp, mixed with ſmall *Acini.*

Several of theſe, by their grateful Taſte, and delicate Flavour, are not only very acceptable to Man; but Providence intended many Species of them to be a proper Food for ſeveral kinds of (1) Birds, which could not ſubſiſt, eſpecially in the Winter Months, without them.

The

(1) Quiſts and Pivets, &c.

The LIGNUM VITÆ, *or* GUAIACUM.

THIS very useful Tree grows here in Plenty, but not to any great Bigness.

Its medicinal Virtues, either by way of Decoction, or the Use of its Gum, are sufficiently described by several Authors.

I shall therefore only observe, that the Berries, which are of a testiculated Form, and of a yellow Colour, inclinable to a Red, make an excellent Diet-drink, greatly assisting to purify the Blood.

The upper Branches are many, cloathed with several Pairs of small roundish pennated Leaves, never having an odd one at the End.

The Flowers are pentapetalous, and of the finest violet Colour that can be imagined; and the Tree is scarce ever to be found without Flowers or Berries.

This is delineated in *Plate* IX.

That called the *White Lignum Vitæ* scarce differs from this in any other Particular, but that it bears white Blossoms.

The BASTARD LIGNUM VITÆ.

THIS grows to above Twenty-five Feet high, covered with a rough brownish Bark.

The Heart of this Tree is almost as hard, and as durable, as Iron-wood.

It is sometimes beautifully clouded with Veins of Yellow, and a dark Red.

The Flowers are succeeded by Berries, each near as big as a small Cherry: These are of a purple Colour, when ripe; and taste very agreeably.

They are much coveted by wild Pigeons, Sparrows, and other Birds.

IRON-WOOD.

THE largest of these in this Island grow not above Thirty Feet high, and the Trunk in proportion somewhat slender.

This is covered with a whitish Bark, whose Surface often slightly pe.. off.

The Bark upon the upper Branches is of a reddish Grey.

The Leaves, which are smooth, and of a yellowish Green, are Two Inches and an half long, and above an Inch broad, narrow near the Footstalks.

The Flowers are many and white: These rise in Groups upon one common Pedicle, and are succeeded by small Berries, whose outward Tegument, as well as inward Pulp, is of the finest Red, interspersed with small Seeds.

It hath the Name of Iron-Wood from its great Weight, its Hardness, and the Closeness of its Grain; for it is proof against all Weather, and is scarce known to suffer any Decay in several Ages: It is so heavy, that it will sink in Water.

The

Plate 9 P.142
To the most rev.d Father in God MATTHEW
Lord Arch Bishop of York.
this Plate is humbly inscribed. &c

Pl. 10 Page 143

To his Grace the DUKE of NEWCASTLE
This Plate is humbly Inscrib'd &c.

The RED-WOOD.

THIS grows, eſpecially if ſheltered from the Wind, to be a very large Tree, affording Boards of a cloſe fine Grain, of a reddiſh Colour, remarkably and beautifully clouded with black Veins.

The Trunk is cloathed with a dark-grey Bark, which is ſomewhat ſulcated.

The Leaves on the upper Side are of a deep Green, and paler on the under: Theſe are ſmooth-edged, and ſharp-pointed, generally about Six Inches long, and Three broad.

The upper Branches bear many red Flowers, which are ſucceeded by middle-ſized Berries, of a dark-purple Colour, when ripe, containing many reddiſh ſmall Seeds.

Theſe Berries are eatable, and looked upon neither unpleaſant nor unwholſome.

It is ſaid, that the young Leaves, applied to the Temples, give great Relief in the moſt violent Head-ach.

The WILD CINNAMON-TREE; *Lat.* CINNAMOMUM SYLVESTRE.

THIS Tree grows to a conſiderable Height.

The Trunk is cloathed with a brown rough ſulcated Bark, which taſtes hot and biting; and hath ſomething of the Fragrancy of the true Cinnamon.

Its Leaves are of the Laurel-kind, ſmooth, green, and ſharp-pointed: Theſe are about Four Inches long, and one and an half broad; and, when bruiſed, yield an agreeable aromatic Smell.

It bears ſmall yellow Flowers: Theſe are ſucceeded by ſmall red Berries.

The LOBLOLLY-TREE.

THIS is a middle-ſized Tree, whoſe Leaves, which are about Three Inches long, are generally in Tufts upon the Extremity of the Branches.

Among theſe riſe a great many Bunches of white Flowers.

Theſe are ſucceeded by ſmall white Berries.

They ſerve as Food for ſeveral Species of Birds.

The BALSAM; *or the* SEA-SIDE SAGE.

THIS is a large buſhy Shrub, growing ſometimes to about Nine Feet high, and its Leaves finely ſerrated.

A ſubtile reſinous Juice perſpires thro' the Leaves, and their ſmall Footſtalks; which, by the Heat of the Sun, is granated, and intirely incruſts them.

This Incrustation, as well as the transparent Liquid, which instantaneously issues from the broken Footstalks of the Leaves, hath a fine aromatic Smell.

A few Drops of this, rubbed with the Finger on the Palm of the Hand, soon thicken to the Consistency of a Salve, tasting hot and bitter: This is excellently good to cure green Wounds.

This is delineated in *Plate* X.

The POISON-TREE; *Lat.* TOXICODENDRON.

THIS is distinguished into Two sorts; the one bearing a small smooth-edged narrow Leaf.

The Trunk of this is very solid, and good Timber for Building.

The other bears a large green smooth-edged sharp-pointed Leaf.

The Timber of the latter is very soft and perishing.

From each of these Trees issues, when wounded, a great Quantity of milky corrosive poisonous Juice.

This, when exposed to the Heat of the Sun, turns so very clammy, that it proves a good Bird-lime, and is with great Success made use of to that Purpose.

Each of these Trees have their Branches near upright and spiral, cloathed with a whitish grey Bark.

The Flowers are succeeded by a greenish-purple Berry, marked with slight indented Seams into Three Parts.

A Cataplasm of the new-extracted Juice, applied to the Feet, is of Service to kill the Vermin called *Chigers*.

The ROD-WOOD.

THIS grows to be a large Timber-tree.

The Bark on the lesser Branches is of a light whitish Grey.

The Leaves, which are set on alternately, are of a yellowish Green, especially the middle Rib.

The Footstalks of these Leaves, at their setting off from these Branches, are remarkably welted with a strong greenish husky Substance, which surrounds the Footstalk, as well as the Branch. I take it that this Bandage is intended to secure and strengthen the Leaves; for, as the least of these are about Ten Inches long, and Five broad, their Weight would otherwise be too heavy for the smaller Limbs to bear.

This Tree bears round white Berries.

The BIRCH-GUM-TREE.

THIS grows to be a middle-sized Tree.

Its Bark is of a reddish Colour, and smooth, somewhat resembling that of a Birch-tree.

From

Plate X *
Page 145
To the Right Honourable
The Earl of Halifax,
First Lord Commissioner of
Trade and Plantations
This Plate is humbly Inscrib'd,
&c.
G. D. Ehret. delin. & sculp.

From thence, and from the great Quantity of tranſparent Gum that its wounded Trunk and Branches afford, it derives its Name.

Its upper Branches ſuſtain a great many ſmall apetalous Flowers.

Theſe are ſucceeded by ſeveral red Berries, near as big as ſmall Cherries.

The Gum is looked upon to be very ſerviceable in all inward Bruiſes, or Strains, of what kind ſoever.

The Branches of this Tree, being ſtaked into the Earth, will grow; and I have known a Branch of it, tho' ſtripped of its Leaves, and expoſed to Wind and Weather (as Part of an Arbour for a Grape-vine), which, in this Situation, budded, and put forth young Shoots and Leaves.

It grows in moſt Parts of the Iſland.

The BAY-BERRY-TREE; *Lat.* MYRTUS arborea aromatica.

THIS is a beautiful upright Tree, its Trunk often of near Two Feet Diameter.

Its Bark, until the Seaſon of its peeling off, is very ſmooth, and ſomewhat reddiſh.

The Trunk divides into numerous Branches, the lower always the longeſt; ſo that towards the Summit it ends pyramidically, being very thickly cloathed with middle-ſized ſharp-pointed ſmooth very green Leaves.

This continual Verdure, added to its manner of growing, makes it appear very beautiful.

Its Berries are well known by the Name of *Jamaica* Pepper.

There is another of the ſame Species called the *Black Bay-tree.*

This differs from the other, chiefly by having its Bark and Leaves of ſomewhat a darker Colour than the above-deſcribed.

The WILD BAY-BERRY-TREE; *Lat.* MYRTUS.

THIS is a far leſs Tree than that laſt-deſcribed.

Its Bark is as ſmooth as, and much redder than, the former.

Its Leaves and Berries are of the ſame Make and Texture with the true Bay-berry-trees.

However, the Berries want the Virtue, as well as the fragrant Smell, of the former.

BLACK-BERRY-HUNTERS.

THIS Shrub hath a great many ſtrong Roots.

Its Trunk is covered with a dark-greyiſh Bark.

The Leaves are ſharp-pointed, and in Length above an Inch and an half, and about an Inch in Breadth.

It bears white ſmall Flowers, ſucceeded by ſmall black Berries.

FIDDLE-

Fiddle-Wood-tree.

THESE are common in this Island, and grow to a considerable Height

They are diſtinguiſhed into the White, the Red, and the Black ſort.

The latter is by far the moſt durable Timber, and not often to be met with.

The Body of the Trunk is covered with a light-brown Bark.

Its upper Branches are thickly cloathed with Leaves of above Three Inches long, and an Inch and an half broad.

The Extremities of the Branches end in ſeveral ſlender pendulous green Strings or *Ligulæ*, which are almoſt ſurrounded with ſmall white pentapetalous ſweet-ſcented Flowers, each ſtanding in a ſmall green *Calyx*.

Theſe are ſucceeded by ſmall pulpy Berries, which are at firſt of a green Colour; afterwards red; and, when ripe, black.

The Bee-Wood-tree.

THIS is a large ſhady Tree, cloathed with numerous ſmooth green Leaves, of Four Inches long, and Three broad.

Theſe have a great many round Excreſcencies growing on them in the Shape of Galls, tho' not much larger than a Grain of black Pepper.

Theſe are Neſts of ſome ſmall Inſects, which make an Hole in each of them; and then either creep, or fly away.

The Top-Branches bear ſeveral Bunches of ſmall white Flowers.

Theſe are ſucceeded by ſmall Berries, whoſe Pulp is of a purple Colour, when ripe.

The Fustic-tree; *Lat.* Morus.

THIS grows to be a large Tree, having generally a ſtrait Trunk.

The Bark is of a greyiſh Brown, ſlightly furrowed.

It bears Katkins; and the Fruit hath many *Acini*, which are as large as a large Strawberry, green within and without.

The Heart of this Tree is very yellow, and is well known for its Quality in Dyeing.

Yellow Hercules.

THIS grows to a middle-ſized Tree, having a great many Roots.

It is chiefly diſtinguiſhed from the White Hercules, by the Yellowneſs of the inner Bark, which, when properly prepared, affords an excellent yellow Dye: Perhaps it might be very uſeful in that Way, if in the Hands of ſkilful Dyers.

The outward Bark is of a reddiſh Grey, thickly covered with ſhort ſtubbed Prickles.

The

The Leaves, which are ſet on alternately, are about Four Inches long, and Two broad, having their middle Ribs ſtrong and prominent.

Its numerous ſmall herbaceous Flowers grow in Cluſters at the Extremities of their Stalks; and are ſucceeded by ſmall round Berries.

The inner yellow Bark of the Root, bruiſed, and ſteeped in boiling Water, tinges it with a yellow Colour.

Waſhing ſore Eyes with this Water hath of late been diſcovered to be a ſovereign Remedy in moſt Diſorders of the Eyes.

The PYE-CRUST.

THIS is a middle-ſized Tree, having many ſtrong Roots.

The Bark of the main Body is of a dark Grey.

The Leaves are very many, thick, ſmooth, and round-pointed, but very narrow towards their Stalks.

Theſe are of a deep ſmutty Green on the upper Side, and ſomewhat lighter on the under, and ſet on the Branches in an irregular manner.

They bear ſeveral apetalous Flowers, which are ſucceeded by yellow Berries, of the Bigneſs of a middle-ſized Olive, whoſe yellowiſh bitter Pulp incloſes Three whitiſh Seeds.

The Texture of the Tree is very brittle, from whence it derived the Name of *Pye-cruſt*.

The DWARF MANGROVE.

THIS Shrub grows by the Sea or Pond-ſides in wet Places.

The Trunk and numerous Branches are cloathed with ſhining green Leaves, Two Inches and an half long, and an Inch and an half broad.

Among the upper Branches are many ſmall tubular Flowers, filled with yellow *Stamina*, ſucceeded by ſmall Berries.

MONTABBA, *alias* TRUBBA.

THIS is a low ſhrubby Buſh, divided into many Branches; the outſide Bark of a whitiſh Grey, covered with an hoary Mealineſs.

The Inſide is of a ligneous Subſtance, which ſurrounds a greeniſh Pith.

The Roots are many and white.

The Leaves are ſet on alternately; and theſe are of a green Colour above, and whitiſh underneath; their Length about Five Inches, and above Three in Breadth, eſpecially at ſuch Places where they are bluntly and irregularly ſcalloped about their Edges.

The Flowers are monopetalous, and of a reddiſh White, expanding, in the Heat of the Day, almoſt horizontally flat; their *Oræ* being ſlightly ſegmented, and divided into Six ſharp-pointed Angles.

From the Centre of the Flower riſes a white Piſtil, ſurrounded by Six yellow thick triangular *Stamina*.

The Flowers are succeeded by a Pumkin-like Fruit, of a fine deep Purple on the Outside, and whitish within; the Pulp interspersed with white flattish Seeds.

This Fruit is boiled and eaten as a Pumkin.

The small RED TRUBBA.

THIS Plant grows to about Four Feet high.

Its Branches, as well as the Leaves, are few and straggling; the latter about an Inch and an half long, and an Inch broad.

Its Flowers are composed of Five white Petals, surrounding several *Stamina*.

These are succeeded by a scarlet Button, whose Inside is full of Juice, and small whitish *Acini* or Seeds.

This Fruit is chiefly used by Negroes, in their Soups.

It hath, when boiled, a bitterish Taste; and is by them looked upon as useful to alleviate any colicky Disorder of the Stomach.

TAMMATAS SHRUB; *Lat.* LYCOPERSICON.

THIS is likewise a shrubby Plant, much resembling, in its Make, that which produces the Bachelor's Pear; with this Difference, that the Prickles on this are not so stiff.

The Fruit is about as big as an Hen's Egg, red when ripe.

They are generally made use of, boiled in Broth.

These were brought hither by his Excellency Governor *Worsley*, from *Portugal.*

The MAIDENHEAD.

THE shrubby Vine, bearing this Fruit, grows very bushy, and about Three Feet high; its Bark bearing a red Fruit, but not eatable.

The BULLY-BERRY-TREE; *Lat.* CAINITO.

THIS grows to be a very large Tree, branching chiefly towards the Top.

It bears a round Fruit, of a russet-yellow Colour, standing upon an Inch and a Quarter Footstalk.

The inside Pulp of the Fruit is milky, and of a soft sweet Taste, not unlike a Sappadillo.

It is about the Bigness of a very large Cherry, but not depressed at each End.

The Pulp incloses a very sharp-pointed oval Seed.

SOAP-

SOAP-BERRY-BUSH, *or* FIRE-BURN-LEAF; *Lat.* SAPINDUS.

THIS is a ſcandent ligneous jointed Vine, whoſe Heart-like Leaves are thin, and ſharp-pointed.

Theſe are, in general, of about Four Inches long, Two and an half broad, and high-ribbed.

Almoſt every Leaf is ſupported by a crooked Tendril, which takes hold of the neighbouring Trees or Shrubs.

The Leaves, when bruiſed, and agitated in Water, will cauſe as ſtrong a Lather as an equal Quantity of Soap,

The Juice is likewiſe very much eſteemed to heal and cure Fire-burns, Scalds, or ſuch-like Sores.

The BASTARD LOCUST, *or* FOREST-TREE.

THIS grows to be a large Tree, cloathed with a whitiſh Bark, red within.

Its Leaves are of a dark Green, about an Inch and an half long, ſmooth, and very blunt-pointed.

The Bark is uſed as an Aſtringent.

The VERVAIN; *Lat.* SHERARDIA SPICATA.

THE Roots of this ſhrubby Plant are very numerous, and white, penetrating not deep into the Earth.

The main Stalk, when full-grown, is as thick as one's Finger, woody, and brittle; ſoon dividing into many leſſer Branches.

Theſe are jointed, putting forth, at every Joint, Duplicates of winged Leaves within one another; whoſe Edges are regularly ſerrated, and of a middle Size, ending very narrow towards the Stalks; the Stalks themſelves terminating in long ſcaly Spikes.

Round the Middle of theſe riſe ſeveral ſmall naked tubular monopetalous Flowers, of a fine blue Colour; whoſe Edges are divided into Five Segments, having likewiſe a whitiſh Spot in the Middle.

The Seeds, which are longiſh, and ſmall, are included in the ſeveral *Laminæ*, or Foldings, of theſe Spikes.

The Juice of this Plant is ſo great a Deobſtruent, that a large Spoonful taken inwardly, for Three or Four Mornings ſucceſſively, hath been more powerful to bring down the *Catamenia*, than either the Uſe of Chalybeats, or any other Method.

This Plant grows in moſt Parts of the Iſland.

The MASTICH-TREE; *Lat.* CALABA.

THIS often grows to a great Height.

The Colour of its Bark is browniſh.

The

The larger Branches divide into a great many leſs ones; and are thickly cloathed with deep-green ſhining Leaves, whoſe middle Ribs are tinged with Yellow.

Theſe are about Three Inches long, and near Two broad.

Their Edges are neatly labiated into ſeveral eaſy *Sinus*'s, and ſet on the Branches alternately, ſupported by Footſtalks of about an Inch long.

The Flowers are yellow, and ſucceeded by a Fruit, which hath a diſtant Reſemblance to a Shrub.

This Tree is of a very cloſe Grain, hard and durable; therefore much made uſe of in Mill and Houſe-building.

The CASSADO-TREE; MANIHOT.

THE Caſſado-trees, or rather Shrubs, are pithy on the Inſide; and ſeldom grow higher than Five Feet.

They are diſtinguiſhed into the White, and Old Sort; the former being ready to be rooted up, and uſed, in about Four Months after it is planted; and its Juice is not ſo ſtrong a Poiſon as the other old Sort.

The Root, which is the Part uſed in both, is large and white, and ſo ſoft, that it may be eaſily grated.

Theſe Shrubs have a few weak Branches.

Theſe, as well as the main Trunk, towards their Extremities, are cloathed with many large digitated Leaves upon reddiſh Footſtalks, of about Six Inches long.

The Flowers are deeply divided into Five Sections, whoſe Edges are tinged with a pale Purple.

Out of the Centre riſe Eight *Stamina*, tipped with yellow farinaceous *Apices*.

Five of theſe *Stamina* fall back, and lean againſt the different Sections.

The Flowers are ſucceeded by many ſmall round green Berries, whoſe Outſide is divided into Five ſelvaged Seams, the Inſide containing Three blackiſh Seeds.

The Method of making Bread from the Roots is, firſt, to grate them very fine, and preſs out the Juice, and dry the mealy Part in the Sun; and then make it into thin Loaves, like thoſe made with Oatmeal in many of the Country Parts of *England*, *Scotland*, and *Wales*.

This Bread is well taſted, and nouriſhing; and, by its abſorbent Quality, it is highly ſerviceable to Perſons of a plethoric Habit, or inclinable to Dropſies.

The Juice is a quick mortal Poiſon to every Animal.

Having tried an Experiment upon Two half-grown Fowls, by pouring down their Throats Two Tea-ſpoonfuls of this Juice, newly expreſſed, it ſoon threw them into Convulſions; and they both died in about half an Hour's time.

LEATHER-

LEATHER-COAT-TREE.

THIS grows to a confiderable Height.

The Bark is of a darkifh Colour.

The Branches are many and ftraggling, their Tops generally bending downwards.

The younger Twigs are feemingly jointed.

Thefe are cloathed with broad Leaves, reddifh when young, whofe Stalks, as well as their Ribs, are then thickly covered with a pale-yellow Pile, or foft-pointed *Setæ*.

The Leaves, as they grow old, turn green, being then in Length about Ten Inches, and Twelve in Breadth, having their middle, as well as their tranfverfe Ribs prominent and ftrong.

From the Bofoms of the Leaves rife many Six-inch long Spikes, furrounded with fmall Berries, red when ripe; the Infide being likewife full of reddifh Pulp, inclofing feveral fmall darkifh Seeds.

CLAMMY CHERRY-TREE; *Lat.* MALPIGHIA.

THIS Tree grows to a confiderable Bignefs, covered with a greyifh Bark.

The Leaves are in Bunches from Three to Five, upon the Tops of the leffer Twigs.

Thefe are about Five Inches long, fharp-pointed, and near Two Inches and an half broad.

The Flowers are of a pale-yellow Colour; and the fucceeding Fruit, when ripe, as big as a middling Cherry.

Thefe grow in Clufters, of a fine red Colour.

The Pulp is very clammy, and furrounds a great many fmall Seeds.

The BLACK CHERRY-TREE.

THIS very much refembles the red Cherry-tree, except that the Leaves of the black Cherry-tree are fomewhat larger, fmoother, and thicker, than the four Cherry.

The Leaves often grow clofe to the Branches; and, when upon the leffer Twigs, they are pennated.

From among thefe rife the Flowers, which (except in Colour, which is red) very much refemble thofe of the Coffee-tree.

Thefe are fucceeded by fmall blackifh Berries.

The RED CHERRY-TREE.

THIS Tree feldom grows above Eighteen Feet high.

Its Branches are numerous, but fmall, and cloathed with a great many deep-green round-pointed Leaves.

From the Boſom of theſe riſe many Footſtalks, ſupporting ſmall Five-leaved Flowers.

Their Petals ſurround ſeveral green *Stamina*, tipped with yellow Summits.

The whole Flower is ſupported by ſeveral ſmall capſular Leaves, and ſucceeded by the Fruit, which is red, when ripe; of about the Bigneſs of a ſmall Cherry, but more compreſſed at both Ends.

Theſe are of an agreeable ſweet, mixed with an acid Taſte.

The Inſide is full of ſmall whitiſh Seeds.

The BELLY-ACH.

THE Roots of this Shrub are few in Number, and white, penetrating deep into the Earth.

The main Stalk, which is covered with a light greyiſh Bark, grows to about Three or Four Feet high, ſoon dividing into ſeveral wide-extended Branches.

The Body of the Shrub, within Three Inches of the Ground, ſwells, or bulges out.

From this Swelling, by Inciſion, is taken out, when ripe, a rough granated Core or Kernel, of a pale Red, and ſometimes near as large as a Pullet's Egg.

This is extraneous to the Nature of the Shrub, and as diſtinctly ſeparable from it, tho' in cloſe Contact with it, as a Peach-ſtone is from the pulpy Part thereof.

It is alſo of the ſame Nature and Quality as Hellebore; and anſwers, to all Intents and Purpoſes, the Uſe of it.

The Branches of the Shrub are neither decorated with Leaves nor Flowers, till near the Top: There the former, which are deeply ſegmented into Three or Four Diviſions, ſurround it.

Theſe, including the Sections, are about Five Inches long, and near as broad.

Their Footſtalks, as well as the young Buds, on the Extremity of the Branches, are guarded round with ſtiff hairy Briſtles, which are always tipped with glutinous liquid Drops.

From among theſe riſe ſeveral ſmall deep-red pentapetalous Flowers; the Piſtil of each being thick-ſet at the Top with yellow farinaceous Duſt, which blows off when ripe.

Theſe Flowers are ſucceeded by hexagonal husky blackiſh Berries; which, when ripe, open, by the Heat of the Sun; emitting a great many ſmall dark-coloured Seeds, which ſerve as Food for Ground-doves.

The Leaves of this Shrub are few, and ſeldom or never drop off, nor are torn or eaten by Vermin of the *Eruca*, or any other kind.

As Nature makes nothing in vain, I beg Leave to conjecture, that this clammy Liquid, like Bird-lime, with which each of the above-mentioned

Setæ

Plate II
Page 152
Belly Ach — Fig. 1.
SAPERE AUDE
To the Right Honourable
Earl of Macclesfield,
This Plate, is humbly Inscrib'd
&c.
Acacee — Fig. 2.

Setæ or Briſtles are tipped, is deſigned to intangle and prevent Caterpillers, or other Vermin, from climbing up to deſtroy the few Leaves, with which Nature hath ſo ſparingly ſupplied this Plant.

As for other Plants, which are not thus guarded, as their Leaves are generally very numerous, if ſome of theſe are deſtroyed by Worms, they ſoon, in ſo warm a Climate, recover this Loſs, and the Plant renews its wonted Verdure

Perhaps, likewiſe, this gluey Subſtance may be of further Uſe, if gathered by Butterflies, or other flying Inſects (eſpecially the former, which are often ſeen hovering about the Flowers), to glutinize their Bags or Webs, before they enter into their *Aurelia*-ſtate.

The Seeds dropping from the ripe Berries are ſo great a Specific againſt Melancholy, that even Doves, that have uſed to feed on them, will not, when confined to a Cage, whoot, if deprived of theſe, and Bird-pepper.

It is ſaid, that Fifteen of the Berries, pulverized, and taken inwardly, prove a good Purge.

It grows in every Part of the Iſland, but moſt luxuriant in a ſandy looſe Soil.

This is delineated in *Plate* XI. *Fig*. 1.

The PRICKLY HOOP, *or the* WHITE THORN.

THIS derives its Name from the Uſe that is ſometimes made of it to hoop Veſſels.

The main Stalk, when full-grown, is generally bigger than one's Arm, dividing, near the Ground, into many prickly Branches.

Theſe grow, eſpecially if ſupported by low Underwood, to ſometimes near Thirty Feet in Length.

The Bark is of a whitiſh Grey.

The ſmall Side-branches, as well as the Leaves, are ſet on alternately.

The latter are, near the Stalk, about Two Inches in Breadth, and Two Inches and an half in Length, ending in a long ſharp Point.

Their Edges are ſlightly ſnipped, and the Colour of the Leaf of a faint Green, inclined to a Yellow.

The middle Rib is ſtrengthened, beſides the tranſverſe ones, with Two, or ſometimes more, longitudinal ſtrong Veins, or Ribs.

The Flowers are ſucceeded by middle-ſized Berries, red when ripe.

Theſe are ſometimes eaten by Men, but chiefly by Birds.

The ALOES PLANT; *Lat.* ALOE vulgaris.

THIS very ſucculent Plant hath one large Root, with a great many other ſmall ſtringy Roots growing from it.

The Number of Leaves are generally about Twenty.

Theſe,

These, near the Roots, are Two Inches broad, and near one thick, gorwing to Eighteen or Twenty Inches high, ending in a sharp Point.

Each Side of the Leaf is guarded, at every Half-inch Distance, with a strong small Prickle.

Out of the Middle of these Leaves rises a slender conic Stalk, sometimes divided into Two Forks, at uncertain Distance from the Root.

This rises about Eighteen Inches above the Top of the Leaves, bearing, near the Summits, several pendulous yellow monopetalous Flowers, whose *Oræ* are divided into Six Segments; these surrounding the Stalk for near Four Inches downwards from the Top, making a very agreeable Appearance.

Each Flower hath a small Aperture at the End; thro' which a great Number of small blackish Seeds drop out, when ripe.

It will not, I hope, be improper, in this Place, to observe the Method of planting, cutting, and boiling of Aloes, since there are many Parts of his Majesty's Plantations, besides these Summer-Islands, that would produce this Plant, with proper Care and Nourishment.

The Land designed for Aloes must be first weeded clean, and then holed Three Inches deep, at about Six Inches Distance.

Then the Suckers from the old Plants, or those produced from the Seeds, are planted in these Holes.

The Land, for the first Year, must be kept clean from Weeds; afterwards, the Shade of the Leaves of the Plant will keep the Weeds under.

Being thus planted, there will be no Occasion to replant it for Seven Years.

It comes to its Perfection in a Year's time.

The Month of *March* is the properest Season to make the Aloes; which is done in the following manner:

Every Slave hath by him Three or Four portable Tubs.

The Leaves, being cut near the Roots, are thrown into these, with their broken Ends downwards; and as the Leaves are full of large longitudinal Veins or Vessels, they yield an easy Passage to the Juice (which is of a greenish-yellow Colour) to drip out.

This being boiled for about Five Hours in a Copper, or Kettle, the watry Particles evaporate; and the Remainder comes to a Consistency, and thickens, as Sugar doth when sufficiently boiled.

The way to know when it is enough boiled is, to dip a Stick in the Liquor, and observe whether the Aloe, sticking to it when cold, breaks short: If it doth, then it is boiled to Perfection, and fit to be poured into Gourds or Calabashes, or other Vessels, for Use.

Aloe is much made use of in Purges, and justly esteemed of great Service in many Cases.

However, Dr. *James*, in his Medicinal Dictionary, says, it ought not to be given to Women with Child, nor to Persons subject to the Piles; for it rarefies the Blood too much, and causes Hemorrhoids.

HEN-

HEN-TURD, *or* DART-WOOD.

THIS grows in ſpiral Branches from the Root, but ſeldom above Twelve Feet high, each ſeemingly jointed at about Twelve Inches aſunder.

The Wood is of a very cloſe Grain, and heavy; very little inferior in Solidity to Box-wood.

It hath the Name of Dart-wood, from the former Uſe made of it in Lances and Darts; with which, by the Permiſſion of our Laws, Negro Watchmen were to be armed, to defend their own, or the Goods of their Maſters, from Robbers. However, a late Cuſtom hath allowed theſe the Uſe of Swords.

The Name of Hen's-turd was likewiſe given it from the blackiſh unſavoury Pulp of the Berry, which it bears.

This ſucceeds a ſmall white Bloſſom.

The Berry itſelf, when ripe, is of a browniſh Colour.

The CRAB-BUSH, *or the* SEA-SIDE LAUREL.

THIS is a ſmall buſhy Shrub, thickly cloathed with green ſtiff ſhineing Leaves, many in Number, and about Three Inches long.

Theſe are narrower at the Stalk than their Extremities.

They bear many apetalous Flowers, ſucceeded by browniſh Berries.

Theſe, when ripe, are greedily eaten by the Land-crabs.

The WHITE SAGE; *Lat.* CAMARA SALVIÆ.

THIS is a thick buſhy Shrub, growing to about Two Feet high.

Its Leaves are ſmall and rough, of a whitiſh Grey, and thick-ſet, generally Three together round the Stalk, in a triangular manner; their Edges being bluntly ſerrated.

The Flowers are of the galeated, monopetalous kind, divided at their Extremities into Four unequal Segments, of a pale Red, altering generally at the *Diſcus* to a bright-yellow Colour.

Theſe Flowers are many in Number, ſupported upon Stalks riſing from the Boſom of the Leaves; the tubulous Part ſerving for Seed-veſſels.

Each Flower is ſucceeded by a ſmall purpliſh Berry.

Theſe ſerve for Food to Sparrows, and other Birds; and the Leaves of the Shrub, being either boiled into a Decoction, or made into Tea, are an excellent Sudorific and Pectoral.

It grows chiefly in dry Places, and flouriſhes all the Year round.

The MISLETOE, *or* BIRD'S TURD.

THE Origin of this Shrub is a small white Berry, containing a very glutinous milky Juice.

These, when ripe, are eaten by Birds; and by them often voided upon other Trees, where they stick very fast, and soon after germinate.

I am apt to believe, that this brownish Juice hath likewise a corrosive Quality, which frets and wears away the outward Bark of the Tree on which it sticks; by this means opening a Passage for the new tubular Roots of the Berry, to penetrate among those larger Vessels of the Tree, thro' which the nutritious Juices are conveyed.

These Roots being thus able to suck Nourishment, the young Sprout soon grows generally at first in Two or Three spiral Convolutions, or Twistings round the Branch: By thus grasping, it not only strengthens its weak Roots, and takes a firmer hold of the Tree; but such a Ligature, as it grows, sinks deep into the Bark, and prevents, in a great measure, the further progressive Motion of the Juices; so that these, receding but a little Way back towards the Root, are absorbed, and sucked up, by the numerous Mouths in the Roots of this fostering Shrub, which daily penetrate more and more into the Tree.

What confirms me in this Opinion, is, that the Branch, from the Place where the Misletoe takes Root, to the Top, is generally in a very decaying Condition.

The Leaves of this parasitical Shrub are of a dark Green, set on in a pennated manner, never terminated by an odd one.

This grows to about Three Feet high, and bushy, bearing a small white Flower, succeeded by a Berry, as above described.

The milky Juice, being squeezed out, is made use of to cure Fluxes and Lasks.

It grows chiefly upon Orange-trees.

The NARROW-LEAV'D MISSLETOE.

THIS grows in Tufts, consisting of Six or Seven narrow Leaves, of about Six Inches long.

They are generally to be found in the Clefts of the Bark of Cedar-trees.

The SPIRIT WEED.

THIS is a strong Shrub, having many substantial Roots.

Its Leaves are pennated about Three Inches long, and sharp-pointed.

The

The Flowers are yellow, standing upon long Footstalks, rising generally from the Bosom of the Leaves, and of the tetrapetalous Kind.

These are succeeded by a small Berry, of a dark-red, when ripe; containing sometimes Two or more blackish Seeds.

A Decoction of this Wood is esteemed to be a strong Diuretic.

The PIGEON-WOOD.

THIS is a shrubby Tree, seldom growing to above Six Feet high.

The Wood is very solid; and the Bark whitish.

Its numerous small Branches are thickly cloathed with small round-pointed winged Leaves.

The small white Blossoms it bears are succeeded by a great many white Berries.

The SWEET WOOD-TREE.

THIS is of Two sorts; the one bearing a Berry like a Damascen.

That which bears the white Bark is the best and most solid Timber.

That with the green Bark, and white within, is of less Value.

The Leaves of the former resemble those of the Bay-tree.

These are set on alternately.

The Berries are supported by small Pedicles, rising from the Bosom of the Leaves.

These, when ripe, are eaten by Pivets, and Ground-doves, *&c.*

The BALSAM-TREE.

THIS is a middle-sized Tree, growing generally in Gullies, and shady Places.

Its Roots are few, but strong.

The Colour of the Bark is of a reddish Grey.

Its Branches are many and spiral, thickly cloathed with very smooth shining succulent Leaves, pale underneath, and green above.

These are narrow at the Stalk, but broad and round-pointed at their Extremities.

Their Edges are smooth, and tinged with a pale Yellow.

The Flowers consist of Six strong white Leaves, surrounding a yellowish Thrum, and supported by Four white capsular Leaves.

The Stalks bearing the Flowers are long, white, and brittle.

The succeeding Berries are of an oblong Shape, about the Bigness of a Nutmeg, and of a deep Red, when ripe.

These are decorated with a *Corona* at the Top.

The

The Berry is likewise marked with Six unindented Seams of a blackish Colour, seemingly dividing it into so many Partitions.

The Leaves, Stalk, and Fruit, when broken, emit a white glutinous Liquid, which will harden by the Heat of the Sun.

The LIGNUM RORUM.

THE Bark of this shrubby Tree is of a dark Green, streaked with White, especially on the younger Branches.

These are covered with smooth deep-green round-pointed Leaves, Five Inches long, and near Three broad.

The smaller Twigs sustain Bunches of white Flowers, each composed of Six Petals surrounding Six slender pale yellowish *Stamina*.

These inclose a whitish Pistil, which is succeeded by a small Berry of a purple Red, when ripe.

The younger Branches are seemingly jointed.

These, and every other Part of this Tree, have so much of a terebinthial Quality in it, that it will, when even half-dried, burn like a Candle.

From this Quality it derives the Name of *Jack-Lantern*; for, being tied into Fagots, and lighted, it serves the Fishermen instead of Torches, to fish in the Night-time for Crabs and Lobsters.

It grows generally in shady Places.

The COFFEE-TREE; *Lat.* JASMINUM ARABICUM.

THE Coffee-tree, if sheltered, grows to be about Fifteen Feet high, tho' its common Growth seldom exceeds Twelve.

It divides into several slender Branches, generally spiral.

The Bark likewise is of a smooth reddish-grey Colour.

The Branches are cloathed with Four-inch long sharp-pointed dark-green Leaves, their Edges being waved or sinuated in an elegant regular manner.

Among these Leaves rise a great many Five-leaved white Flowers, each Petal being round-pointed; the Whole surrounding several short white *Stamina*, loaded with *Apices*.

These likewise surround the Pistil, which is the Rudiment of the succeeding Berry.

This, even when ripe, is covered with a thin husky *Capsula*; which, when sufficiently dried in the Sun, easily cracks, and discovers the Berry, which is too well known to want a further Description.

Box-

Box-Wood.

THIS Shrub is covered with a whitiſh-grey Bark.

Its Leaves are about Two Inches long, and One broad; and are generally ſharp-pointed.

Theſe are ſet on the Branches alternately.

The Flowers are of the tubular ſtellated Kind, ſucceeded by ſmall Berries, which are eaten by Birds of different kinds.

This Shrub generally grows in the Shade: Several of theſe are to be found on the Eſtate of *John Colliton*, Eſq; in *St. Lucy*'s Pariſh.

The Wild Honey-tree.

THIS Tree grows to a conſiderable Bigneſs.

The Bark is ſomewhat ſulcated, and of a reddiſh-grey Colour.

The Leaves are of a deep Green, ſmooth, and winged, exactly reſembling Ivy-Leaves.

Its white Flowers are ſucceeded by Berries, ſomewhat round, as big as a Cherry.

Theſe are of a ſnowy White without; and the Pulp within (which is as ſweet as Honey, and as yellow as Gold) is divided into Partitions, including many ſmall Seeds.

I found this Tree on a ſmall Grove belonging to Mr. *Joſeph Jordan*, jun. in *St. Lucy's* Pariſh.

The Black Cherry-tree.

THIS Tree ſeldom riſes above Eighteen or Twenty Feet high.

The Bark is of a ruſſet Grey, and the Branches thickly cloathed with dark-green Leaves of about Two Inches long, and one and an half broad.

The Flowers are white, and of the roſaceous Kind.

Theſe are ſucceeded by ſmall black Cherries, ſomewhat larger, and more compreſſed at both Ends, than the *Engliſh* Black-cherry.

Their ſweetiſh Pulp ſurrounds a ſoft Kernel.

They are generally ripe in *Auguſt* and *September*, and are eaten by Men as well as Birds.

This grows in almoſt every Pariſh in the Iſland.

The Black Sage-bush.

THIS hath a great many ſtrong Roots.

The main Stalk is cloathed with a blackiſh Bark.

The Leaves, which are high-ribbed, and corrugated, are about Two Inches and an half long, and above One broad.

These are set on alternately.

The Top-branches sustain a Spike, thickly studded with very small stellated Flowers, which are succeeded by as many small sharp-pointed Berries, red when ripe.

The Leaves, as well as the whole Shrub, have a strong Sage-like Smell.

A Decoction of these is justly looked upon to be an excellent Sudorific.

It grows in most Parts of the Island.

The INK-VINE.

THIS is a creeping weak Vine, supported by the neighbouring Rocks, or Underwood, *&c.*

Its Leaves are of a middle Size, divided into Three deep Sections.

It bears many blue Flowers.

These are succeeded by small Berries, very black within.

This Vine grows in dry shady Places.

The Juice of this Vine is so black, that it resembles Ink, from whence it derives its Name.

CHIGERY-BUSH.

THERE is scarce any Difference between this and what we have already called *Soldier-bush.*

And, as the latter, in some Parts of the Island, go under the Name of the former, I take the only Distinction between these seemingly different Plants to be more owing to either the Male or Female kind, or to the more luxuriant Growth of this same Plant in various Soils, than to any other real Difference.

It grows generally in shady Places.

The WILD CALABASH.

THIS is a Shrub, growing about Ten Feet high, the Branches being generally very strait, thickly covered with long green Leaves; in Shape not ill resembling those of a Laurel.

From the Bosom of the Leaves, on short Footstalks, arise a great Number of small Berries, black when ripe.

These are eaten by Pivets, and other Birds.

CAT'S-BLOOD.

THIS is a small weak Shrub, cloathed with a Bark of a reddish Grey.

The lesser Branches are geniculated.

The

The Leaves which thickly cover these, are about an Inch and a Quarter long, and about half an Inch broad, and sharp-pointed, bearing, on slender Spikes, several very small Four-leav'd Flowers.

The Pistil, which grows from the Centre of each, is tipped with an *Apex*, and succeeded by a small red Berry.

A Cataplasm of this Plant, bruised, is very justly looked upon to be a good Vulnerary.

It grows chiefly in shady Places; and blooms in *June* and *July*.

I have found this Plant in great Plenty in the Parishes of *St. Peter* and *St. Lucy*.

The SOLDIER'S-BUSH.

THIS Plant is jointed, at different Distances, from Three to Eight Inches asunder.

Its Roots are strong and fibrous; the main Stalk, and its divided Branches, growing often above Four Feet high.

The Leaves are near Six Inches long, and Four broad.

Its upper Side is thinly beset with very short hairy Bristles.

From the Top of the Branches rise small white Flowers.

These are succeeded by small white Berries, spotted with Black.

There often grows at the Root of this Plant a fungous Excrescence, which answers the End of Touchwood.

It is justly esteemed an excellent Vulnerary, and with great Success, made a Part of most kinds of healing Salves.

POPS; *Lat.* ALKEKENGI Indicum majus.

THIS Plant hath a fistular quadrangular Stalk, growing to about Two Feet high, cloathed with thin slender Leaves, Four Inches long, and Two broad, having their Edges, at uncertain Distances, neatly snipped.

Upon the Top of the Stalks appear several yellow apetalous Flowers.

These are succeeded by thin bluish capsular Pods, which inclose a round green Fruit of about the Bigness of a small Cherry.

As this ripens, and turns yellow, the outward Pod, or Covering, which is of a blunt conic Form, withers, and drops off.

The Fruit is full of small Seeds, not disagreeable to the Taste.

They are justly looked upon to be very diuretic.

There is another Plant, which bears the same kind of Fruit.

This differs from that already described, by being a creeping scandent Plant, and its Leaves shorter and thicker than the former.

This is called the *Pop-Vine*, and grows in most Parts of the Island; especially under the Shelter of Hills.

The

The Berry-bush, *or* Hedge-bush.

THIS often grows to about Seventeen Feet high, cloathed with green Leaves, Six Inches long, and an Inch and an half broad.

These are set on the Branches alternately, having their middle Ribs thickly guarded with sharp reddish Prickles.

The Ends of the lesser Branches sustain a great Number of white Five-leaved Flowers.

These Petals surround a yellow Pistil, divided at the Top.

This is succeeded by a small red Berry, which is looked upon to be very wholsome: However, its Taste is not very agreeable, and therefore seldom eaten by Men.

The Wild Pepper-grass.

THIS hath a great many strong stringy Roots, penetrating about Five Inches into the Earth.

The Bark of the main Stalk is greyish, soon dividing into numerous small Branches, thickly covered with very small sharp-pointed Leaves.

From among these rise very many Pedicles, supporting several small Star-like Flowers, succeeded by very many capsular round Pods, inclosing a great many very minute russet Seeds.

The whole Plant seldom grows above Two Feet high.

It is to be found in almost every Part of the Island.

The Fire-burn Weed.

THE main Stalk of this Plant grows to be about Three Feet high. It is generally of a green Colour, chanelled and pithy.

The Leaves are about Three Inches long, and Two in Breadth, having their middle, as well as their transverse Ribs strong and prominent.

From among these rise a great many Footstalks, or Pedicles, of a pale Red, supporting, upon a Spire, a great many small white-leaved Flowers, each consisting of Four Petals, surrounding a white Pistil.

These are succeeded by many small Berrries, red when ripe; whose Pulp (which inclose a great many blackish Seeds) is likewise as red as Blood.

Gooseberry Shrub.

THE main Stalk of this scandent Shrub divides near the Earth into many lesser ones.

The Bark of these is of a dark-yellowish Green.

The Leaves which grow in Pairs upon one common Footstalk, are small, smooth, and succulent, having their Edges somewhat tinged with Yellow.

The

The Flowers are ſucceeded by Berries about as big as a middling Gooſeberry.

Theſe are coronated at the Top; the Inſide containing, in an agreeable ſweetiſh Pulp, ſeveral ſmall *Acini*, like thoſe of Gooſeberries.

This Berry is, above all others, remarkable, by a great many ſmall Leaves, which grow upon the Outſide of it.

The ſeveral Branches of this Shrub are likewiſe guarded with very ſharp-pointed Prickles.

I found this growing in great Plenty on the Eſtate of the Honourable *John Maycock*, Eſq; in *St. Lucy's* Pariſh.

The WILD CATERPILLER; *Lat.* BLITUM Americanum ſpinoſum.

THIS chiefly differs from the eatable Sort, by the Smallneſs of its Leaves, as well as that the ſeveral Tufts of Leaves are guarded near the main Stalks with Two or Three ſharp-pointed Prickles.

Theſe are likewiſe of the red and white Sort, differing ſcarce in any thing but Colour.

They grow chiefly in open Fields; eſpecially at the Plantation of the Honourable *William Maynard*, Eſq; in *St. Thomas's* Pariſh.

The MILK-WEED.

THIS lacteſcent Plant is diſtinguiſhed into the red and white Sort.

Theſe ſeldom riſe above Fifteen Inches high.

The main Stalk, with its ſlender Side-branches, is thickly covered with ſeveral Pairs of ſmall pennated Leaves, ſomething inclinable to an oval Shape, having their Edges finely ſerrated.

From the Boſoms of the Leaves riſe a great many ſmall Pedicles, ſuſtaining on their Summits a Group of very ſmall, ſcarce perceivable, white tubular Flowers.

Theſe are ſucceeded by ſmall Berries

The Roots or the Stalks of this Plant, when wounded, emit a milky Juice.

It grows in moſt Parts of the Iſland, and is made uſe of as Part of the Ingredients in Diet-drinks, to cleanſe the Blood.

DOVE-WEED.

THIS ſmall Plant ſeldom riſes above a Foot high.

Its Roots are fibrous and many.

Its main Stalk divides into ſeveral Side-branches.

These are very thickly cloathed in an alternate Order, with small oval Leaves, not ill resembling those of Tamarinds.

Its Flowers, which are monopetalous, whose *Oræ* are divided into Five Sections, are succeeded by small *Capsulæ*, somewhat resembling Berries.

The small Doves feed on these; and from thence this Plant derives its Name.

This Plant, reduced to Powder, is a great Drier-up of old Sores.

The milky Juice, which issues out of the Stalk, is looked upon to be good to destroy Warts.

It grows in most Parts of the Island, especially in rocky or gravelly Soil.

Gumma-Bush.

THIS grows to be about Eighteen Inches high, the Stalk very green and succulent.

The Leaves are about Two Inches long, and above an Inch broad; their Edges being irregularly and deeply serrated.

From among the Leaves arise a great many white small Five-leaved Flowers, inclosing broad close yellow *Stamina*, surrounding a green *Stylus*.

The Flowers are succeeded by very round small Berries, black when ripe.

The Leaves of the Plant are made use of by the Negroes, as boiled Sallad; but seldom, if ever, made use of by the White Inhabitants.

The White Hercules.

THIS differs chiefly from the other by the Colour of its inner Bark, which is white.

The Flowers are very small, and of the herbaceous Kind.

These are succeeded by a capsular Bunch, full of small black oval Seeds.

The Buds of this Tree, boiled into a Decoction, are very good to cure a Distemper called the *Red Water* in Cattle.

This grows best under the Shade of an Hill.

Boumbo Bush.

THIS Bush hath a very offensive strong Smell.

It grows to about Two Feet high, thickly cloathed with Leaves.

It grows in most Parts of the Island.

Purslain,

PURSLAIN, WHITE *and* RED.

THESE differ very little, if any thing, from thoſe in *England*, except in the Rankneſs of their Growth here.

They are generally boiled and uſed, when more valuable Greens are not in Plenty.

The ARABIAN JESSAMMY; *Lat.* JASMINUM ſive Sambac Arabum.

THIS is a woody creeping Vine, cloathed with dark-green ſharp-pointed Leaves, of about Two Inches long, and one broad, bearing Seven-leaved white fragrant Flowers.

Theſe are ſupported by Seven very ſharp-pointed graminous Leaves.

This is planted chiefly in Gardens, or near Dwelling-houſes.

It yields a very agreeable fragrant Smell, and keeps its Verdure all the Year.

TITHYMALOIDES.

THE Roots of this are very numerous and ſtrong.

The main Stalk branches near the Earth, into many ſpiral ſtrong Twigs.

The Bark of the Trunk, near the Roots, is ſmooth and whitiſh; but all its Branches of a deep ſhining Green, keeping their grateful Verdure undiminiſhed thro' every Seaſon of the Year.

Theſe are cloathed with thick ſucculent green round-pointed Leaves, of about Four Inches long, and Two in Breadth, having their ſmooth Edges a little tinged with Yellow.

Theſe are likewiſe ſet on the Branches alternately.

The Flower, which is of a fine deep Red, cannot be reduced to any Claſs that hath hitherto been deſcribed by any Botanical Writer, it being a triangular ſharp-pointed ſmall horizontal Sheath; the Top of the Piſtil jutting out, making the Extremity of the longeſt Angle: This, on the Inſide of the Flower, is covered with a yellow farinaceous Duſt.

From the ſame narrow Aperture appear ſeveral ſmall *Stamina*, tipped with green *Apices*.

Theſe Flowers are ſucceeded by triangular ſmall Berries, of a dark Red when ripe.

Each of theſe incloſes Three angular Seeds.

The Berries, Branches, and Leaves, are full of a thick glutinous milky Juice.

The only Place that I have ſeen this was at *Cluff's Bay*, in *St. Lucy's* Pariſh.

This very ſeldom grows in any other Part of the Iſland.

The

The ſmall WILD CUCUMBER.

THIS is a ſmall Vine, creeping generally along old Walls.

Its Leaves are thin, and ſharp-pointed, of about an Inch and an half long, and as broad near the Stalk, where they grow out, as it were, into Ears.

The Flowers are ſucceeded by a ſmall ſmooth Fruit, of about the Bigneſs and Form of a Thimble.

Theſe, when ripe, are of a bluiſh black Colour, and are eatable.

The POISON-WYTH.

THIS is a large ſcandent ligneous ſhrubby Vine, whoſe Roots are very many, and ſtrong; and the main Body of the Wyth, near the Ground, as large as one's Arm, and ſomewhat flattiſh.

This climbs to a great Height, and is cloathed with ſharp-pointed green Leaves of above Three Inches long, and near Two broad.

If the main Stalk is cut off near the Ground, or at any Height, the upper Part, covering the Trees or Rocks, will ſtill ſurvive, and in a ſhort time ſend down, from ſeveral Parts, long ſtringy Filaments; which, growing downward, take Root, and ſupply the Place of the broken-off Stalk.

The Flowers are, in Appearance, like a Bunch of red Coral, ſucceeded by ſmall Berries, black when ripe.

TREES, SHRUBS, *and* PLANTS,

OF THE

Capsule-bearing FRUITS.

BOOK VI.

The PRICKLE-YELLOW-WOOD; *Lat.* Xanthoxylum.

THIS is a Timber-Tree, growing large enough to be saw'd into Planks, or Boards, for Tables, and other Uses. Its Roots are many and strong. The Wood is of a very fine close Grain, ponderous, and of a yellow Colour; and the Bark of a reddish Grey. The main Branches are many, shooting out into a great Number of lesser ones: These are beautifully decorated with a great many winged Leaves. The Edges of these are regularly indented; and between each Segment the Leaf is generally waved, or labiated. Its Length is about two, and its Breadth about one Inch. The whole Tree, particularly when young, is guarded with short Prickles. The Flowers are succeeded by a Group of black shining flat Seeds, inclos'd in small *Capsulæ:* These are very small, and of a shining Black, affording by Expression a great Quantity of Oil.

WILD HEMP.

THERE are two Sorts of this Plant, the White and Red: The latter never grows above two Feet high: The other, which is most useful, to above five Feet. The outward Coat of this, when ripe, affords numerous ſtrong Filaments, ſomewhat coarſer than what is peeled off from our *Engliſh* Hemp-plant: This ſerves to make Ropes, and ſuch Uſes. The Flowers are yellow, ſucceeded by ſmall *Capſulæ*, incloſing a great Number of ſmall black Seeds. The Stalks and Leaves are thickly covered with ſtiff Hairs, ſharp even to a Pricklineſs. The Leaf is bluntly ſerrated, and high-rib'd; in Length about two Inches, and in Breadth about an Inch and an half.

The Barbados Pink.

THIS is a very ſlender, buſhy, creeping Vine, full of ſmall Side-branches, cloathed with numerous very narrow graſſy Leaves, generally ſet on alternately. The Flowers, which are tubular and monopetalous, are of a deep Scarlet. Their *Oræ* are divided into five Segments, the intermediate Spaces forming as many ſharp-pointed Angles. The Inſide of the tubular Part of the Flower appears whitiſh. From this Part riſe ſeveral *Stamina*, tipped with whitiſh *Apices*. The Flowers are ſucceeded by ſmall *Capſulæ*, each containing a blackiſh Seed.

The Hop-Shrub; *Lat.* Meliſſa altiſſima globularia.

THIS Shrub grows to be about five Feet high. The Leaves are ſmooth, and round-pointed, and of about three Inches long, and above one broad. Theſe are ſet on the Branches in an alternate Order. The Flowers, which ſtand on long ligneous Pedicles, are each ſucceeded by very thin browniſh *Capſulæ*, reſembling at a Diſtance Hops, and of a triangular Shape. This is blown out like a Bladder between each Angle, as well as every Angle terminating in a thin Border, or *Ala*.

Monkey-Vine.

THIS is a long creeping Vine, of a dark-reddiſh Colour, and hairy. The Leaves are of an Heart-faſhion, but ſomewhat auriculated near the Stalk. From the Bottom of the Leaves riſe many long tubular white Flowers, whoſe *Oræ* are much expanded, and ting'd with Purple. From the Centre of the Flower riſe ſeveral white *Stamina*, tipp'd with long white *Apices*. The Flowers are ſucceeded by a ſmall round whitiſh *Capſula*, each incloſing ſix black angular Seeds, not much larger nor unlike Onion-ſeeds.

Plantain-Shot; *Lat.* Cannacorus.

THIS Plant grows generally in ſhady Places, and often to five Feet high. Its main Stalk is tubular, cloathed at different Diſtances with very ſmooth green Leaves, in Make and Subſtance very much reſembling a Plantain-Leaf. Theſe are about nine Inches long, and five broad. The

Flowers ſtand upon ſtrong Footſtalks, each having at firſt the Appearance of a ſcarlet Pod of an Inch and an half long: When this opens, it diſcovers three ſcarlet ſpoon-like Leaves. The Inſide of one of theſe, when it deflects back, appears ſtained with Yellow and Red. From the Centre of the Leaves riſes a reddiſh flat Pointal, incircled by another reddiſh Leaf, ſpotted with Yellow; to whoſe Edge, near the Top, ſticks a long whitiſh *Apex*. The Flowers are ſucceeded by ſmall *Capſulæ*, each incloſing a round black hard Seed, as big as Swan-ſhot: From theſe, and the Make of its Leaves, they derive the Name of Plantain-ſhot.

LOGWOOD; *Lat.* Campechiana.

WHAT is called here the Logwood-tree hath a dark-coloured Bark, the Branches ſpreading open and wide: Theſe are here-and-there guarded with Prickles. The Leaves are many, and very ſmall.

WILD CLARY; *Lat.* Heliotropium Americanum.

THIS Plant grows to about eighteen Inches high; the Stalk green and hairy. The Leaves are ſet on alternately, ſurrounding the Stalk: Theſe are of a middle Size, rough and crumpled, having their Edges much ſinuated. From the Top riſes a long gramineous Spike, ſtudded with very ſmall monopetalous whitiſh Flowers: Theſe are ſucceeded by many ſmall Seeds.

HOP-WEED.

THERE are two Species of this Plant, diſtinguiſhed by the Name of White, and Red: The Red grows to about eighteen Inches high. The Leaves are rough, and about two Inches long, and above an Inch broad; their Edges being irregularly ſerrated. From the Boſom of the Leaves riſe many Footſtalks, thick-ſet with blue tubular Flowers, ſucceeded by ſmall Seeds. A Decoction of this Plant is made uſe of, as a gentle Gargle, to cure ſore Mouths.

MUSKETO-BUSH, *or the* WHITE HOP-WEED.

THIS derives its Name from its either real or pretended Service in driving away, by its Smell, Muſketoes from Bed-chambers, or elſewhere, by having a Buſh or Bough of it hanging in the Room. Its Roots are many; and the Stalk is four-ſquare, and every Square chanelled. It grows in rich Land, to often above four Feet high. This is ſurrounded by two-inch long ſharp-pointed Leaves, whoſe Edges are irregularly ſerrated. The Top of the Stalk ſupports a blunt-pointed conic pluſhy Tuft, diſcovering ſeveral ſmall tubular blue Flowers, ſucceeded by many ſmall Seeds.

The

The WILD PENY-ROYAL.

THIS is of two Sorts, the Red, and the White, diſtinguiſhed by the Colour of the Stalks. It is, in general, a low groveling Plant, jointed at every Inch and an half Diſtance: From theſe Joints iſſue a Pair of winged Leaves. From the Boſom of the Leaves riſe ſeveral ſmall Tufts of whitiſh ſtamineous Flowers, ſucceeded by a Pair of winged Leaves, extending each ſide the Stalk: Theſe are ſomewhat ſharp-pointed, of about an Inch long, and their Edges ſomewhat ſinuated. From the middle of theſe Bunches of Flowers riſe ſeveral leſſer Leaves. The *Capſulæ* that ſucceed the Flowers, produce ſmall Seeds.

The WORM-SEED WEED.

THIS, it is ſaid, very much reſembles the true *Semen Santonicum*, which is exported to *England* from *Alexandria:* It is here to be found in great Plenty, upon the ſandy Intrenchments near *Bridge-Town.*

CATERPILLERS. *Lat.* Blitum Braſilianum.

OF theſe there are two Species, the White, and the Red, both eatable, when boiled as Spinage; and they are, in my Opinion, very little inferior to the latter in their agreeable Taſte, and their wholſome Qualities. The red Sort grows often to above two Feet high, eſpecially if cultivated in Gardens. Its Roots are many, white, and ſtringy. The main Stalk, as well as the leſſer ſide ones, are reddiſh, thick, and ſucculent: Even theſe, eſpecially the leſſer ones, when boiled, are very delicious. The Leaves are ſomewhat ſharp-pointed, and of a pale Green, of about two Inches long, and one and a quarter broad. The ſeveral Stalks end in long Spikes: Theſe are covered with ſmall farinaceous green Seeds. It grows to Perfection chiefly in the Months of *July* and *Auguſt*; though it may be raiſed at any time of the Year, if the Seaſon proves wet, or if they are watered. There is a third wild Sort, called the prickly Caterpiller.

BLOODWORTH.

THIS is a bulbous Plant, its Roots much reſembling that of an Onion, conſiſting of ſeveral Coats or *Laminæ.* It divides near the Ground into ſeveral long green Leaves. Theſe are kept very upright by many longitudinal ſtiff Ribs. Among the Leaves riſes a ſlender Stalk, which, near the Summit, bears a ſmall tubular white Flower, which is ſucceeded by a ſmall conic *Capſulæ*, containing ſeveral ſmall Seeds. The Root of this Plant, pounded, is an excellent Antidote againſt Poiſon.

The WILD-MARIGOLD; *Lat.* Asteriscus frutescens.

THIS grows to about two Feet high, the main Stalk is square, and the Leaves broad, and rough; the Top, of the Stalks support yellow Flowers like those of Marigold; the Petals are succeeded by a hard Button-like *Capsula*, which incloses several flat small Seeds.

POND-WEED.

THIS Plant grows best in wet moist Land, where it often rises three Feet high; its Roots are many and fibrous; the main Stem and Branches are of a dark Green, and hairy, cloathed with Leaves four Inches long, and one broad; these are likewise high-ribbed, and have their Edges finely denticulated. From the Bosom of the Leaves rise many long Spikes thickly covered on the upper Side with very small, white, pentapetalous Flowers, and succeeded by several minute *Capsulæ*, inclosed in a green *Calyx*; the former are succeeded by several small black Seeds.

The HOG-SLIP.

THIS is a trailing herbaceous Vine, cloathed with sharp-pointed Leaves; from the Bosom of the Leaves rise several two-inch Footstalks, supporting an umbilical Group of monopetalous yellow Flowers, whose Edges are slightly segmented. From the Centre of the Flower rise several broad short *Stamina*, surrounding a Pistil tipp'd with bluish *Apices*; these are succeeded by several *Capsulæ*, inclosing, when ripe, several black Seeds. The Vine is of the herbaceous Kind, and much coveted by Hogs. From thence it derives this Name.

The BROOM-WEED.

THE Species of this Plant is divided into three Sorts, the White, the Red, and the large crumple-leaved Broom-weed; a Decoction of each Kind proves to be a strong Diuretic; the Leaf of the white sort exactly resembles that of the Green-Tea.

TOBACCO; *Lat.* Nicotiana.

THIS Plant hath been so often and so well described, that I need only observe, that it is here planted but very sparingly, and that chiefly by the Slaves, and the poorer Sort of White Inhabitants, but none for Exportation. It is very liable to be destroyed at the Roots by a Grub, or large Worm, called by the common People *Kitisonia*; the Leaves are likewise often destroyed by a small green Worm of the Eruca Kind.

The THISTLE, *or* RABBIT-WEED.

THIS hath its prickly Leaves in common with most other Thistles. From the Centre of these Leaves rises a green Spire, whose Extremity supports a Flower something like a Marigold, whose whitish Petals surround a yellow Thrum.

TUBEROSE; *Lat.* Hyacinthus Indicus Tuberosus.

THIS is so well known here, and in *England*, that it would be Loss of Time to afford it a particular Description.

The WILD-SLIP.

THIS Vine is generally supported by the neighbouring low Shrubs; its Leaves are of a middle Size, and of a slender Texture, and are set on alternately; from among these rise on short Footstalks several Bell-fashion Flowers, of a pale White, on the Outside and the Inside, beautifully stained with a fine purple Colour.

BACHELOR BUTTON, *or* EVERLASTING; *Lat.* Amaranthoides Lichnidis.

THIS Plant hath many white Roots, the Stalk succulent, large and jointed, of a fine deep scarlet Colour; from each Joint rise two opposite Footstalks, of about an Inch long, somewhat hairy like the main Stalk: These support Leaves of four Inches long, and an Inch and an half broad, round-pointed and smooth edged; the Inside of the middle Rib somewhat reddish; the Flower somewhat resembles that of Clover.

The FIRE-BURN BUSH.

THIS is a scandent Shrub, having a middle-sized sharp-pointed Leaf, bluntly jagged about the Edges; the Middle, as well as the transverse Ribs, are prominent, and very neatly disposed: The Seed-vessels are of a triangular Shape, and are set, many in Number, round lo ng slender Stalks, rising from the Bosom of the Leaves; the Juice of these boiled, and brought to a Consistency with Hog's Lard, and other Ingredients, are said to be of great Service to cure a Fire-burn.

The DIALTHEA; *Lat.* Abutilon Indicum.

THE Roots of this Plant are few, penetrating not deep into the Ground; the Stalk, which grows to three Feet high, is woody and brittle, divideing often into many Branches: The Bark of a full-grown Plant is of a reddish Brown, the Leaves, which are about two Inches long, one broad, and

and blunt pointed, are ſoft, and downy, and of a hoary Green, having their Edges irregularly ſerrated; their middle Ribs and tranſverſe ones are ſtrong and prominent. They are ſet on the Stalk alternately, and from the ſetting on of theſe iſſues a Group of very ſmall, fiſtular, one-leaved, yellow Flowers, whoſe Edges are divided into five Segments; theſe are ſucceeded by ſmall black Seeds in form of ſhort blunt Cones, every *Capſula* having one.

A Cataplaſm of the pounded Leaves of this Plant is looked upon as an excellent Vulnerary.

It grows generally in dry Places.

TREES,

Of TREES, SHRUBS, *and* PLANTS,

OF THE

Pruniferous K I N D.

BOOK VII.

The Bearded Fig-Tree, Ficus Bengalienſis.

THE Miſtakes of many Authors, in their Deſcription of this Tree, are ſo many, that it would be endleſs to particularize them; at leaſt, it would be an ungrateful Taſk to one who wiſhes they had left no Room for Correction. The Roots of it are many and large: The Body of the Tree, when old, is rather a Compoſition of many joined irregularly together, than one undivided Trunk. This is occaſioned by the ſurpriſing Manner that this Tree is naturally propagated; for from its Branches iſſue, at unequal Diſtances, ſeveral Tufts of ſtringy Filaments, each in Subſtance and Make about as thick as the Baſe-ſtring of a Violin. Theſe grow always downwards, perpendicular to the Earth, and when they are ſome Feet in Length, the Part next the Branch conſolidates into firm Wood, whilſt the lower Parts hang wavingly downwards in ſo many looſe-twiſted Shreds or Filaments; but when theſe grow to reach the Ground, they take Root, unite into firm Wood as the upper Part; and in a ſhort time grow to a conſiderable Bulk, affording great Increaſe of Nouriſhment to its once Parent-Branch; by which means this grows large, and produces other progreſſive, lateral, as well as many upright Branches; theſe, in like manner, eſpecially, the progreſſive Branches, by

 their

their descending Filaments produce new Trunks, and so on successively.

When these Filaments grow from Branches near to, and take Root within a small Distance from the original Parent Tree, many of them by their quick Growth are joined sometimes partially, sometimes totally to it, and so in time form one irregular enormous Trunk.

I am of Opinion, that if one of these Trees was carefully and properly cultivated by Art, and Cattle of all kind kept from browsing upon it; if the bearded Filaments were suffered to insinuate themselves, especially those from near the Extremities of its longest lowermost Branches, or were they guided to take Root in the Earth, one single Tree might be brought to make a considerable Grove. Its Leaves are of a smooth shining Green, in length about five Inches, and above two broad, smooth-edged, and sharp-pointed. The Flowers, as in others of the like Kind, are contained in the Embryo Figs; these latter stand upon short Footstalks, and are about the Bigness of small Cherries, the inside full of small *Acini* resembling eatable Figs; which are Food only for Birds, especially Pivets, which come in great Number to feed upon them when ripe. The tender Buds and Leaves of this Tree afford, when bruised, a milky Juice very much resembling in Colour and Consistence that which issues from the Manchaneel Tree, but differing in its Quality, the former being, as it is said, a sovereign Remedy against the Poison of the latter.

The most remarkable of these Trees for Bigness grows near *Codrington* College. This, about six Feet from the Ground, divides into five Branches, each equal to a large Tree, some of them girting round about eight Feet. The most lofty of these Branches is by moderate Computation above forty Feet high. But as there is nothing that can be called great or little but by Comparison; let us examine, as far as any credible History hath informed us, whether any Part of the World affords of this, or any other kind of Tree, a larger. What bid fairest for Superiority are the Cedars of *Libanus*, which they will always deserve in regard to the Goodness and Solidity of their Timber, but not in Bulk; for, according to the Reverend Mr. *Maundrel*'s Account, the utmost Extent of the Branches of the largest Cedar upon Mount *Libanus*, from one Side of the Tree to the other, did not spread above a hundred and eleven Feet; whereas the Branches of this reaches above an hundred and twenty-seven; the Circumference of the Body of the former was but twelve Yards, of this eighteen; therefore we may pronounce it to be the largest Tree that hath been taken notice of in any well attested History. The next to this in Bigness in this Island, is in *St. James*'s Church-Yard, which spreads a Shade (very near circular) of eighty-five Feet in Diameter; and its Height, by Computation, is at least seventy Feet. The great Mr. *Milton* was of Opinion, that this was the Tree with whose Leaves our first Parents made to themselves Aprons: for in describeing their Fall he says;

——And both together went
Into the thickest Wood: There soon they chose
The Fig-tree, not that Kind for Fruit renown'd;
But such as at this Day to *Indians* known
In *Malabar* or *Decan*, spreads her Arms
Branching so broad and long, that in the Ground
The bearded Twigs take Root, and Daughters grow
About the Mother Tree; a pillar'd Shade
High over-arch'd, and echoing Walks between:
There oft the *Indian* Herdsman, shunning Heat,
Shelters in cool, and tends his pasturing Herds,
At *Loopholes* cut through thickest Shade. These Leaves
They gather'd broad as *Amazonian* Targe;
And, with what Skill they had, together sew'd
To gird their Waist.

However noble this poetical Description is; yet, as to the Matter of Fact, it wants even Probability to countenance it; for the Leaves of this Tree are so far from being of the Bigness of an *Amazonian* Target, that they seldom or never exceed five Inches long, and not quite three broad; therefore we must look for another of the Fig-Tree kind, that better answers the Character given of this Tree by *Moses:* And as the Fruit of the Banana-tree is often by the most antient Authors called a Fig, I may, I hope, without Presumption add my own to the already numberless Conjectures of others, and look upon the Fig-tree in Paradise to be no other than the Banana-tree; for when *Pliny* describes the *Asian* Fig-tree, he says its Leaf is *maximum umbrosissimumque*; and as the Leaves of these are three Feet long, and about two broad, they may be deemed more proper than any for a Covering; especially since they might be easily joined together, with the numerous thread-like Filaments, which may without Labour be peeled from the Body of this Tree. These Fig-trees grow in the *East* as well as the *West-Indies*. *Quintus Curtius*, in his Account of *Alexander*'s Expedition to the *Indies*, excellently describes it. Had both the *Indies* been so well known some Centuries ago as they are now, that Author would not have been (at least in this Instance) thought fabulous. His Words are these:

Sylvæ erant prope immensum spatium diffusæ, proceri sque & in eximiam altitudinem editis arboribus umbrosæ. Plerique rami instar ingentium stipitum flexi in humum, rursus, qua se curvaverant, erigebantur, adeo ut species esset non rami resurgentis, sed arboris ex sua radice generatæ.

Quint. Curt. Lib. IX.

The Stopper-Berry *Tree*; *Lat.* Malpighia.

THIS grows to be a considerable large Tree. Its Bark is of a whitish-red, and scaly, often dropping off in Flakes; the upper Branches are thickly cloathed with deep-green smooth shining Leaves, of about three Inches

Plate 12.

Page 177

To the most Rev.d Father in GOD, THOMAS Lord Bishop of LONDON.

Inches long, and near two broad; theſe are generally ſet on in Pairs. From the Boſoms of the Leaves riſe a great many Pedicles ſupporting ſmall white Flowers; theſe are ſucceeded by Berries black when ripe; of about the Bigneſs of a black Cherry, or rather larger and flatter; the Pulp is of a ſweetiſh Taſte; this ſurrounds a greeniſh ſoft Kernel; the Fruit is eaten by Men as well as by ſeveral Kinds of Birds.

The BULLY-BERRY *Tree.*

THIS is a very durable Timber-tree, growing ſo large as often to exceed ſeventy Feet in Height; the Bark is of a rough rugoſe Texture; the Leaves are ſmooth, thick, and of a ſhining-dark Green. The Fruit, which ſucceeds a ſmall white Flower, is round, and of a golden Colour, having a ſmall *Corona* at the Top: The Inſide is of a milky, clammy Subſtance, and very ſweet. This thick Pulp ſurrounds two dark, reddiſh, flat Stones, very much reſembling the Seed of a Sappadillo; the greateſt Number of theſe Trees are in *Scotland*, at the Eſtate of *Thomas Alleyne*, Eſq;

The Eaſt-India MANGO *Tree.*

THIS Tree, or its Seed, was lately brought from the *Rio Janiero*, and grows only at the *Guiney* Plantation belonging to *Edward Laſcelles*, Eſq; Its preſent Height is about twelve Feet; its Leaves are narrow and ſharp-pointed, being at leaſt ſeven Inches long, and one and an half broad: The middle Rib ſtrong and prominent. Theſe are ſet on the Branches in an alternate Order. Tho' this is called the *Mango* Tree, yet the Leaves do not anſwer the Deſcription given by others of the real *Eaſt-India Mango* Tree.

The OLIVE-TREE.

I Have ſeen ſeveral of what are here called the true Olive-Trees; but as they never bore any Fruit, we have only a traditional Certainty of their being the real Trees, which, in other Parts, bear Olives; however, I ſhall inſert its Character. The Trunk is conſiderably large, cloathed with a whitiſh-grey Bark; the Height of the whole Tree is often above twenty-five Feet; its Leaves, which are many, are ſet on in an alternate Order. Theſe are about two Inches long, half an Inch broad, and ſharp-pointed; the upper Side green, and the under covered with a hoary Mealineſs.

The Wild OLIVE-TREE

THIS grows to the Height of a common Willow. The Bark of the main Trunk is much ſulcated, and of a dark Colour; its Top, by its many ſmall Branches, and numerous ſharp-pointed Leaves, is very buſhy; theſe Leaves are generally four Inches long, and one broad, ſet

on

on the Branches in an alternate Order. The firſt Appearance of the Flower is a Tube of a browniſh-white Colour, about the Thickneſs of a Quill, and in Length three-quarters of an Inch; this, near the Top, firſt bulges out, and then ſplits into two Parts; the upper, which I call the Creſt, being excavated Spoon-like; the under Lip deflects very much back, ending in three diſtinct Diviſions, each Point curling inward; the middlemoſt having from the Socket of the Flower, to its Extremity, a narrow Liſt of fine purple Fur, or ſoft *Villi*. As the Flower comes to its Perfection, the Creſt, or the upper Part, deflects back from the intermediate Space; between the upper and the lower Lip riſe four *Stamina*, tipp'd with *Apices*, ſurrounding a naked Piſtil, which, when the Flower drops, becomes the Rudiment of the Fruit, which is a ſmall Olive, Yellow when ripe. Theſe Trees are valued for their quick Growth, and good Shelter, and they are chiefly planted near the Sea-ſide, to ſhelter the neighbouring Fields of Corn and Cotton from being blaſted by the ſalt Spry of the Sea, which too often, after all Precaution, is very deſtructive to thoſe Plantations that adjoin to the Sea to the Eaſt and North-Eaſt of the Iſland. This is delineated in Plate XII.

The FUSTIC-TREE; *Lat.* Morus.

THERE are two Species of Fuſtic-Trees, the Green and the Yellow; the former is very ſcarce in this Iſland; the latter is diſtinguiſhed into the Male and Female, the Male bearing a two-inch long Spike, or Katkin; the Female bears on the under Part of the Leaves, on the middle Rib, ſeveral ſquammous ſweetiſh Berries, much reſembling Mulberries; theſe are eaten by Birds whilſt upon the Tree, and much coveted by Dogs when ripe, and fallen off. The Heart of this Tree is very yellow, and much made uſe of by the Dyers; it is likewiſe very ſolid and durable Timber, which makes it uſeful to make Cart-wheels, *&c.* Theſe Trees grow beſt in Gullies, and ſuch Places as are ſhaded from the Wind.

The Gully PLUM-TREE; *Lat.* Mombin.

THIS Tree, eſpecially if ſhaded from the Wind, grows conſiderably large, both in Bulk and Height; the Bark upon the Trunk is of a very dark Grey, ſulcous, and very rough; its Branches are many, and generally crooked. Theſe are cloathed with unequally pennated Leaves. The Lobes are about three Inches long, and ſomewhat above an Inch in Breadth; the Leaf conſiſting generally of four Pair of Lobes, with an odd one at the End. The upper Twigs ſupport pyramidal Cluſters of ſmall granulated herbaceous Flowers, which are followed by Plums of oblong Shape, yellow when ripe, and ſomewhat bigger than large Nutmegs; theſe in Bunches hang downwards, ſupported by two-inch-long Footſtalks. Their pulpy Part is of an agreeable Taſte; this is

Pl. 18 Page 178
To the Rt. Honble George Littleton, Esqr
one of his Majestys Commissioners for executing the
Office of Lord High Treasurer:
This Plate is humbly inscrib'd &c.

not above one ſeventh of an Inch thick, covering a rough, ſulcous, ſtony Kernel. This is delineated in Plate XIII.

The WHITE PLUM-TREE.

THIS is a middle-ſized Tree, the Body of it cover'd with a whitiſh Bark; the Branches are cloath'd with Leaves of about four Inches long, and two broad, ending in a round Point; the ſlender top Branches ſuſtain Groups of white ſmall Flowers, which are ſucceeded by a great Number of ſmall Plums, green even when ripe; theſe by the Heat of the Sun open into three thick Partitions, emitting a great many ſcarlet ſtringy Seeds.

The Jamaica PLUM-TREE.

THIS grows to a conſiderable Bigneſs; the Branches are cloathed with pennated Leaves; the Flowers are ſucceeded by oval Plums, ſomewhat depreſſed at both Ends, having their Surface here-and there indented. Theſe are of a purple Colour when ripe, and of a ſharp agreeable Taſte.

The DAMASCEN, *or the* CALLIMATO-TREE; *Lat.* Icaco.

THIS Tree grows to about twenty Feet high. The Leaves are long, ſharp-pointed, and ſmooth-edged; very green on the upper, and pale on the under-ſide; they are ſet on the Branches alternately. It bears about *April* a great many Flowers, very much reſembling thoſe of a Sugar-apple; theſe are ſucceeded by black, longiſh, very clammy Plums, having one, ſometimes two Stones or Kernels in them, the pulpy Part being very ſweet, and agreeable. The great Turtle-Doves, Pivets, and Thruſhes, feed greedily upon them.

The CHIGERY-GRAPE-TREE.

THIS is a middle-ſized Tree, having ſtrong numerous Roots; the Bark of the Trunk is of a dark Grey; the Branches are thickly covered with Leaves; theſe are about five Inches in Length, and two in Breadth. The Flowers, which are ſmall, five-leaved, and of a browniſh Colour, in great Number ſurround ſeveral ruſſet ſtrong Spikes, of about two Inches long; theſe Flowers are ſucceeded by Berries ſomewhat ſmaller than white Currans, each compoſed of ſeveral thick juicy *Laminæ*, incloſing a blackiſh ſharp-pointed Stone or Kernel; theſe are red when half-ripe, and white when full-ripe: They have an acid, ſweet, agreeable Taſte, like white Currans; but if eaten to Exceſs, they cauſe a tingling Itching in the Skin. They are generally ripe in *Auguſt* and *September*.

The Bay-Grape-Tree.

THE Body of this Tree is conſiderably large, though ſeldom growing to any great Height, unleſs ſhelter'd from the Wind; the Branches are many, but crooked and ſtraggling: The Bark is of a whitiſh Grey; from the Extremities of the Branches hang pendulous Spikes, thickly covered with very ſmall Bloſſoms, which are ſucceeded by a great Number of Grapes in Cluſters: Theſe are round, and ſmaller than the leaſt red Cherry, and of a ruſſet Purple when ripe; they have an agreeable Mixture in their Taſte of both ſweet and poignant Acid; their Pulp, which is but ſmall in Proportion to their Bigneſs, ſurrounds a Stone or Kernel; theſe, if bruiſed and ſteeped in Wine, are looked upon to be a good Reſtringent. The Tree grows beſt in a looſe ſandy Soil; their Leaves are very broad and ſleek, cloſely embracing the Stalks, and neatly ſtrewed with purple or red Veins; and even the whole Leaf, whilſt young, is red. The Grapes ſteeped in Water, and worked up with Sugar, make a very agreeable Wine.

The Fat Pork-Tree.

THIS chiefly grows near the Sea-ſide, in the Pariſhes of *St. Joſeph* and *St. Andrew*. The Fruit is diſtinguiſhed into the white and red Sort. The Trees bearing each differ very little, if any thing, in their Texture, Leaves, or Flowers; the main Body of the largeſt exceeds not five or ſix Inches in Circumference; the Leaves are of a middle Size, round-pointed, and ſet on the Branches alternately; the Flowers are white, and very ſmall; the Petals are ſurrounded with a green ſtiff *Calyx:* Theſe are ſucceeded by Plums called here *Fat Pork*. The outſide Skin of one Sort is, when ripe, of a fine ſcarlet Colour, faintly clouded over with a purpliſh glaucous Mealineſs, like that which covers Plums in *England*. The white Sort differs only in Colour; for the pulpy Part of each cuts ſtringy, or rather wooly, and is of a white Colour, not ill reſembling, though leſs firm than, the Fat of Pork whilſt warm. This Pulp hath a ſweetiſh Taſte, but to a great many diſagreeable; this eatable Part covers a pretty large Stone, which incloſes a white Kernel, no-way diſagreeable to the Taſte; the whole is full as large as our common black Plums in *England*.

The Black Wood.

THIS grows to be a large Tree, of an hard ſolid Texture; it derives its Name from the blackiſh Colour of its Bark and Leaves; it ſheds yearly its Bark, which is very bitter; its Flowers are ſucceeded by ſmall Plums; the pulpy Part ſurrounding a Stone or Kernel, or ſometimes two.

The

To His most serene Highness
The Duke of Orleans
This Plate is humbly Inscrib'd
&c.
G. D. Ehret delin. & sculp.

Plate 15
Page 181
HONI SOIT QUI MAL Y PENSE
To His most serene Highness
The Prince of Hesse,
This Plate is humbly Inscrib'd
&c.
G.D. Ehret delin. & sculp.

The MALE POPAW TREE; Papaya.

THIS Tree is generally of an undivided Trunk, and diftinguifhed into the Male and Female; as well as the Fruit into the long *Mango Popo*, and the round Sort.

It's Roots are many and ftrong, penetrating not deep into the Earth, but extending feveral Yards in a circular Compafs. The Bark, which is of a whitifh Colour, is marked with the *Vestigia* of the fallen off large Footftalks of the Leaves. The Body of the Tree grows tapering to the Top; it is often at the Root of about a Foot Diameter, and generally from fifteen to twenty Feet high. The moft fubftantial Part of the Tree is a thick reticulated Web compofed of feveral complicated Divifions, fomething like the *Tunica reticularis* in the Skins of Animals; it is in this that the Strength of it confifts. The Infide of all the young, efpecially towards the Top, as well as moft old Trees, are hollow: by this we hierographically reprefent a Perfon of no Sincerity, and from hence in thefe Parts comes that Proverb to that Purpofe, *As hollow as a Popo.*

Within three Feet to the Top, the Leaves begin to furround the Tree in a circular and regular Manner, and confequently very beautiful. Thefe Leaves, efpecially the lowermoft, which are the largeft, being often above a Foot and an half long, are neatly divided into eight or nine large Sections, each again fubdivided into leffer ones, and ending in a Point; the middle Ribs are ftrong and prominent; the whole Leaf appears very beautiful, being on the upper Side of a deep Green, and beneath covered with a very vifible, yet very fhort hoary Down. The largeft Leaves, which are lowermoft, ftand upon green hollow Footftalks, often two Feet long; however thefe, as well as the Leaves, gradually grow lefs toward the Extremity of the Tree. The Stalks, when bruifed, yield a difagreable Hemlock-like Smell.

The Male Tree bears feveral fmall pentapetalous white Flowers upon pendulous Stalks of near a Foot and an half long, but was never here known to produce Fruit. This Tree is delineated in Plate XIV.

The Female bears fomewhat fimilar, but larger Flowers, of a yellowifh Colour, growing on very fhort Footftalks, arifing from, and furrounding the Tree among the lower Leaves chiefly; thefe in ftill Evenings and Mornings afford a very fragrant and grateful Smell. The Female Flowers and tender Buds of thefe are preferved into Sweetmeats, and the long *Mango Popo* into Pickles, the latter being very little inferior to an *East India Mango.* Both thefe Fruits, efpecially the round Sort, are likewife, when near ripe, boiled and eaten with any kind of Flefh-meat, and efteemed wholfome, if they are cleanfed of the milky corrofive Juice they contain, and eaten but feldom. This Juice is of fo penetrating a Nature, that if this unripe

ripe Fruit, when unpeeled, is boiled with the toughest old salt Meat, it will soon make it soft and tender; and if Hogs are for any considerable Time fed with it, especially raw, it is said that it will wear off all the mucous slimy Matter, which covers the Inside of the Guts, and would in time, if not prevented by a Change of Food, intirely lacerate them.

I know of no physical Virtue in any Part of this Tree, unless that the milky Juice of the *Popo* is sometimes made use of to cure Ring-worms, and such cutaneous Eruptions.

It grows best in shady Places.

This Tree is delineated in Plate XV.

The PLANTAIN-TREE.

IT is the Opinion of many Writers, that this Tree was formerly peculiar to *Ethiopia* only, though now very common in all the hot Parts of *Asia*, *Africa*, and *America*; especially in *Guiney*, and the *West Indies*. Its Height, great Bulk, and large Leaves, claim a Place among Trees; but its soft bulbous Roots, the pulpy Texture of the Trunk, which is so herbaceous, that it is often sliced and given by way of Fodder to Cattle, seem to partake of a liliaceous Plant, more than a Tree: but as it hath generally been classed among the latter, I shall treat of it in the same Light. Its Roots are numerous, white, and spongy; the Trunk near the Earth is about thirty Inches in Circumference, round, tapering, and undivided, till about nine or ten Feet high, at which Height it puts forth several large green Leaves in an alternate Order: These are often five Feet in Length, and near two and an half in Breadth, of a delightful shining Sea-green Colour, and of a long oval Shape; these stand upon long tapering Footstalks, the middle Rib in each Leaf is very prominent, and deeply chanelled on the upper Side. This serves as a Gutter to convey the Water that falls upon the Leaf, to the main Trunk, where it is soon absorbed by so soft and porous a Body; for the Trunk of the Tree is composed of several *Laminæ* upon *Laminæ* of large longitudinal Veins, or Vessels horizontally crossed at about one tenth of an Inch Distance, with very thin membranaceous Filaments. These last prevent both the copious Juices from the Roots, or the Dew and Rain descending from the Leaves, to penetrate through the other perpendicular Vessels, till each Part is saturated with its proper nutricious Juice. From the quick Growth, and great Bulk, of such succulent Plants, spongy Shrubs, and Trees which have their Vessels so much distended, we may perhaps account for the far slower Growth of more durable Timber, both here and elsewhere: For the Closeness of the Grain of the latter having their Vessels very fine in close Contact, the annual *Laminæ* of these, when succeeded by exterior new ones, close and consolidate together, and so add to the Bulk of the Tree: Yet such an Addition will be no more, when compared in Quantity to the gross *Laminæ* of succulent Plants, spongy Shrubs or Trees, than so many Layers of Muslin compared

in

in Bulk to an equal Number of coarſe Bays. But to return to my Subject: From the Top of this Tree (iſſuing from among the upper Leaves) at about ten Months Growth, riſes a tough ligneous Stalk, about three Feet long, bending downwards, and bearing on its Extremity a conic purple *Spatha.* " The Flowers (which ſurround this in three or four Rows) " are monopetalous, irregular, incomplete, and Hermaphrodite, compoſed of a Tube which fills the Ovary, and a Pavilion divided into four " Lobes, and forming a kind of a Mouth. The Ovary, which adheres " ſtrongly to the Tube, is triangular, and crowned with five Chives, which " grow from the Side of the Flower. The Style, which is alſo terminated " by a little Head, afterwards becomes a ſoft ſomewhat angular Fruit, " whoſe outward huſky Tegument is very ſmooth, and yellow when ripe." This is from five to nine Inches long, and near an Inch Diameter, growing ſmaller, and by Degrees a little crooked, at each End: The outward Coat or Rind eaſily peels off when ripe. The inſide eatable Part is of a gold Colour, and of a ſweetiſh Taſte. The whole Bunch, which generally contains ſome Scores of theſe Plantains, often weighs forty, fifty, or ſixty Pounds Weight. The moſt common Method of uſing this Fruit, when deſigned to ſupply the Place of Bread, is, to take them when green, though-full grown, and bake them in the Embers, or boil them.

Vide Philoſoph. Tranſ.

In a ſhort time after the Plantain-Tree hath borne its Bunch, it decays near the Root, and falls proſtrate to the Ground, and periſheth: However, the Planter's Hope periſheth not with it; for long before the Mother Tree decays, two or three large Suckers or young Trees grow up from the Root of the old one. The largeſt of theſe, in about a Twelvemonth's Time, bears ſuch another Bunch of Plantains as the above deſcrib'd; and as this Tree likewiſe dies, after it hath produced Fruit, there ſpring up from the Root freſh young Shoots; ſo that there is an annual Succeſſion of Trees without any Trouble to the Planter. However, it is thought the moſt prudent Method is to replant them once in ſeven or eight Years; in doing this to the greateſt Advantage, the Situation muſt be rich, and ſheltered from the Wind; and the Land intended for this Purpoſe muſt be dug in Holes two Feet deep, one and an half broad, and twelve Feet aſunder: Theſe being well manur'd, large Roots of ſuperfluous Plantain-Trees are cut through in two or three Pieces; one of theſe put in every Hole, ſlightly covering it with Earth, in a ſhort time ſprings up. Another common Way of propagating theſe Trees is, to dig up other young ones, which in great Number are to be found growing about the Roots of old decaying Trees, and cutting off the Top of theſe within three Feet to the Root, and ſo tranſplant them into Holes prepared for that Purpoſe. Having cut one of theſe young Trees horizontally in the middle, the remaining Stump vegetated ſo ſtrong from the Centre, that it thruſt out a ſmall ſlender Shoot near a Quarter, or above * an Inch long in ſeven Hours time.

* However ſurprizing this may appear, it is not more extraordinary than the quick Growth of Aſparagus in *England*, a far colder Climate.

The Wild PLANTAIN-TREE.

THIS hath a great Resemblance, in its Trunk and Leaves, with the Plantain-Tree already describ'd, differing chiefly, that though it blossoms, it never bears any Fruit.

The BANANA-TREE.

THIS differs in its general Make from the Plantain-Tree, only by the greater Height of its Trunk, which is likewise here-and-there spotted with Black; the Leaves, as well as the Fruit, are somewhat smaller than those of the Plantain-Tree, the Fruit being not above five Inches long, of a fine yellow Colour when ripe, and of a fragrant Smell, and sweet Taste. These are eaten raw by way of Dessert. The learned Doctor *Derham* observes, that the finest Needle, the Work of Art, appears, when viewed through a Microscope, rough and unhewn, when compared to the Sting of a Bee, or even the meanest Blade of Grass, the Work of Nature: so there is no less distinguishing Delicacy in the Taste of those large Drops of natural Honey found in the Blossoms of this Tree, when compared with what our Hive produces; the former intirely void of that Harshness in the Throat, from which the best of the latter, by its essential Salts, is not free. This was a symbolical Tree in *Egypt*; for, among other Hieroglyphics, we often find the Head of *Osiris* adorn'd with *Banana* Leaves, and the Child *Orus* sitting on her Lap. It grows best under the Shelter of a Hill, or some such shady Place. As these are justly reckoned among the most delicious Fruits that *Arabia* produces, *Ludolphus*, in his *Ethiopic* History, conjectures that these Fruits were the Mandrakes which *Jacob*'s Wives contended for. This Tree is described in Plate XVI.

The WATER LEMON-VINE.

THIS is a very large ligneous Vine, creeping, if supported by neighbouring Trees, to a very great Length. The Leaves are of a dark Green, and sharp-pointed. The Flowers very much resemble those of the *Granadilla* Vine; these are succeeded by a Fruit as big as an uncased Walnut, of an irregular roundish Shape, the Outside of a yellowish skinny Substance, the Inside much resembling that of a Gooseberry, being full of a sweet gelly-like Substance, mixed with a great many small *Acini*: This justly claims a Place among the most delicious Fruits.

The WILD WATER LEMON-VINE, *or* LOVE IN A MIST.

THIS is a sort of a Passion-Flower, and called by BOERHAAVE *Granadilla flore albo, fructu reticulato.* This is a trailing gramineous hairy Vine, provided with a great many Tendrils, or Claspers, which

Plate 16
Page 184
To His Grace the
Duke of Chaulnes
President of the Royal Academy of Sciences
at PARIS &c.
G. D. Ehret. delin. & sculp.

which twist round the neighbouring Trees or Bushes, and so supports itself from the Ground. The Leaves, which are likewise somewhat hairy, and cut into two shallow Sections, are set upon the Vine upon an Inch Footstalk, and about two Inches asunder. From the Bosoms of the Leaves rise Stalks of about an Inch long, supporting many beautiful Flowers, composed of fistulous *Radii*, round each Border, purple at the Bottom, and whitish at the Top: These are almost inlaid in an outward Covering of very thin white membranaceous Leaves, which are much crumpled and labiated bending inwardly; a double *Perianthium* divided into five deep Sections. The Ovary, which is the Rudiment of the Fruit, rises from the Bottom of the Flowers, from the middle of which Ovary come five short Chives, bending downward, and tipp d with large oval Summits: Between this and the Top is the Fruit seen in Miniature, from whose Top come three Styles tipp'd with round Buttons. These likewise bend downwards. As the Flower decays, the *Perianthium* closes together, forming a green somewhat conic Fruit, about the Bigness of a Cherry; the outward Skin being seemingly divided, or rather mark'd with six unindented Seams. The Fruit, when ripe, is yellow on the Outside, includeing a great many very agreeable pulpy Seeds, and Juice, of the same Taste as the Water-lemon already described. What is most remarkable in this Fruit is, that it is intirely covered or surrounded with a bluish white reticulated Web, composed of innumerable small Strings, and soft Bristles, whose Tops are tipp'd with a glutinous Substance. Both this Web, and this clammy Matter, are designed by Nature to preserve the inclosed Fruit from being destroyed by Vermin, such as Ants and others.

PIMPLOES; *Lat.* Opuntia.

THESE are likewise called the wild prickly Pear; their Roots are many, extending several Feet round: It hath scarce the Appearance of any Stalk near the Earth. The whole Tree or Bush is composed of numerous Leaves of a round oval Shape: These are about nine or ten Inches long, about seven broad, and three-quarters of an Inch thick. The Inside of these contains a great Quantity of mucilaginous Juice; the Surfaces of them are thickly beset with Tufts of whitish Prickles, each Tuft consisting generally of four of these Prickles. The Body of this Bush, after the green outward *Lamina* is scrap'd off, is a reticulated Substance, strongly wreathed in several Folds one among another. The Flowers come out of the Sides of these Leaves, and are compos'd of a great many roundpointed Petals of a yellowish Red. The Chives are very many, tipp'd with yellowish Summits. The Pistil is red, and surrounded with these Chives. These Leaves are supported and spring from a conic green husky Pod-like scaly Substance, and are succeeded by a Fruit whose Inside is full of small black Seeds.

The

The Cochineal.

THIS hath at a Distance the Resemblance of the Pimploe; but it generally grows higher; and the Leaves are smooth, bearing many rosaceous red Flowers, which are succeeded by somewhat round scaly Fruit, whose Inside is full of small *Acini*, and purple Juice.

The Prickly Pear-Vine; *Lat.* Cereus scandens minor.

THIS likewise is of the same Texture as the three last already described, each of them, in my Opinion, being a strange Mixture between a Plant and a Shrub. This, if supported by a Wall, will grow above fifteen Feet high, composed of several thickish Leaves of a triangular Form, almost as deeply indented between every Angle as the Blade of a three-edged Sword. These Leaves are of a strong ligneous Substance, of about two Feet long, the Extremity of one giving Birth to another. The Edges of every Angle are, at certain Distances, thickly covered with Tufts of very sharp-pointed Prickles. From these Eminences likewise proceed the Flowers: These are of the rosaceous Kind, standing upon longish green bulbous Stalks. The Petals of each Flower are sixteen in Number, two Inches long, and blunt-pointed; these are so very thin, that when laid over the smallest Print, it may be seen and read through the Leaves. These membranaceous Petals are supported by other green capsular Leaves; the Style is large and strong, surrounded by numerous white Chives, tipp'd with whitish Summits. The Fruit, when ripe, is of an oval Form, and often as large as a Turkey-egg: The outside Skin or Rind is, when full-ripe, of a dark-purple Colour, and scaly at about every Half-inch Distance, each Scale being of a triangular Shape, and sharp-pointed. The whole Rind is likewise here-and-there studded with Tufts of very small sharp Prickles; the Inside is full of purplish solid Pulp, intermixed with small whitish Specks; the whole somewhat juicy, and well tasted.

The Wild Cucumber-Vine; *Lat.* Anguria.

IT is called by Father Plumier, *Anguria fructu echinato eduli.* This quadrangular rough hairy Vine hath a long Tap-root, very little divided, even at its Extremity; it creeps generally upon the Ground, and by its numerous twisting Claspers takes hold on every Side of the neighbouring Bushes: Its Leaves, which stand upon two-inch long Foot-stalks, are deeply divided into three, sometimes more, round-pointed Sections; its Flower, which is yellow, is succeeded by a small Cucumber, whose Surface is covered with many soft-pointed Prickles; they are sometimes eaten, but are esteemed to be of too cold a Nature to be wholsome.

The GRANADILLA-VINE.

THIS is the *Granadilla latifolia fructu maliformi* of Father PLUMIER. The Stalk of this Vine is quadrangular, bordered upon every Angle: Its Claſpers, or Tendrils, by which it ſupports itſelf, are very numerous, as well as the Leaves many; theſe are ſix Inches long, and five in Breadth, ſtanding upon four-corner'd Footſtalks. The Flowers are of the ſame Make with the Paſſion-Flower. The Fruit is of a long oval Sh ape, of a yellowiſh Green when ripe; containing a great Number of ſmall Seeds or *Acini*, covered with a gelly-like Juice, or ſweetiſh Pulp. This Vine grows beſt in ſhady Places, and muſt be ſupported with an Arbour.

The FIG-TREE.

AS neither this Tree, nor its Fruit the eatable Fig, differ from that of the ſame Species in *England* and elſewhere, it will be needleſs to give it a particular Deſcription.

Here ends the *Pruniferous* Kind of Fruit.

OF

TREES, SHRUBS, *and* PLANTS,

OF THE

SILIQUOSE KIND.

BOOK VIII.

BY Trees, Shrubs, and Plants, of the ſiliquoſe Kind, are to be underſtood all thoſe whoſe Fruits are incloſed in Pods.

It is obſervable, that very few Trees bear Pods in cold Climates, in Compariſon to the great Number which warm Countries abound with.

And, as moſt of the ſiliquoſe herbaceous Plants are ſcandent, it is remarkable, that as they are weak, and, as it were, helpleſs of themſelves; therefore indulgent Nature hath provided them with either Tendrils or Claſpers to intwine about others, and be ſupported by them; and often with even ſharp-pointed Hooks at their Extremities to faſten upon the neighbouring Trees or Rocks: And thus, like helpleſs Orphans, by this foſtering Aſſiſtance, they not only ſubſiſt, climb up, and flouriſh; but alſo help to adorn the Face of Nature with their varied Beauties.

It is likewiſe worthy our Notice to obſerve, that theſe are not only in greater Plenty, but that there is likewiſe a greater Variety of them, in warm Climates: And indeed they are, by their Qualities, better adapted for

 the

Plate 17
Page 189
QUO VIRTUS
To the Right Honourable
Earl of Bath,
This Plate, is humbly Inscrib'd
&c.
G. D. Ehret. delin. & sculp.

the Conſtitution of Perſons in hot Countries; for, by the ſmall Quantity of Oil and Salts they contain, they are leſs liable to produce groſs Humours than animal Food.

The TAMARIND-TREES; *Lat.* Tamarindus.

THESE are of two Sorts; at leaſt different in the Taſte of their Fruit, the one bearing a ſour Fruit, the other of a Mixture of a ſweet and acid Taſte.

The Trees, though they thus differ in their Fruit, are however of the ſame Bulk, Make, and Texture. Each having numerous large Roots, and Branches, the latter by their many Subdiviſions, and being thickly cloathed with very many pennated Leaves, afford an agreeable Shade.

The Flowers are yellow, ſomewhat ſpeckled with Purple; and of the papilionaceous Kind.

Theſe are ſucceeded by falcated Pods, conſiſting of four or five Cells, each including a flattiſh oval Stone incloſed in a Snuff-coloured Pulp.

This Pod is of a browniſh grey Colour; when ripe, eaſily broken, and ſeparated from the Pulp.

This, and the incloſed Stones, are faſtened together by a great many ſmall ſlender Fibres from the woody Stalk, which run through the Pod.

This Pulp is preſerved in Sugar, and ſent to *England*, or elſewhere; and it is looked upon, taken inwardly, to be a good Cooler in Fevers.

Theſe Trees are not peculiar to our Climates; for they thrive not only in *Egypt*, but likewiſe in the Land of *Palæſtine*, in *Arabia*, and in other Parts of *Aſia*.

The TREE JESSAMIN.

THIS grows to be a conſiderably large ſtraggling Tree, cloathed with many Leaves generally ſet in Groups on the blunt Extremities of the Branches.

The Flowers are pentapetalous, and are ſucceeded by long Pods.

This Tree is deſcribed in Plate XVII.

The SPANISH ASH.

THIS Tree hath an Aſh-coloured Bark: the Trunk of it, in rich ſhady Land, grows to ſo great a Bulk, that the Heart alone is of a conſiderable Bigneſs.

Its Branches, which are clothed with oblong, pointed Leaves, placed in Pairs, each Leaf about three Inches long, and one broad, grow generally very upright, till near their Extremities, where they are ſubdivided into many leſſer; and then bend wavingly downwards, clothed with many pendulous Groups of papilionaceous Flowers, of a fine violet Colour.

Theſe

These are succeeded by broad flattish Pods; each inclosing two or three Seeds, somewhat resembling, but flatter than, a Kidney-Bean.

Each Seed hath its partitional Cell, in which, when the Seed is ripe, is always found a small Quantity of a glutinous Substance, of the Consistence of a very soft Gum.

The WHITE SPANISH ASH.

THIS differs chiefly from the *Spanish* Ash, already described, by the Colour of its Flowers; those of the former being of a beautiful Purple mixt with White, of the latter intirely white, and succeeded by a flattish small Pod, containing three flattish Seeds or Beans.

The SILK-COTTON-TREE; *Lat.* Ceiba.

THE Roots of this Tree are very numerous, jutting out above-ground like so many Buttresses, supporting the Trunk; which, when full-grown, is often sixty Feet high before it branches, and, near the Ground, from three to four Yards in Circumference.

The whole main Body of the Tree is almost intirely covered with short stubbed Prickles.

At the above-mentioned Height it divides into several Branches, cloathed with long narrow Leaves set orbicularly on their Extremities.

They are so neatly joined to one another, that the whole Group, consisting of seven Leaves, have the Appearance of one digitated Leaf standing upon one common Footstalk of about four Inches long.

The Extremities of these Branches sustain likewise semicircular Bunches of rosaceous Flowers; each Flower consisting of five middle-sized Petals of a Cream-colour.

These are placed in a circular Order, inclosing the Pointal, which is likewise surrounded with five *Stamina* tipp'd with *Apices* of a yellow-reddish Colour.

The Whole is incircled with the *Calyx*, or a green husky Cup, resembling that of a Pomgranate-Flower, but less regularly segmented about the Edge.

The Petals have a weak faint Smell.

The Pointal becomes the Rudiment of the Pod, which is of a blunt conic Form of about four Inches long. The Inside is full of a greyish silky Down, intermixt with small blackish Seeds.

When this Pod is ripe, it opens; and the Down and Seed are carried off by the Wind to the adjacent Parts.

This Tree seldom bears more than once in three Years.

The

The Small Cotton Tree; *Lat.* Xylon.

THOUGH the Cotton-Wool is diſtinguiſhed into the great, the Ravellin, the Vine and Flying-fiſh Sort; yet the Trees, bearing this Variety of Wool, differ very little from one another.

The Leaves of each are ſcalloped, of thoſe bearing the great Sort, very deeply, not ill reſembling a Curran-Tree-Leaf in *England.* Thoſe of the Flying-fiſh Cotton-Tree are likewiſe deeply ſegmented; having their middle and tranſverſe Ribs of a reddiſh Colour.

The Leaves of the Ravellin and the Vine are leſs ſcalloped, and more blunt-pointed.

Moſt of theſe Trees, or rather Shrubs, if permitted to grow to their own natural Height, would riſe to about fifteen Feet high; but, as ſuch a luxuriant Growth would prevent their bearing the Number of Pods, they would otherwiſe do, as well as ſhade the Corn and Pulſe planted generally among them, the main Branches are therefore yearly lopp'd off.

The Flowers are compoſed of five large yellow Leaves, each ſtained at the Bottom with a purple Spot.

The Piſtil is ſtrong and large, ſurrounded at, and near, the Top with a yellow farinaceous Duſt, which when ripe falls into the *Matrix* of the Piſtil.

This is likewiſe ſurrounded, when the Petals of the Flowers drop, with a capſular Pod, ſupported by three almoſt triangular green Leaves, deeply jagged at their Ends.

The incloſed Pod, which is rather of a conic than a round Shape, opens, when ripe, into three or four Partitions, diſcovering the Cotton in as many white Locks, as there are Partitions in the Pod. In theſe Locks are interſperſed the Seeds, which are blackiſh and ſmall.

The Cotton-Wool (of which in the *Eaſt Indies* they make their fineſt Callico) is too well known to want a farther Deſcription: Yet it would not be amiſs here to obſerve, that, as the Inhabitants of the warmeſt Climates want cloathing, eſpecially in the wet Seaſons of the Year, indulgent Providence hath ſufficiently ſupplied the want of Wool, here denied to Sheep, by cauſing a Vegetable to bear the fineſt Wool in the World.

However the Certainty of gathering a good Crop of this Kind is very precarious; ſince we may almoſt literally ſay of this Shrub, that in the Morning it is green and flouriſheth, and almoſt in the ſame Evening it decays and withers: For when the Worms begin to prey upon a whole Field of Cotton-Trees, though they are at firſt ſcarce perceptible to the naked Eye; yet in three Days they will grow to a conſiderable Bigneſs, and ſo devouring in that ſhort time, that they will reduce the moſt verdant Field, thickly and beautifully cloathed with Leaves and Flowers, into almoſt as deſolate and naked a Condition as Trees are in the Month of *December* in *England*; leaving often not a whole Leaf remaining; by this means, eſpecially if they come late in the Year, they greatly prejudice if not intirely destroy, the ensuing Crop; and sometimes the very young Trees are by this means killed.

These Worms are of three Sorts, all of the Caterpiller Kind, and distinguished by the Names of the Black-Backs, the Streaked-Back, and the Fire-Worm. The last is of a russet Colour, and the least; but yet the most destructive.

When these grow to their destined Bulk, they spin and inwrap themselves in a Bag or Web, like Silkworms, in the few remaining Leaves, or any other Covering; after a few Days Rest in this their Aurelia-state, they turn into dark-coloured Moths, and fly away.

These Worms are observed most generally to make their Appearance after sultry Weather, especially if it thunders and lightens, and as the Weather is then more than ordinary sultry.

The WHITE-WOOD-TREE.

THIS Tree grows to a considerable Bulk; its Branches are numerous, and thickly cloathed with Leaves.

These in Make, Bigness, and Colour, very much resemble those of a Laurel.

The Flowers are of the tubular Kind, of a pale White without, and yellowish within. Their Extremities are much expanded and crumpled, and generally divided into five Sections.

These are succeeded by long greenish Pods, containing a great Number of long flattish Seeds.

The LOCUST-TREE; *Lat.* Siliqua edulis.

THIS grows to be a large Timber-Tree, very full of Branches.

These are thickly cloathed with Leaves of about three Inches long, of a deep-green Colour, and smooth-edged; and always set on in Pairs upon one common Footstalk; differing from all other Leaves, by having always one Part of the Leaf divided by the middle Rib far larger than the other. The Extremities of the upper Branches have many papilionaceous Flowers.

These are succeeded by a light-snuff-coloured rough Pod, of about three Inches long, and near two broad; inclosing a whitish spongy Pith; which covers two or three hard blackish Kernels or Stones.

The pithy Part hath a sweetish Taste, and is sometimes eaten; but the Whole hath a very disagreeable stinking Smell.

The CASSIA-FISTULA-TREE.

THIS Tree grows in the *East* and *West Indies*, as well as in *Egypt*, and several Parts of *Asia*. Its Height, when full-grown, is often no less than

than forty Feet. The Bark, especially upon the Trunk, is very much sulcated and cracked; it generally branches pretty near the Top, bearing several middle-sized sharp-pointed green Leaves.

The Flowers, which are pentapetalous and many, are of a yellow Colour, hanging in Clusters upon several small Twigs.

These are succeeded by blackish Pods, from ten to twenty Inches long, and about three-quarters of an Inch Diameter; "having a Seam running "the whole Length on the one side; and another less visible on the "other."

The Inside is divided into a great many Cells, separated from each other by thin brittle Plates or Partitions covered with a black sweet Pulp.

Between these are the Seeds which are small, flat, and smooth.

The Pulp, taken by way of a Purge, is too well known to want farther Explanation.

This Tree grows generally in a dry rich Soil.

The MOABITE; alias, *the* MANGROVE-BEARD-TREE.

THIS Tree never grows to be fit for Timber.

Its Branches are very thick, covered with small roundish green Leaves, set on alternately.

The Flowers are tetrapetalous and white.

These are succeeded by long falcated twisting Pods, containing eight or nine shining-black flattish Seeds; which are half-sheathed in a pulpy, Snow-white, and sometimes scarlet, Pith. These Pods open, when ripe; the inner as well as the outward Side being then of a fine Maidens-blush; which Diversity of Colours in the Seeds, Leaves, and Pods, makes it then appear very agreeable.

The white Pith is sometimes eaten; but looked upon to be very unwholsome. The Seeds are strung upon Silk, and made use of by the Negro Women for Bracelets.

This Tree is more planted for its thick Shade, to keep off the salt Spry of the Sea from Corn or Cotton, than any other Use.

The GARDEN MANGROVE.

THIS is perhaps, above all Vegetables, the most beautiful Evergreen; keeping, without the least Decay, or Withering, it's grateful Verdure all the Year round. The Roots of this Tree are strong and many, penetrating deep into the Earth.

If suffered to grow to its natural Height, it often rises above forty Feet high, branching very thick on every Side, and consequently affording a delightful Shade.

The Leaves are many, very thick, and of a deep-green Colour, sharp-pointed, and smooth-edged; in Length about two Inches, and the largest about an Inch broad.

The most common Method of propagating this Tree is by laying the small lower Branches in Baskets of Mould, or Earth, till they take Root; but the most natural Way, as this is of the Fig-tree kind, is to suffer the several slender small Filaments, which issue from the main Branches, to take Root in the Earth.

The RED MANGROVE-TREE.

THIS generally grows about the Edges of Ponds, and often several Feet in Water. Its Roots divide into several Branches, and join together above the Surface of the Water, where each straggling Division meets, uniting into, and forming the Body of the Tree; which, from that Place to the Water, seems to be supported by numerous Stilts.

What *Pliny*, speaking of the Palm-tree, says, *Gaudet riguis, & toto anno bibere amat*, may be literally said of this, which very seldom grows, except in, or very near, the Water.

The first Appearance of the Flower is a small conic husky *Calyx:* This opens into four downy whitish-brown Petals, surrounding the Pistil.

When the Petals drop, the Pistil becomes the Rudiment of the Fruit.

The Flower-Cup, or *Calyx*, opens, and expands horizontally into four Parts.

The HOLY-THORN, *or* ROYAL-CASHIAW.

THE Trunk of a grown Tree of this Kind is above eight or nine Inches Diameter; and grows to above twenty Feet high.

The Bark is rough, somewhat sulcated, and of a dark-brown Colour. The Branches are numerous, and very prickly, especially towards the Top. From these extend very long flat slender gramineous Twigs; having on each side a great Number of very small Leaves, set on alternately.

From the Tops of the woody Branches, upon long green Footstalks, rise many yellow pentapetalous Flowers, out of whose *Discus* come several short *Stamina*, whose *Apices* are of a dark Brown. The Leaves in general are somewhat wrinkled; and one above the rest is much deflected, black, and appears stained with red Spots.

From strong Footstalks hang a great many Pods, black when ripe, including four, five or six blackish oval Seeds, separated from each other by long Partitions.

It is the Opinion of some *Roman* Catholics, that our Saviour's Crown of Thorns was made with the Branches of this Tree.

The BEAN-TREE, *or the* SHROVE-TUESDAY; Corallioden-dron.

THIS Tree is generally larger than most Apple-Trees; widely spreading its Branches on every Hand.

Plate 15.
P. 195
To His most serene Highness
The Hereditary Prince of Saxegotha,
This Plate is humbly Inscrib'd.
&c.
Erbets de: sc.

It is called *Shrove-Tuesday* from its never wanting Flowers by that Day. These are of a flaming Purple, and as they very thickly cover the Tree, they appear very beautiful, especially at that time, when the Tree is intirely void of Leaves.

The Make of these Flowers is not less remarkable; for they cannot well be registred under any Botanical Class.

Their first Appearance being an intire falcated small Pod, when this unfolds, it discovers several long purple *Stamina*, all joined to one common broad Footstalk.

These are succeeded by an husky Pod, containing one, two, or more red small Beans, with a black Eye to them.

The COCOA-TREE.

THIS seldom grows to any great Height; and never thrives, unless much sheltered from the Wind.

Its Roots are many, and the Bark of the Trunk, as well as its Branches, is of a whitish Green.

The Leaves are set on alternately, and are about eight Inches long, and near three broad, highly ribb'd on the lower Side.

The HORSE-NICKER, *or* CHICK-STONE-TREE.

THIS is a small groveling Tree, growing chiefly in a loose, marly, or sandy Soil.

Its Trunk soon divides into many Branches, cloathed with unequal pennated Leaves, consisting of six or seven Pairs, of an oval Shape.

The small, as well as the large Branches, are likewise thickly covered with strong crooked sharp Prickles. These are so very numerous, and so sharp-pointed, as well as so variously inclined by the twining different Position of the Branches, that it is next to an Impossibility, without the greatest Deliberation, to extricate one's self when once entangled in them: They are therefore very properly planted upon the Borders of our Intrenchments; for they are the best Defence of this Kind that hath been hitherto known.

The Tops of the Branches end in an upright Spike, surrounded with small yellow Flowers.

From the Flower-cup, or *Calyx*, rise three small green Leaves, which deflect very much back.

These inclose five yellow spoon-like Petals, which surround several short *Stamina*, tipp'd with *Apices*.

The Flower is succeeded by a brownish echinated Pod, of about two Inches long, and one broad.

This incloses several roundish Seeds or Nuts, of the Size, Colour, and Shape of Marbles, which Children play with.

This is described in Plate XVIII.

SPANISH OAK.

THIS grows to be a large Tree, covered with a ſmooth greyiſh Bark. From the upper Branches riſe many long Pedicles or Footſtalks. Theſe ſuſtain ſometimes one, but oftener two Pair of ſharp-pointed ſmooth Leaves, four Inches long, and two and an half broad.

From different Parts of the Branches riſe many purple white papilionaceous Flowers. Theſe are ſucceeded by four-inch-long falcated Pods, incloſing ſeveral oblong dark-colour'd Seeds, covered with a fine white ſoft ſweet Subſtance; which by a great many People is eaten.

The RED WILLOW.

THIS, as well as every other Species of Willows in this Iſland, grow to be large and very ſhady Trees; for which Purpoſe they are generally planted near Dwelling and Out-houſes.

The main Trunk hath a blackiſh ſmooth Bark. The upper Branches are long and ſlender: Thoſe eſpecially near the Top are thickly cloathed with Leaves of above three Inches long, and one broad, ſharp-pointed at both Ends, and of a greeniſh Yellow. When at full Growth, the Weight of ſo many Leaves upon the Tops (chiefly) of ſuch a great Number of ſlender long Branches makes them bow wavingly downward: By this means each Tree affords an agreeable imbrowned Shade.

The WHITE WILLOW.

THE Bark of this Tree is of a whitiſh-grey. The Leaves, which are four Inches long, and one and an half broad, are of the ſame Colour. Theſe are ſet on the Branches alternately. The upper Twigs bear many white Bloſſoms. Theſe are ſucceeded by ſeven-inch-long Pods, containing a great many greeniſh Seeds.

The BLACK WILLOW.

THIS Tree grows to a conſiderable Height, ſoon dividing into many Branches. The Subdiviſion of theſe into leſſer ones is always thickly covered with Leaves. Theſe are about three Inches long, and above an Inch and an half broad; their under Sides of a Pale-white, and the upper of a Dark-green, ſmooth-edged, and ſharp-pointed. There is ſomething very remarkable in the Make of theſe Leaves, when young; for, at their firſt Appearance, they are cloſely folded, or doubled together lengthways, from the Stalk to the Point, ſo that they appear like Half-leaves divided along the middle Ribs: When they grow near their deſtined Bigneſs, theſe ſeeming Half-leaves unfold into perfect regular whole Leaves. The Flowers, which are of two Sorts, both purple and white, ſtand upon ſtrong Pedicles,

Pedicles, and are composed of four capsularLeaves. When these open, they discover several *Stamina* tipp'd with white *Apices*, surrounding a snuff-coloured Pistil, which is succeeded by a round long Pod, red within, containing several black Seeds in partitional Cells; these are no bigger than Grains of black Pepper. The Pods are about three Inches long, and of a russet Colour. This Tree, by its numerous waving Branches, affords a delightful Shade.

The DOWN-TREE.

THIS grows to a considerable Height, cloathed with large, roundish, scalloped Leaves.

The Extremities of the Branches sustain a great many Flowers.

These are succeeded by seven-inch-long blackish Pods, which are lengthways neatly gouged into seven regular Chanels.

The Inside consists of a greyish fine silky Down, much resembling that of the Silk-cotton-tree-down.

The only one that I know of in this Island, is at the Estate of —— *Evans*, Esq; in *St. Joseph*'s Parish.

The FINGRIGO; *or the* SAVINE-TREE.

THIS shrubby Tree seldom grows above twelve Feet high, being very bushy, especially towards the Top.

Its numerous Branches are thickly guarded with small crooked Prickles; and cloathed with very small Leaves of a surprising Make, being rather a Continuation of many Leaves, than distinct ones; for what in other Trees and Plants we call Footstalks, or Pedicles, are in this but so many narrow Leaves, supporting others somewhat broader.

They bear almost an innumerable Quantity of small Flowers.

These are succeeded by as many brownish very small *Capsulæ*, each inclosing a black somewhat flattish shining Seed.

A Decoction of the pounded Leaves answers the End of *English* Savine.

They have likewise this additional Virtue; that if dried and powdered, and mixed with Corn, and given to Horses. it frees their Maw from the Bots, as well as all other Worms.

The PAPAW-OCKROE.

THIS Plant grows about a Foot high, having one strong Taproot, besides several small Side-ones. The Leaves surround the Stalks and are very neatly serrated, very much resembling those of Nettles.

The Tops of the main Branches or Stalks support each a pentapetalous yellow Flower; the Petals small, and spoon-like: These are supported by five capsular sharp-pointed Leaves, yellow above, and of a russet Colour underneath.

 The

The Petals ſurround a great many yellow *Stamina*.

Each Flower is ſucceeded by a ſmall ſharp-pointed Pod, incloſing a great many ſmall Seeds.

The only Uſe that I know of this Plant is, that it is boiled as Sallad, and eaten by moſt Negroes.

The LARGEST SENSIBLE PLANT.

THIS grows to be about four Feet high. The Branches are thickly cloathed with a great Number of oval ſmall pennated Leaves. From among theſe riſe ſeveral three-inch-long Stalks, bearing each of them, on their Extremities, an uncommon ſtamineous Flower. The greateſt Part of this is yellow; but on one Side of it appears a Tuſt of pale-white *Stamina*, tipp'd with browniſh *Apices*.

When theſe *Stamina*, as well as the yellow Part of the Flower, drop off, there appears an horizontal Group of ſmall Pods of about an Inch and a quarter long, when ripe, and of a blackiſh-brown Colour, incloſing ſeveral ſmall flattiſh black Seeds.

CRABS-EYE-VINE.

THIS is a long ſcandent Vine, cloathed with ſmall pennated Leaves, each conſiſting of eleven Pair, with an odd one at the End.

The Flowers, being papilionaceous, ſmall, and white, are ſucceeded by Pods, each incloſing two or three ſcarlet Peas.

Every one of theſe, at its Extremity, has a very black Spot, which makes it much reſemble the Eye of a Crab. From thence it derives its Name.

The LEAST WILD PEA-VINE.

THIS is a ſlender reddiſh hairy Vine; cloathed at uncertain Diſtances with ſmall green Leaves, generally three in Number, upon one common Footſtalk.

From the Boſom of the Leaves riſe many yellow papilionaceous Flowers, each ſucceeded by half-inch-long Pods, containing two ſmall Peas.

The WILD POTATO-VINE.

THIS is a weak ſlender Vine, creeping upon the next Supporter it meets with; but moſt commonly it twines itſelf about *Guiney*-Corn-ſtalks, and bloſſoms about *Chriſtmas*. Theſe are of the papilionaceous Kind, and of a beautiful Mixture of the moſt ſnowy-white and deep-ſcarlet Colours.

Plate. 19.
p. 199.
NON SIBI SED TOTI
To Richard Mead M.D.
Physician in Ordinary to the King;
THIS PLATE
Is most humbly Inscribed
&c.

The ROPE-MANGROVE.

THIS is a groveling Tree, ſeldom riſing above eighteen Feet high: The Bark is of a light-ruſſet Colour; the Leaves are high-ribbed; their Length about ſeven, and their Breadth near ſix Inches, ſupported each by a ſix-inch-long Footſtalk. The Extremities of the Branches are beautifully decorated with large yellow Flowers, very much reſembling thoſe of the Cotton-Flower, except that they want the purple Spot at the Bottom. The Petals of this Flower, which are five in Number, are ſupported by a double Row of green capſular Leaves neatly jagged. The Piſtil is ſtrong, and divides into four knobbed Diviſions: From the Side of it riſe many yellow *Stamina*, tipped with the ſame colour'd farinaceous *Apices*. The Flower is ſucceeded by a roundiſh ſharp-pointed Pod, divided into five Partitions, incloſing each a black kidney-like Seed. The moſt common Method of cultivating theſe Trees is by cutting the younger Branches into many Pieces, and ſticking them in the Ground in wet Weather. It is called the Rope-Mangrove, from the Uſe that is made of the Bark of it to make Ropes or Halters for Cattle. Theſe Trees grow moſt commonly in low moiſt Land, near the Water: They flower about *June* or *July*.

The JUNCTION-VINE.

THIS is a long trailing Vine, bearing a ſharp-pointed heart-like Leaf. The Flower is monopetalous, very much repreſenting, in its whole Shape, the *Placenta Uteri*. The *Labia* of the Flower are freckled with purple and white Spots. The Root of the Vine is exceeding bitter, and very uſeful to make bitter Diet-Drink. The Flowers are ſucceeded by a conic blackiſh Pod, which is very much ſulcated on the Outſide, and the Inſide divided into partitional Cells, incloſing ſeveral ſmall black Seeds.

The PIGEON-PEA-TREE; *Lat.* Cytiſus.

THIS grows to about nine Feet high: Its Roots are few, penetrating not deep into the Earth. Its main Stalk, or Stem, at about three Years Growth, is commonly near as big as a Man's Wriſt.

I know of no Part of this Shrub but what is of ſome Uſe. The Wood is good for Fuel; and by the often falling of its numerous Leaves, the Land it grows upon is very much enriched; and its Fruit is of great Service, by affording hearty nouriſhing Food to Man and Beaſt. The main Stalk is generally divided into ſeveral buſhy Branches: Theſe are cloathed with long ſharp-pointed ſmooth Leaves, green above, and covered with an hoary Mealineſs underneath: It bears, upon two-inch-long Spikes, very many yellow papilionaceous Flowers, ray'd with purple Veins. Theſe are ſucceeded by Pods of a ruſſet Colour, when ripe; containing three, four, or five, ſomewhat flat roundiſh Peas, ſeparated from one another by a ſlender Partition. Theſe Peas, green or dry, are boiled and eaten, and eſteemed very wholſome, eſpecially if uſed in the wet time of the Year; for, being of a binding Quality, they prevent Diarrhœas and Dyſenteries, ſo common in wet Seaſons. I attribute their Reſtringency to a very ſmall Quantity of a gummy reſinous Subſtance, which is generally found, more or leſs, in each Pod.

Theſe Trees are produced from the dry Peas, planted about an Inch deep. They grow to a conſiderable Bigneſs even the firſt Year, each bearing ſome Hundreds of Pods; the ſecond and third ſtill more numerous; the fourth and fifth they bear but very ſparingly; in two Years more the Tree gradually decays and dies. They thrive beſt in a dry Soil, and ſoon periſh in a wet one.

Thus " is much Moiſture hurtful to ſome, whilſt kindly to others; " thus ſome Plants require a ſtrong and rich, others a poor and ſandy " Soil; ſome do beſt in the Shade, others in the Sun."

This is delineated in Plate XIX.

NEMNEM, *or* TOOTH-ACH-TREE.

THIS Tree was firſt brought hither by a *Portugueſe*, about fifteen Years ago; it takes its Name of the Tooth-ach-Tree from its rather ſuppoſed, than real Quality of curing the Tooth-ach. It differs very little, if any thing, except in its greater Bulk, the Length of its Prickles, and the more falcated Form of its Pod, from the common *Akaſee*, already deſcribed.

The AKASEE *or* SWEET-BRIER.

THIS Shrub is of the thorny Kind, growing to about ten Feet high: Its Roots are ſtrong and many, and penetrate very deep into the Earth. The main Stem, as well as the whole Shrub, is cloathed with a reddiſh-grey Bark; the Branches are ſomewhat geniculated backward and forward, alternately. From each of theſe grow ſeveral Side-twigs; on theſe are ſeveral Pair of very ſmall oval pennated Leaves. The Flowers, which are of a globoſe Form, yellow, and of the ſtamineous Kind, riſe from the Boſom of the Leaves on Stalks, or Pedicles, of about an Inch long, guarded at the Bottom with two ſharp Prickles: Theſe Flowers are ſucceeded by Pods of about four Inches long, black when ripe, containing eight or nine oval Seeds, ſeparated from each other by fungous Partitions. The Roots, when bruiſed, yield an offenſive Smell; and, if boiled to a Decoction, and drunk, prove mortally poiſonous to Man, or Beaſt. The Pod, when half-ripe, affords ſo glutinous a Jelly, that it is made uſe of, inſtead of Cement, to join together broken *China* Ware. If the main Stem is wounded, there ouſes out, in few Days, a tranſparent Gum like Gum-Arabic. The Deſcription which Mr. *Lemery* gives of the

Shrub which produces Gum-Arabic, anſwers very near to this. This is delineated in Plate XI. Fig. 2.

The FLOWER-FENCE, *or* SPANISH CARNATION; *Lat.* Ponciana.

THIS Shrub is diſtinguiſhed into two Sorts, the one bearing a yellow Flower, the other a red and yellow mixt; the former is very ſeldom to be met with. The Shrub bearing both is of the ſame Species, and grows to be about ſix Feet high, ſending out near the Top ſeveral ſlender Branches on every Side. Theſe are thickly cloathed with many *Pennæ* of ſmall oblong winged Leaves; the main Stalk is covered with a whitiſh grey Bark, the upper Branches ending in green long Spikes. Theſe are ſpirally ſurrounded, upon long green Footſtalks, with very beautiful Flowers, each compoſed of five Petals, which are generally red near the Bottom of the Leaves; and their Edges, which are finely waved, or ſinuated, are deeply bordered and tinged with Yellow; and, where the Yellow predominates, it is generally rayed with fine Streaks of Red. Theſe Petals are ſupported by five under-capſular, reddiſh, ſpoon-like Leaves. From the Centre of the Flower riſe nine two-inch-long *Stamina*, of a flaming Purple, tipp'd with reddiſh-brown *Apices*, ſurrounding a reddiſh Style, which is the Rudiment of the Pod. This, when ripe, is about five Inches long, flattiſh and of a purple Black, divided into ſeveral partitional Cells; each Cell including a flat blackiſh Seed, not ill-reſembling the Kernel of an Apple. Lye-water, made from the Aſhes of the Roots of this Shrub, is looked upon to be good to bring down the *Catamenia*; and one of the Flowers, bruiſed and ſteeped in Breaſt-milk, is a gentle Anodyne; for which Purpoſe it is often given to quiet very young Children. Theſe flouriſh all the Year round.

The DOG-WILLOW-RATS-BANE, *or* RAW-HEAD *and* BLOODY-BONES.

THIS is a ſcandent Shrub, cloathed with green ſmooth-edged round-pointed Leaves of about two Inches long, and an Inch broad. Theſe are ſet on the Branches alternately. From the Extremity of the latter grow ſeveral roundiſh Pods, compoſed of four ſpoon-like Leaves, ſupported near the Stalk with four ſmaller capſular Leaves. When theſe Pods open, they diſcover a Taſſel of long white *Stamina*, tipp'd with *Apices*; from the Middle of theſe riſes the Piſtil, which is near three Inches long, bearing upon its Top the Rudiment of the Fruit. This Piſtil ſoon grows ligneous, and, at its full Growth, produces a five-inch-long Pod of a whitiſh-yellow Colour, ſtreaked with two oppoſite ſcarlet Seams. The Inſide is full of Cells, containing ſeveral ſmall green Seeds. This Shrub grows chiefly upon rocky Places.

The

The French Guava.

THIS is a ſhrubby Plant, whoſe main Stalk hath a ſtrong ligneous Texture. Its Side-branches are cloathed with ſeveral Pair of large oval winged Leaves, the upper Stalks ending in upright Spikes, which are covered for three Inches in Length with pendulous yellow Flowers, not ill-reſembling thoſe of the Aloe Plant. Theſe are ſucceeded by ſeveral blackiſh long Pods, whoſe ſeveral partitional Cells incloſe a great many round blackiſh Seeds. A Decoction of this Plant, or an Ointment made of a Mixture of its Juice, is looked upon to be of great Uſe to cure and dry up any cutaneous Eruptions. This flouriſhes about *Chriſtmas*, and loves a rich Soil, as well as a ſhady Place to grow in.

The Wild French Guava, *or* Stinking-Bush.

THE main Stalk of this Plant riſes above four Feet high, bearing on its Side-branches ſeveral Pair of ſharp-pointed winged Leaves. The Flowers, which are yellow, and much reſembling thoſe of the Tamarinds, are ſucceeded by five-inch-long blackiſh Pods, containing many ſmall blackiſh Seeds. This derives its Name from the Likeneſs it hath to the *French Guava* Shrub, and from its ſtinking Smell.

The Arnotta Shrub.

THIS, even at its greateſt Growth, exceeds not twelve Feet, nor the Trunk, where largeſt, above three Inches in Diameter. It branches pretty much towards the Top, and is cloathed with middle-ſized ſharp-pointed green Leaves, whoſe middle and tranſverſe Ribs are ſomewhat reddiſh. Theſe are diſpoſed on the Branches alternately. The Flower is pentapetalous, and of a bluiſh Yellow, compoſed of ſpoon-like Petals. The incloſed yellowiſh *Stamina* are tipped with purpliſh *Apices*. The Style, which riſes from the Centre, proves the Rudiment of the ſucceeding echinated Pod, which is of a conic Shape. Its Inſide is divided into ſeveral partitional Cells, incloſing a great many ſmall Seeds, covered on the Surface with a red Subſtance; which the *Indians* on the Main dye their Bodies with. As the Method of preparing this Dye is not well known here, I ſhall not attempt to give any Account of it.

The Lucern.

THIS Kind of Graſs hath been but very lately cultivated in this Iſland, and even now but by ſome curious Perſons. It is of a very quick Growth in wet Weather; but ſo ſoon dies in dry Seaſons, that it is not worth cultivating here, however valuable it may be where there fall more conſtant Rains.

The WOOD-SORREL.

THIS differs very little, if any thing, from the ſame Species that grows in *England*: And its Virtue in Ptiſans or Sallad is equally the ſame.

The JALLOP, *or the* FOUR-o'CLOCK-FLOWER; *Lat.* Jallapa.

THIS ſhrubby Plant grows to be about four Feet high, and is divided into numerous Branches, as well as by the different Colours of its Flowers diſtinguiſhed into ſeveral Sorts. However, the Root of that, which bears a brown yellowiſh Flower, comes neareſt in phyſical Goodneſs to the true Jallop. Its Leaves are winged, and the Flowers of the tubular monopetalous Kind, whoſe *Oræ* are divided into five Sections. The *Stamina* are tipped with five *Apices* of a ſcarlet Colour. Theſe Flowers, though diſtinguiſhed into the Purple, the Red, and the Yellow already deſcribed, have all their Footſtalks covered over with ſoft *Villi* or Hair; and the Bloſſoms of each Kind are ſucceeded by ſmall *Capſulæ*, containing in each a ſmall black, huſky Seed, incloſing a white Kernel.

I ſhould have added before, that the Leaves are ſharp-pointed, and above two Inches long, and an Inch and an half broad. The Flower expands early in the Morning, and generally continues open till Eight o'Clock; and then ſhuts, and ſeems almoſt wither'd, till about Four in the Afternoon; at which Time it blows anew, and continues open till Night.

The CRAB-TREE.

THIS tall Shrub hath a dark-colour'd reddiſh Bark. The Branches are many and crooked. The Leaves, which are generally in Pairs, are likewiſe unequal and ragged; the largeſt not exceeding two Inches long, and the ſmalleſt not an Inch. The Branches are guarded with very ſmall ſcarce-perceptible Prickles. The Flowers are of the papilionaceous Kind, and ſucceeded by many ſmall Pods.

The INDIGO-WEED.

IN the Infancy of Trade in this Iſland, Indigo, which is produced from this Weed, was one of its ſtaple Commodities: But as the Improvement of Sugar became more beneficial, or rather, when our Neighbours the *French*, by their ſeveral Indulgences in Trade, as well as the cheap Purchaſe of their Land, were able to underſell us at foreign Markets, this Branch of Trade was ſo intirely ſwallowed up by them, that we have had no Indigo manufactured here for above forty Years paſt. Therefore it would be needleſs in me either to deſcribe the Plant, if I could meet with any in the Iſland, or to trouble my Reader with the Method of making Indigo from it.

The Jallop-Vine.

WHAT is here called the Jallop-Vine, grows near the Houſe of the Honourable *Samuel Rouſe*, Eſq; Its Stalk yields a milky Juice. It bears an heart-like Leaf, and a Bell-flower.

Sorrel.

WE have here, beſides the Wood and the Garden-Sorrel, another tall ſhrubby Sort, diſtinguiſhed by their Colour into the White, and the Red. Theſe grow ſtrong, and are ſeemingly ligneous, though they are of an annual Growth, produced from the Seed. Their Roots are many and fibrous.

The main Stalk grows often to be above three Feet high, red and hollow. The Leaves, which are deeply digitated, and their middle Ribs and Stalks reddiſh, are ſet on alternately. The Flower conſiſts of five pale-white Leaves, gyrally incircling one another, and each deeply ſtain'd with a purple Spot at the Bottom, and incloſing a ſhort ſtrong Piſtil ſurrounded with farinaceous Duſt.

The Petals are likewiſe incircled with five ſharp-pointed capſular Leaves, which, when the Petals drop off, cloſe together into a conic-ſhap'd Pod. This, when ripe, opens into five Sections, diſcovering a ruſſet Berry, which likewiſe, when ripe, opens into five Sections, and diſcovers a great many blackiſh Seeds.

Theſe Leaves, as well as the Berry, are decorated as well as ſupported by a ſtrong *Calyx*, which on the Outſide is divided into nine ſharp-pointed Sections.

The red Leaves, and Flower-cup, being thick and very juicy, are, when ripe, ſeethed in boiling Water, which in a few Hours extracts both their Colour and Strength. This Decoction, work'd with Sugar, makes a very ſtrong reddiſh heady Wine; and as theſe Plants are of two Kinds, the Red and the White, their reſpective Wines will be of theſe different Colours, but of the ſame Taſte.

The red capſular Leaves, when young (firſt ſtew'd,) make excellent Tarts.

The Fruit is generally ripe about *November* and *December*.

The Wild Sena, *or the* Wild Cassia Fistula; *Lat.* Colutea.

THIS buſhy Shrub grows to about four Feet high; the main Stalk pithy and brittle; the Branches thickly cloathed with round deep-green

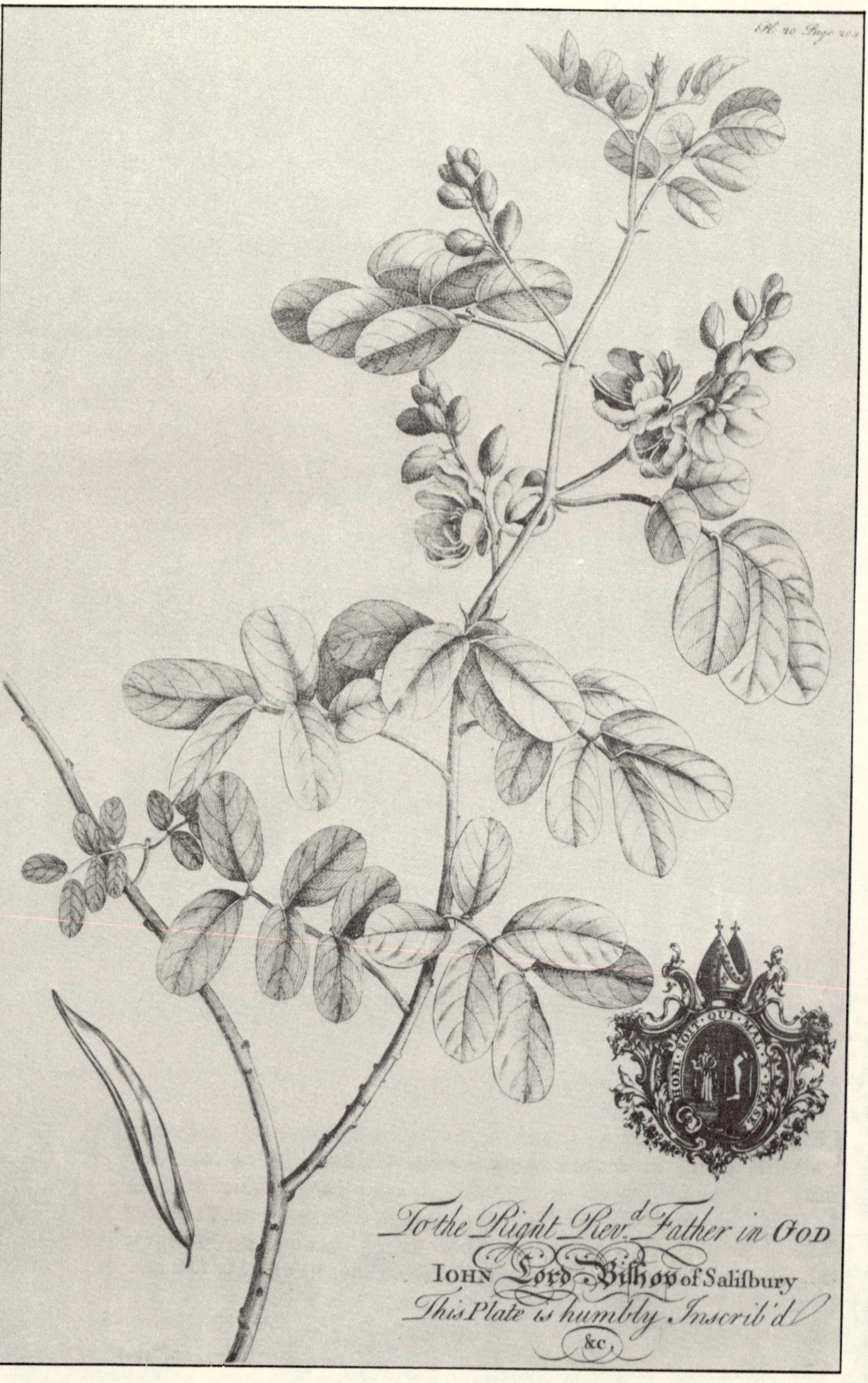
Pl. 20 Page
HONI · SOIT · QUI · MAL · Y · PENSE
To the Right Revd Father in GOD
IOHN Lord Bishop of Salisbury
This Plate is humbly Inscrib'd
&c,

green winged Leaves, ſet upon the Twigs in three or four Pair. The Flowers are yellow and pentapetalous.

Theſe are followed by five-inch-long browniſh round Pods, ſomewhat falcated when ripe.

The Inſides of theſe are divided, like the large *Caſſia-Fiſtula*, into very many ſmall Cells, ſeparated from one another by thin pulpy Partitions of a ſweetiſh Taſte, including, in the intermediate Spaces, ſmall blackiſh flat Seeds.

This is delineated in Plate XX.

The CHRISTMAS-BUSH.

THIS Plant grows to about three Feet high. Its Roots are many, but weak, always blooming in *December* ; and from thence it derives the Name of *Chriſtmas-Buſh*.

It continues to bear Bloſſoms till the middle of *March*, and then gradually decays.

The main Stalk and Branches are of a ſtrong gramineous Texture, and its Leaves, which are about two Inches long, and near as broad, end in a ſharp Point.

Their Edges are ſnipped in three or four oppoſite Places, their upper Sides deeply furrow'd, and by far greener than the under, which, when young, are whitiſh, ſoft, and downy. On the Top of each Branch (generally ſpeaking) ſtands a Group of very ſmall white tubular monopetalous Flowers ; their Tops are but barely to be ſeen, peeping out of long ſcaly *Calyces* ; and from each Flower riſe two very fine white *Stamina* with pendulous *Apices*. Theſe numerous thready *Stamina*, ſo cloſe together, give it, at a Diſtance, the Appearance of an intire ſtamineous Flower. When theſe drop, their *Calyces* become pappous, and full of long ſmall blackiſh Seeds, each ſurrounded, like the Feathers in a Shuttlecock, with very fine white Down. A Cataplaſm of this bruiſed Plant is eſteemed not only an excellent Vulnerary, but likewiſe ſo great a Detergent, that it will not ſuffer the leaſt proud Fleſh to grow where it is applied. It grows in every Soil all over the Iſland.

The GULLY-ROOT.

IF the Virtue of *Aſa-Fœtida*, in nervous and hyſterical Diſorders, conſiſts chiefly in its efficacious Power of diſperſing, by its ſtrong Smell, the animal Spirits to their ſeveral proper and diſtinct Parts of the Body, the Roots of this Plant may perhaps claim an equal Degree of Virtue ; for its Smell is ſo very offenſively ſtrong, that a ſmall Piece of it will, in a ſhort time, fill the largeſt Room with its diſagreeable Odour.

Pops.

THIS is a ſucculent perennial Plant, having many fibrous Roots. The main Stalk is hollow, and the Outſide divided into many irregular Angles. It widely branches towards the Top, and is thickly covered with thin ſharp-pointed Leaves, of about three Inches and an half long, and about two Inches broad. The Edges of theſe are widely and irregularly ſerrated; the whole Plant growing to be about two Feet and an half high.

The Flowers, which are monopetalous and yellow, are divided into five Angles in a Star-faſhion. The Inſide of each Angle, near the *Diſcus*, or Bottom of the Flower, is ſtained with a browniſh purple Spot. The whole Flower ſurrounds five *Stamina*, tipped with pale bluiſh *Apices*. Theſe likewiſe ſurround a whitiſh Piſtil. The Flowers are ſucceeded by ſmall round bluiſh Berries, full of an agreeable Pulp, mixed with many whitiſh Seeds.

Theſe are looked upon to be good Diuretics: The Berries, till ripe, are ſurrounded with a thin green *Capſula*, ſomewhat of a conic Shape: which is of a very ſlender Texture, and divided principally into five high convex purple Veins, interſperſed with a great many leſſer longitudinal as well as tranſverſe Ribs or Veins. When the incloſed Fruit is ripe, this *Capſula* opens, and withers away.

Foster's Plant.

AS this Shrub hath been hitherto nameleſs, I have preſumed to call it after the Perſon's Name who diſcovered to me, not only this, but ſeveral others. This ſhrubby Plant grows to about four Feet high. Its Roots are ſtrong, penetrating deep into the Earth. The inſide Bark of theſe, as well as the Pith in the main Stalk, is of a yellowiſh Colour; the latter ſomewhat inclinable to a red. Its ſlender Side-branches are cloathed with five Pair of pinnated Leaves without, an odd one at the End; theſe are about an Inch and an half long, and near an Inch broad. From among theſe riſe a great many papilionaceous Flowers; which are ſucceeded by long flattiſh Pods, incloſing a great Number of flat Seeds of a ruſſet Colour.

The Leaves of this Plant, if bruiſed, yield a very offenſive Smell. It grows chiefly in rich Land, and blooms in *June* and *July*.

The Cow-Itch Vine.

THIS is a long ſcandent Vine, creeping often to above twenty Feet high, if ſupported by neighbouring Trees or Rocks. Its Leaves are of a middle Size, ſharp-pointed, and very thin, covered with ſoft *Villi* or Down, and ſet on the Vine in a Triparture-order. The Flowers of this Vine grow in Bunches.

The

The Appearance of each, at firſt, is an inch-long purple petalous Pod; which, when it blows, divides into two narrow purple Petals, diſcovering in the Middle, between both, a white long Pointal, hooked even to a Prickline∫s at its Extremity. This is likewiſe ſheath'd with a thin white Membrane. When this, as the Flower grows, ſplits open, it diſcovers the Pointal to be fringed or divided near the Top into ſeveral *Stamina*, tipp'd with *Apices*. The two above-mentioned Petals are, near the Bottom, incircled with another purple Petal, not ill reſembling the *Carina* of a papilionaceous Flower. The Whole is ſucceeded by a Pod near of the Shape and Size of a Pod of a full-grown *Engliſh* Field-Pea. This incloſes ſeveral blackiſh ſmall Beans. The Outſides of theſe Pods are thickly covered with very fine ſhort *Setæ*, or ſtiff Hairs. If theſe, either blown by the Wind, or by any other means, touch the Skin, they will inflame the Part far worſe than if touched with Nettles.

The ITALIAN SENA.

WHAT is called here the *Italian Sena*, hath many white ſtringy Roots ſpreading on every Hand. Its main Stalk ſoon divides into leſſer Branches; theſe are cloathed with ſeveral Pair of winged pale-green Leaves of an oblong Shape. The Branches likewiſe ſuſtain, on two-inch-long Footſtalks, ſeveral five-leaved yellow Flowers; theſe are ſucceeded by falcated broad Pods, divided by ſeveral Indentings into many Partitions, which incloſe dark-colour'd Seeds.

This, with a great many other curious medicinal Plants, is to be found in the Garden of Doctor *John Douglaſs*, near *Speight*'s-*Town*, as well as in the Garden of Doctor *Reynold Alleyne* in *St. James*'s Pariſh.

The MUSK-BUSH, *or* WILD-OCKRO.

THIS buſhy Plant bears many ſmall Branches, cloath'd with very hairy Leaves, divided into three deep Sections, forming as many partitional ſharp-pointed Angles. The Ribs on the upper Side are ſomewhat reddiſh, on the under green. The Edges of theſe partitional Sections are irregularly indented: The upper Branches ſuſtain a beautiful yellow Flower, compoſed of five round-pointed large Petals, each Petal above two Inches long, and ſtained at the Bottom with a purple Spot. Theſe ſurround a large Piſtil cover'd with yellow farinaceous Duſt. The Top of this is almoſt ſurrounded, on ſhort Footſtalks, with blackiſh-purple *Apices*. The Flower is ſucceeded by a multangular fleſhy Pod, every way reſembling that of an Ockro. They are ſometimes eaten, eſpecially when very young; otherwiſe they taſte muſky.

The HORSE-VINE.

THIS Vine is ſmall and creeping; taken notice of chiefly becauſe it is looked upon to be good Feeding for Horſes: Its green round-pointed Leaves are ſet on an inch Footſtalk in a Triparture-order. The Flowers are ſucceeded by ſmall Pods, incloſing a great many Seeds, or Peas, ſeparated from each other by partitional Cells. It grows chiefly in ſhady Places.

The WILD-BASIL; *Lat.* Acinus.

THIS hath very ſmall fibrous Roots: Its Stalk and Branches are ſquare and deeply chanell'd. The Leaf is about an Inch and an half long, ſharp-pointed, and ſomewhat indented about the Edges. It bears ſmall fiſtular blue ſtar-like Flowers, ſtanding in Groups upon conic Tufts, whoſe Apertures are guarded with ſhort hairy Spikes. In theſe are included the Seeds, which are very ſmall, and of a blackiſh Colour. This Plant is juſtly look'd upon to be an excellent Vulnerary.

The HOLLY-HOCK; *Lat.* Malva Arborea.

THIS Plant much reſembles the *Gundelia Orientalis, Acanthi aculeati Folio, Capite glabro*, growing in the *Levant*, and deſcribed by *Tournefort*. Its Leaves are very deeply laciniated, as all of the Thiſtle-kind are. The middle as well as the leſſer Ribs are ſtain'd with blue milky Veins. The Stalks, near or at the Top, bear many yellow ſtamineous Flowers. Theſe are ſucceeded by echinated browniſh Pods, incloſing a great Number of browniſh ſmall Seeds: A Thimble-full of theſe, taken inwardly, proves an excellent Purge; and the yellow Juice, that, when broken, plentifully diſtils from the Stalk, is a ſovereign Remedy to cure old Ulcers.

The CONEY-FLOWER-VINE.

THIS Vine creeps upon and ſupports itſelf by its numerous Claſpers, twining round the Branches of neighbouring Trees. It bears a great many ſmall heart-like Leaves; and at different Diſtances is decorated with large papilionaceous Flowers, the Extremities of whoſe *Alæ* are white, terminating near the *Carina* in a moſt deep beautiful Purple.

The *Carina* itſelf is white, and its Orifice cover'd with two ſmall membranaceous Petals, having their Riſe from the Footſtalk-Part of the *Carina*; and, meeting on each ſide at the Orifice, they cloſely join together to cover it. Theſe Leaves have their Parts, which thus join together, beautifully ſtain'd with Purple. The Piſtil takes its Riſe at the Bottom of the *Carina*; and, as it grows ſtronger and larger, its Point becomes divided into ſeveral *Stamina*, tipp'd with *Apices*, which, when ripe, fall into

into the Ovary; and then the Piſtil, which becomes the Rudiment of the future Pod, puſhes open the Orifice of the *Carina*, as well as burſts through the two membranaceous Leaves, that covered it. Soon after the Expanſion of theſe the Flower withers, and drops off; and is ſucceeded by a four-inch long Pod, chanelled in the middle on each Side with a deep Furrow. The Pod contains ſeveral ſmall Peas or Seeds.

The SENSITIVE PLANT; *Lat.* Mimoſa.

THIS is diſtinguiſhed into the large and ſmall Sort, the main Stalks ſomewhat hairy; each bearing proportionable ſmall longiſh oval-winged Leaves. The Flowers, which ſtand upon three-inch-long Footſtalks, are ſtamineous. Theſe are ſucceeded by ſmall browniſh Pods of an oblong Shape, and of about an Inch long. The Edges of theſe Pods are thin, and through their Middle, on each Side, riſes a Ridge, which, on the Inſide, contains a great Number of blackiſh ſmall Seeds. The winged Leaves upon this Plant ſhrink very much upon the leaſt Touch. The Pedicles, which ſupport the Flowers, and the ſucceeding Pod, are, near their Setting off from the main Stalks, embraced with two heart-like ſharp-pointed Leaves, of near half an Inch long. Theſe intirely differ from all the reſt.

The RABBET-WEED.

THIS is of the Thiſtle-kind. Its main Stalk grows to be three Feet high, and hollow like a Pawpaw-Shank. It is thinly covered with ſmall narrow Leaves, which are prickly about their Edges. The upper Branches ſuſtain a great many Flowers compoſed of a Border of pale-white ſmall Petals, ſurrounding a yellowiſh long Thrum, which, when ripe, proves pappous, and flies away into whitiſh Down. This Plant derives its Name from the Uſe that is made of it to feed Rabbets with.

The INDIAN ROOT.

T E Seed of this, ſome few Years ago, was ſent hither from *Jamaica*, as a very valuable Plant. If the chimerical Gentleman, who ſent it, had but had Coolneſs of Thought enough firſt, fully to diſcover its pretended Virtues, before he had recommended it, the Planters of this Iſland would have been much happier without it; for, as its Seeds are of the pappous Kind, they were ſoon carried over all the Iſland; and it too unluckily flouriſhes in every Soil, though uſeful to neither Man nor Beaſt. Its Roots are white, fibrous, and many. The main Stalk riſes about eighteen Inches high.

The

The OCKRA; Ockro.

THIS excellent shrubby Plant is plainly a Species of Mallows. The Stalk is like that of a Mallow, and rises to about four or five Feet high, bearing, on and near the Summit, many large yellow Flowers succeeded by green conic fleshy Pods chanelled into several Grooves.

Each Pod, when ripe, is from an Inch to two Inches and an half long, full of as many longitudinal Cells, as there are Chanels of small round black Seeds. The long Ockro differs from that already described, chiefly by having its Stalk twice as high as the other, and its Fruit often four Inches long, and the Pod smaller. This likewise, with Care, may be brought to bear all the Year round. The Fruit is taken, when young and tender, and boiled and eaten with Butter.

They are of a very mucilaginous Nature; a great Restorative, very wholsome, and extremely lubricative. These are undoubtedly the Plants which the *Romans* so much valued, and which *Horace* so well describes by calling them *leves malvæ*.

COYERS.

THIS shrubby Plant hath several whitish Roots, smelling not unlike a Radish. The main Stalk, which is of a purplish Colour, branches very much near the Ground: From the several Side-branches issue a great many Footstalks, whose respective Tops from one common Centre sustain seven sharp-pointed Leaves, being almost equally sharp near their common Footstalk, where they all join; at which Juncture there is a yellowish Spot. The Flower very much resembles that of a Garlick-Pear-tree, consisting of four small Spoon-like Petals.

From the Centre of these rises a dark-coloured Pistil, from whose Sides, somewhat higher, issue six purplish *Stamina*, tipped with brownish *Apices*; the Pistil in the Middle still continuing larger than the *Stamina*, bearing upon it the Rudiment of the future Pod, which, when ripe, is of a flattish Shape, of about six Inches long, inclosing a great many small Seeds. The Juice of this Plant, mixed with sweet Oil, is looked upon to be a sovereign Remedy against the Pain in the Ear, if poured into it warm.

The MANY-ROOTS.

THIS Plant derives its Name from the great Number of its Roots. The several Stalks, which grow up from these, are about seventeen Inches high, and hairy. Their Leaves, which are crumpled, and have very turgid Veins, are about four Inches long, and three broad. From the Bosom of the Leaves rise many Pedicles sustaining large dark-blue Flowers. These are succeeded by very small brownish Pods.

A De-

A Decoction of this Plant, drank warm, is looked upon to be very good to heal sore Throats. It grows to equal Perfection exposed to the Wind, or in the Shade; and generally blooms in *June* and *July*.

NIGHTSHADE; *Lat.* Solanum.

THE Roots of this Plant are many, stringy, and white. The main Stalk, as well as the lesser side-ones, are of a deep-purple Colour, and herbaceous Texture, rising often to near two Feet. The Leaves, which are about five Inches long, and about four broad, are irregularly serrated into shallow as well as deep Sections; the Extremity of each being sharp-pointed. The Flowers are of the tubular Kind, having their Borders or Extremities much expanded, and divided into five equal Parts, ending in corniculated Points. This Border or Extremity of the Flower is of a pale White. Its long tubular Shank is finely rayed, the Inside with purple-bluish Veins. From its Centre rise five two-inch-long *Stamina*, tipped with grey-bearded *Apices*. The Pistil, being of equal Length, is tipped with the Rudiment of the Fruit, which, when ripe, is an echinated Pod of about an Inch long, inclosing a great many small blackish Seeds. The whole Flower seems withered, and closed up, in the Day-time; at least, till about half an Hour past Five in the Evening: Then it opens, and fully expands itself into a fine beautiful Flower.

This is no sooner in Bloom than the Fly-bats come from their lurking Holes, and, flying from Flower to Flower, dart into these, severally, their long twirling Tongue or *Proboscis*, with which they suck out either the Honey-dew, or some other Moisture from the Bottom of the Flower.

The Leaves are generally looked upon to be of great Efficacy to cure the Head-ach, by tying them to the Temples of the Person afflicted.

RATA-PEPER, *or* RATA OCKRO; *alias* Hoho.

THIS Plant hath a great many white stringy Roots. The main Stalk is green, herbaceous, and hairy, rising to about fourteen Inches. This and its Side-branches are cloathed with green Leaves. These are two Inches long, and near as broad; having several high Veins on their under Sides, and their Edges irregularly serrated. Among these rise several Footstalks or Pedicles, sustaining a great many yellow Flowers succeeded by long slender Pods.

This Plant is much made use of by our Slaves in their Soups and Broths; who esteem it, and not unjustly, a very wholsome boiled Sallad.

Many-Seed.

I HAVE given this Plant a Name from its many Seeds; which, from reckoning the Number in one ſmall capſular Pod, and multiplying that one by the Number of remaining ſmall round capſular Pods, the whole Plant, which ſprung from one Seed, contained twelve thouſand eight hundred and eighty Seeds. It ſeldom grows above three Feet high, and is forked.

Coot-Weed.

THIS, whilſt young, very much reſembles *Aſtragalus Orientalis*, mentioned by *Tournefort* as growing in the *Levant*; with this only Difference, that the Flowers of the Eaſtern Plant are white, and of this yellow. It affords, when broken, a rank fetid Smell; which ſhews that the Oil and Salts are in a volatile State.

Rattle-Bush, *or* Shake-Shake.

THIS Plant hath a great many white matted ſtringy Roots; the main Stalk dividing, almoſt as ſoon as it comes out of the Ground, into many green ſucculent Branches; theſe different Stalks, at different Diſtances, bearing, on Pedicles of about two Inches long, three blunt-pointed inch-long green Leaves. Theſe are ſet on at the End of every Stalk in a Triparture-order. The Flower, which is of the papilionaceous Kind, is ſucceeded by a ſmall Pod. The incloſed Peas, when ripe, make a rattling Noiſe when ſhaken by the Wind: From hence they derive the Name of *Rattle-Buſh*, or *Shake-Shake*. They grow chiefly, or at leaſt beſt, under the Shelter of an Hill.

The Red Dialthæa.

THIS is a low ſhrubby Plant; its Bark of a reddiſh Grey. The Branches are thickly cover'd with ſharp-pointed Leaves. Theſe are about two Inches and an half long, and near two broad. Their Edges are finely ſerrated: The Flowers, which are of the herbaceous Kind, almoſt ſurround the Stalk in ſeveral Bunches. Theſe, when ripe, are ſucceeded by ſeveral very ſmall Huſks or *Capſulæ*, incloſing ſeveral browniſh triangular Seeds. They grow chiefly in ſhady Places.

Wild-Parsley.

THIS derives its Name from the great Reſemblance its Leaf hath to that of Parſley. The Flowers are ſucceeded by a ſmall triangular *Capſula*, each Angle containing, in its Partition, one round ſmooth Seed, of of about the Bignefs of large black Pepper. This, when ripe, is neatly marbled with black and white Spots.

The only Ufe that I know is made of this Plant, is, that the good Houfewives make the Nefts of their fetting Hens of the Leaves and Branches of this, in order to deftroy the Lice upon the Hens. The Seeds likewife are made ufe of, being ftrung upon Silk, inftead of more valuable Beads, for Bracelets.

It flourifhes about *June*, *July*, and *Auguft*.

PEPPER.

AS the Species of thefe are many, I fhall firft give a Lift of their Names; and then proceed to a particular Defcription of fuch as are any-ways remarkable: Bonnet-pepper, ribb'd. Bonnet-pepper, plain. Negro-pepper. Bird-pepper. Cherry-pepper. Long pepper. White-pepper. Olive-pepper. Spur-pepper. Rofe-pepper. Ifaac-Cape-pepper. Bell-pepper, and Turbilo-Pepper. The Pepper moft commonly ufed, and moft approved of, is what is called Negro-pepper. The Shrub that bears this feldom rifes to above three Feet. It is very thickly cloathed with fmall green fharp-pointed Leaves, of about an Inch and an half long, and about one broad. Its Flowers, which are white, and five-leav'd, are likewife very numerous. It is obferved, that, even before they blow, each of thefe hath a fmall wriggling Worm in its *Difcus* or Socket. The fucceeding Fruit is of a fmall oblong fharp-pointed Pepper, red when ripe, of near one third of an Inch long. This hath a very hot poignant Tafte, attended with an agreeable Flavour.

SWEET-HEART.

THE Roots of this fmall Plant are many, penetrating deep into the Earth. The main Stalks are jointed, and are no thicker than Packthread. Thefe rife feparately from the Root, four or five in Number, growing to about fifteen Inches high. From each Joint of thefe main Stems, rifes a flender Footftalk of an Inch long, fupporting three fmall Leaves, fet on in a triangular manner; the Top of each Stalk ending in a gramineous Spike, furrounded with many purple papilionaceous Flowers. Thefe are fucceeded by fmall Pods of about two Thirds of an Inch long, including five fmall Peas, or flattifh Seeds. Thefe are feparated from each other by Partitions; and the Outfide of the Pod is divided or fegmented into fo many femicircular Sections as there are Peas. The Pod is intirely incrufted with fmall *Setæ*, or hooked Briftles, imperceptible to the naked Eye; by which means they tenacioufly ftick to the Cloaths of thofe who walk among them.

A Decoction of the Roots, boiled in Milk with the Bark of a *Guava* Tree, is look'd upon to be good to cure Fluxes.

The

The SPANISH-NEEDLE.

THIS under ſhrubby Plant grows in moſt Parts of the Iſland. It has many ſmall fibrous Roots. On every Footſtalk ſtand three ſharp-pointed ſerrated Leaves, ſet on in a triangular manner, the oppoſite to the Stalk being the longeſt.

The Flowers are compoſed of a fiſtular yellow Thrum, partly ſurrounded with three white Petals. Theſe are divided at the Top into two deep narrow Sections, which gives each Leaf the Appearance of three. The Seed-veſſels are very ſmall long blackiſh Spikes, ſurrounding the Top in a Whorl-faſhion, having their Ends very much bearded and jagged, incloſing very ſmall longiſh black Seeds.

This Plant is looked upon to be a good Vulnerary.

IRON-VINE.

THIS is a ſmall creeping Vine, deriving its Name from the ferruginеous Colour of its Stalk and Leaves. The Roots penetrate deep into the Earth: The ſlender Side-branches bear upon half-inch Footſtalks three ſharp-pointed Leaves; the oppoſite to the Centre being the longeſt. This is about an Inch and an half long, and an Inch broad; the two other Leaves ſomewhat leſs. The Flower is very ſmall, of a bluiſh Colour, and of the papilionaceous Kind.

The whole Plant, when bruiſed, yields a reddiſh rough Juice; which, if boiled into a Decoction, and drunk, is looked upon to be of great Service in Diarrhœas and Dyſenteries. This Plant flouriſhes in moſt Places, eſpecially in the Summer Months, when theſe are moſt common.

DOWN-VINE.

THIS is a creeping Vine, bearing a great many heart-like Leaves as well as ſeveral Flowers, which are ſucceeded by ſmall Pods, filled with Down intermixed with ſmall Seeds; the former ſerving as Wings to convey the latter over the adjacent Places. This Vine grows chiefly in Gullies, and ſuch ſhady Places.

The WILD DOLLY.

THIS Plant riſes about twenty Inches. Its Roots are ſmall and fibrous. The Leaves are ſet three upon a Footſtalk in Triparture-diſpoſition. Theſe are of a very deep-green Colour, and moderately ſharp-pointed. From the Boſom of the Leaves riſe many ſix-inch-long Spikes or Footſtalks: Theſe are reſpectively decorated at the Top with a beautiful purple papilionaceous Flower, and ſucceeded by a three-inch-long narrow Pod, incloſing in ſeveral partitional Cells ſmall blackiſh Seeds, or rather Peas.

The

The WILD-TAMARINDS.

THIS very ſmall woody perennial Plant reſembles in Miniature, eſpecially in its Leaves, Flowers, and Fruit, the large Tamarind-Tree already deſcribed.

The HORSE-EYE-BEAN VINE.

THIS, if ſupported by any tall Tree or Rock, will grow to above fifty Feet high. Its Leaves are large, and ſet on five-inch Stalks, three in Number, in a triangular Manner. Its Bloſſoms are of the papilionaceous Kind; which are ſucceeded by a large downy Pod, incloſing from one to three Beans, called *Horſe-Eyes.*

This Name is given to them from their Likeneſs to the Eye of an Horſe in Shape and Colour.

The HORSE-BEAN.

THIS is a long trailing Vine, with large beautiful green Leaves. Its Flowers, which are of the papilionaceous Kind, are ſucceeded by Pods often ten Inches or a Foot in Length, which incloſe from nine to one-and-twenty large Beans.

The BEAN-VINE.

THE many different Sorts of eatable Beans, planted in this Iſland, are generally diſtinguiſhed by the Names of Arbour-Beans, ſpeckled Beans, *Carolina* Beans, the Sugar Bean, and the ſmall Six-weeks Beans. The Arbour-Bean, which is by far the largeſt, takes its Name from its being often planted near Arbours; on which, if guided, it will creep, and afford an agreeable Shade. This Vine bears a Pod near ſix Inches long.

The WILD-PEA.

THIS is a ſmall weak creeping Vine, which by its numerous Tendrils climbs up the neighbouring Trees, Shrubs, or Rocks. It is thinly cloathed with ſmall ſharp-pointed Leaves; each Pedicle ſuſtaining three, ſet on in a Triparture-order. The Flowers are of the papilionaceous Kind, and white, except that near the Centre. They are very ſlightly ſtained with red. Theſe are ſucceeded by a three-inch-long Pod, containing in partitional Cells about fifteen ſmall Peas.

This Vine grows chiefly in ſhady Places.

The Eatable Wild-Pea.

THIS is a ſlender Vine, ſupported by any neighbouring Buſhes. It bears on every Footſtalk three Leaves, ſet on in a Triparture-order; as well as, upon two-inch-long Pedicles, yellow papilionaceous Flowers.

Theſe are ſucceeded by near three-inch-long Pods, incloſing ſeveral green Peas, which are often made uſe of by the poorer Sort.

The Cuckolds Increase.

THESE grow upon ſhort upright Stalks; and are diſtinguiſhed into the large and ſmall Sort. The Leaves are ſharp-pointed.

The Flowers are likewiſe of the papilionaceous Kind, and ſucceeded by eight-inch-long roundiſh Pods. Theſe, in their ſeveral partitional Cells, contain nine, ten, or eleven longiſh Peas, which, thoug hſomewhat windy, are generally liked.

The Buona Vista, *commonly called* Bonny-vis.

THE Species of theſe are diſtinguiſhed into the Moon-ſhine, the Green, the White, the Red-and-black, and the Small Sort. The Vines and Leaves of theſe are of a greyiſh Green, round-pointed and broad, and the Flowers papilionaceous and white; except the black Sort, which have purple Bloſſoms. What are moſt commonly planted are the Six-weeks Bonny-vis. Theſe are ſo called from their bearing Pods in about that time: Theſe Pods are ſomewhat flat, of about two Inches long, incloſing three, four, or five Peas or Beans in partitional Cells. Theſe are univerſally made uſe of, either green or dry; being looked upon (though ſomewhat windy, yet) a wholſome Grain.

The old Sort generally bears about *Chriſtmas*; and if the Vine is ſuffered to grow, it will in two Years time, if ſupported, prove ſo ligneous and ſtrong, that it will be no leſs than an Inch or more in Circumference. The Six-weeks Kind hath a more tender Vine; and though it bears ſooner, it will neither thrive, nor laſt ſo long.

OF

OF

TREES, SHRUBS, *and* PLANTS,

OF THE

ANOMALOUS KIND.

BOOK VIII.

GRAPPLE-WOOD.

THIS ſhrubby Tree hath a reddiſh-grey Back. Its numerous Branches are ſpindling and weak. Its Leaves are ſmooth and green, ſomewhat above an Inch long, and three-quarters of an Inch broad, ſmooth-edg'd and blunt-pointed. I have never known this Tree to bear either Flowers or Fruit.

The FIVE-SPRIG-TREE.

THIS Tree derives its Name from the general Number of Branches, which horizontally ſurround the Stalk at ſeeming Joints, from one to three Inches aſunder; the main Stem ſtill growing ſtrait upwards, ſurrounded at ſuch Diſtances with the above-deſcrib'd circular Branches. The Wood, as well as the Leaves, have an agreeable Smell: The latter are round-pointed; and what is moſt remarkable in theſe Leaves, is, that when upon the Tree, they are very green; but when dry, inſtead of turning to a ruſſet Colour, they bleach into an almoſt perfect White.

The

The CORN-TREE.

THE Tree call'd by this Name was brought hither about twelve Years ago from *Guiney*, and now grows at the Estate of Mrs. *Warren*, at the *Black-Rock* Plantation. It is said, that it bears, in its native Soil, several long cylindrical Ears, not ill resembling a large Mold-candle, round which Stalk the Grains are disposed; but as this never bore any, and as we have no authentic Writers mentioning the Growth of such a Tree in any Part of *Africa*, I much doubt of the real Existence of such an one: However, it may not be amiss to describe it. Its present Growth is about sixteen Feet high, branching chiefly towards the Top; the Bark is of a greyish White; and its lesser Branches at the Top cloath'd with five green Leaves, surrounding one common Centre; these are two Inches and an half long, and an Inch and an half broad.

The SOUTH-SEA-ROSE; Nerium Indicum.

THIS is a very tall Shrub, whose Branches are many and slender, generally bending wavingly downwards; these are cloath'd, especially near their Extremities, with a great many sharp-pointed yellowish-green Leaves, near six Inches long, and half an Inch broad, set on three in Number round the Stalk; this Disposition is preserved at irregular Distances to the Tops of the Branches, which sustain, upon longish Footstalks, several five-leav'd red Roses. From the Centre of these first Border of Leaves rise three or four lesser red Leaves streak'd with White. The Colour of the Bark is of a darkish Red: The Roots, boil'd to a Decoction, prove a strong Poison, if drunk by Man or Beast. It grows chiefly in shady Places.

SOBER'S-PLANT.

THIS is a small straggling Shrub, divided in many Branches, cloathed with many small round-pointed Leaves of a Liver-colour. The Footstalks and middle Spine of each Leaf are prickly, and the Leaves, when bruised, smell very strong and disagreeable.

The FRENCH ROSE-TREE.

THIS shrubby Tree grows to about twelve Feet high, discovering in most Seasons of the Year, upon the Extremity of the Branches, several beautiful Roses, each consisting of five large white Leaves, whose bottom Parts are slightly ting'd with Red: By Noon these Leaves are of a blushing florid Red, retaining some small Appearance of their morning Whiteness; at Night, which is its last Stage, its Colour is of deep putrid Purple; then it withers, and its Leaves shrivel up. The *Stylus*, which these Leaves inclose, is divided at the Top into three fornicated Branches, and is surrounded from Top to Bottom with much yellow farinaceous Dust:

from theſe, in time, come a great many ſmall blackiſh Seeds, which are lodged in a pappous Down, in theſe crumpled Leaves.

The Plant called PATIENCE.

THIS is a low creeping Plant, whoſe Leaves very much reſemble thoſe of the Plant called *Bachelor's Breeches*. A Decoction made from this Plant is looked upon to be of great Service to alleviate the Pain in the Stomach.

The TURKY-WEED.

THIS Plant grows to about two Feet high; it bears, at about two Inches Diſtance from each other, ſeveral Pair of ſharp-pointed winged Leaves, in Length near three Inches, and half an Inch broad; the Tops of the Branches ſuſtain ſeveral ſmall round capſular Seed-veſſels. This Plant derives its Name from the Uſe that is made of it to feed Turkeys. It grows almoſt in any Soil.

GOATS-RUE.

THIS is a dark moſs-like Plant, growing on the Inſide of old Wells and Walls, from whoſe compact Bottom riſe ſeveral pluſhy *Villi* or ſmall brown ſtiff Hairs, of about an Inch long; theſe are tipp'd with ſlender *Apices*, affording numerous ſmall Seeds.

The WILD BASIL.

THIS woody Plant grows to the Height of about three Feet; its Bark of a black Ruſſet: The Branches are cloathed with Leaves of a Silver-white beneath, and green above. Theſe are high-ribb'd, and ſharp-pointed. At every Inch Diſtance riſes a ſtiff Footſtalk, ſurrounded at the Top with a burry-ruſſet Button, whoſe numerous *Villi* or *Setæ* are prickly. From among theſe are ſeen ſmall tubular Flowers. The whole Plant is of a detergent Quality, and therefore often made uſe of to cure old Ulcers.

ROCK-BUSH.

THIS Plant hath many ſtringy white Roots. The main Stalk is of a dark Green, growing often to be four Feet high, and jointed at about every ſix Inches Diſtance. The Leaves are large and ſharp-pointed, being about five Inches long, and three Inches in Breadth, and of a dark-green Colour. This Plant grows in great Plenty at *Cluff's Bay*, in *St. Lucy*'s Pariſh.

NETTLES.

THESE are of three Sorts, the White, the Red, and the Vine-nettle: The two former grow into upright Stalks, the latter a creeping Vine. The Red, whose Leaves are elegantly serrated, resemble the *Roman* Nettle, with this Difference, that the Pedicles of these, as well as the middle and transverse Ribs of the Leaves, are of a fine purplish Red. The Stalks and Leaves of each Sort are thickly cover'd with stinging hairy Down. From the Bosoms of some of the upper Leaves rise small Pedicles, supporting an horizontal Group of very small gramineous Flowers, succeeded by very many small flattish Seeds.

The WILD LAVENDER.

THIS is a shrubby perennial Plant, growing in sandy Places near the Sea; and hath in general (except that it wants its fragrant Smell) a great Resemblance to the Garden Lavender: It bears upon the Summit of its Branches many white small monopetalous Flowers, whose *Oræ* are divided into five Segments.

The ANTEGOA ROSE-TREE.

THIS grows to about fifteen Feet high. Its Branches are cloathed with many Leaves, of about two Inches long, and somewhat above an Inch broad. From among these, especially at the Extremity of the Branches, rise a great many pale-white monopetalous Flowers, whose *Oræ* are divided into five deep blunt Segments. These Flowers have a weak faint odoriferous Smell, and blow chiefly in the Months of *June* and *July*.

The SEA-SIDE SAMPHIRE.

THIS is a very succulent Plant, and differs from the *English* Samphire by the less Number of its digitated Sections, as well as by its more luxuriant Growth, being every Way larger; it grows upon the Rocks and grassy Banks near the Sea. It is generally distinguished into two Sorts, the Green and the Red. The Blossoms of each are composed of five pale-red Petals, supported by an equal Number of green capsular Leaves; the Petals surround the *Stamina*, which are of a pale Purple.

The

The WILD PENY-ROYAL.

THESE are of three Sorts: The largeſt hath a whitiſh green oblong Leaf; the ſecond of ſomewhat a deeper Green, with a leſs and rounder Leaf; the third hath likewiſe very ſmall Leaves, which, as well as the Stalks, are of a dull Purple. As the Roots, the Leaves, and Flowers of all theſe have a great Reſemblance of Peny-Royal, it is from hence that they derive their Name, though they want intirely the ſtrong Smell of that Plant.

INDIAN CALE, *or* SEVEN YEARS CABBAGE.

THIS is ſo called from the Term of Years that its Stalks will remain in the Ground, not only unperiſhed, but will yearly produce very ſmall Heads of Cabbage, beſides leſſer Sprouts; though the former are not near as cloſe, nor ſo large, as our annual Garden-Cabbage.

The WILD CLOVE.

THIS Plant hath many white fibrous Roots: The main Stalk riſes to about two Feet high, being cloathed with a great Number of narrow long ſharp-pointed Leaves. The Flowers are ſmall and white, ſucceeded by the blackiſh ſharp-pointed Pods, which exactly reſemble Cloves, from whence it derives its Name. Theſe Plants grow chiefly in wet ſwampy Land.

The ROCK BALSAM.

THIS is diſtinguiſhed into two Sorts, differing only in Colour, the White and the Red, each having thick ſucculent Leaves and Stalks. They grow chiefly upon Rocks, and old Walls; the Leaves are ſleek, blunt-pointed, and roundiſh. From the Boſom of the Leaves riſe long conic rough dentated Spikes, very much reſembling ſmall round ſteel Files; in the ſeveral *Laminæ* or Foldings of theſe are contained the Seeds. The Juice of this Plant is very mucilaginous, and looked upon as an excellent Vulnerary.

The ARROW-ROOT; *Lat.* Maranta.

THIS is a very uſeful Plant, both phyſically and otherwiſe. Its Root is long, white, jointed, and mealy: The Juice of this is exceeding cold, and, being mixed with Water, and drunk, is looked upon to be a Preſervative againſt any Poiſon of an hot Nature. Out of this Root is made likewiſe the fineſt Starch, far excelling any made with Wheat. The main Stalk riſes about two Feet and an half high, ſurrounded at unequal Diſtances with ſmooth ſhining deep-green Leaves of about ten Inches long, and five broad; these somewhat resemble the Leaves of Plantain-shot. The Flowers, which are white and monopetalous, and of the labiated Kind, stand upon long Footstalks. It is generally dug up for Use in the Month of *February*. Of this there are two Sorts, the long and the short-jointed.

The LONG-LEAV'D MISLETOE.

THESE grow in Bunches, containing seven or eight Leaves; each Leaf being sharp-pointed, five Inches long, and near half an Inch broad. They are generally to be found growing in the Clefts of the Bark of Cedar-Trees.

ANTEGOA BALSAM.

THIS is a low groveling Plant, growing very thick, and close to the Earth. The Leaves are longish and sharp-pointed. The Flowers are white, and very small. A Decoction made of this is looked upon to be an excellent Healer of all inward Bruises, as well as a good Vulnerary and Styptic in external Wounds.

The SILVER-LEAV'D PLANT, *or* PISS-A-BED.

THE Leaves of this Plant spread near the Earth like the Leaves of young Lettuce, the upper Side green, and the under of a white hoary Mealiness, and feels plushy. From the Centre of these Leaves rises a Stalk of about two Feet long, whose Top supports a white downy Flower, somewhat like a Sun-Flower, which, when ripe, is carried about by the Wind.

The BROAD POND DUCK-WEED; *Lat.* Nymphæa Indica.

THE Leaf of this exactly resembles the white Water-lily, describ'd in *Gerard*'s Herbal. Its Form is that of a Colt's-foot, green above, and whitish underneath. The Flower consists of five Leaves, and every way answers the Make of the same Species in *England*; but more especially their Roots, which are of a brownish-red Colour, and somewhat hollow within. These Roots are always fix'd in the Bottom of the Pond.

The BLACK-THORN; Pisonia.

THIS is what Sir *Hans Sloane* calls the *Fingrigo*, and under that Name he hath justly described it, excepting that the Colour of the Bark is always rather of a leaden Dark than a light Brown.

The Trunk of a full-grown Tree is about six Inches Diameter: The Tops, by the Weight of its numerous Branches, soon inclines downwards; and if supported by neighbouring Underwood, or even upon the Ground,

They will grow, trailing along the Surface of the Earth, to often above twenty Feet in Length. The main Branches are thickly cover'd with two regular Sets of leſſer Side-ones: Theſe, as well as the Prickles they are guarded with, grow always oppoſite one to another, and by their contrary Poſition cutting, as it were, the larger Branches at right Angles; by which means, look which way you will, theſe leſſer Branches make the Appearance of a Croſs. Its Leaves are very ſlender, ariſing in Tufts three, four, or five, in Number; theſe are generally much torn and eaten by Vermin. It bears upon ſhort Footſtalks Groups of very ſmall monopetalous bell-faſhion'd Flowers, whoſe *Stamina* are tipp'd with *Apices:* Theſe, which are of a dark-yellow Colour, have a faint weak Smell, and are ſucceeded by burry roundiſh Katkins, which are ſo very clammy, and full of very ſmall ſharp-hooked Prickles, that if a Bird alights upon them, they ſo entangle in its Feathers, that it will not be able to fly away.

The SNAKE-WOOD.

THO' this is but a ſlender Tree, ſeldom above ſix Inches Diameter, even near the Ground; yet it ſometimes grows to be above forty Feet high. It hath no Leaves, nor ſeldom Branches, till near the Top: There it is ſurrounded with Leaves ſimilar to thoſe of the *Popo-Tree.* The Trunk appears very knotty, if not jointed. The Inſide is hollow, and the Whole ſo light, that a weak Man may eaſily brandiſh a Piece as big as *Goliah*'s Beam, or *Hercules*'s Club. I have never known it to produce either Seed, Flowers, or Fruit. This is delineated in Plate X. Fig. 2.

The MAY-POLE; *Lat.* Aloe Americana muricata.

THIS very remarkable Tree hath a great many ſtrong ſtringy Roots. The Trunk, which is very ſtrait and tapering, is always green; its Bark, very neatly divided into ſeveral cloſe alternate Scales, or *Laminæ*, of a triangular Shape, ſharp-pointed at their Extremities. Each of theſe, as well as the Branches of the Flowers, leſſen in Bulk, as they draw near to the Summit of the Tree, which is often above thirty-five Feet high, and three Feet in Circumference near the Ground; yet this ſurpriſing Magnitude is but the Growth of three Months time. The green woody Leaves, which ſurround it at the Bottom, are many in Number, each being from three to four Feet long, about ſeven Inches broad, and three thick, ending always in a black horny Point: One of theſe Leaves often weighs ſix Pounds. The Trunk of the Tree, about twelve Feet from its Summit, thruſts out a great Number of ſtrong green ſhort Branches in an alternate Order: Theſe different Branches, with their reſpective Flowers, have been always thought (and not unjuſtly) to reſemble the Branches of the Candleſtick in the Temple of *Solomon:* Each Candleſtick or Branch ſuſtains an horizontal Group of Flowers near twenty-five Inches in Circumference; the Whole, which at a Distance seems to be but one, is composed of several Scores of single yellow Flowers, each of these standing upon a longish pod-like Stalk of a yellowish Green, terminating in six sharp-pointed thick yellow Petals. From the Centre of these rises the Pistil, which is strong and blunt-pointed; the Petals surround six *Stamina* of about two Inches long; these are tipped with large falcated *Apices*, which, when ripe, are covered with yellow farinaceous Dust. The Socket of the Flower is thickly besprinkled with a very sweet Honey-dew: This draws to it Abundance of Humming-birds and Bees. The Inside of the Trunk is a snow-white Pith, spongy and porous: When the Blossoms are dropped, their husky long Pedicles shoot out into many sharp-pointed Leaves of about two Inches long, and near as broad: These, growing close upon one another, are somewhat, at their Extremities, expanded, and form a Sucker not very unlike a Pine-sucker. When these grow ripe, they fall down to the Ground, and take Root.

In a short time after it hath produced Flowers, the Body of the Tree, which so late was tall and flourishing, falls prostrate to the Earth, and perisheth; and the very large Leaves, which surrounded it at the Bottom, in like manner wither and die. Having cut down one of these at the End, as I was informed, of three Months Growth, however surprising it may appear to some, yet it measured full nine-and-twenty Feet in Height, and very near three Feet in Circumference near the Earth; and, having carefully weighed it, its Weight amounted to two hundred and ten Pounds; so that its additional Weight each Day was far above two Pounds: Or if we consider its Height, and its Number of Days in growing, we shall find, that it grew three Inches and about three Quarters of an Inch in every four-and-twenty Hours. This very extraordinary Growth far surpasses any other Computation hitherto taken notice of, being far superior to the Remarks of the ingenious Mr. *Helmont* upon the Willow he planted, after five Years Growth of which, the Leaves, Roots, main Stalk, and Branches did not exceed five hundred Weight.

FOREST-BARK, *or* BASTARD LOCUST.

THIS grows to be a large Tree cloathed with longish green Leaves; the Bark is much sulcated, and made use of as a good Restringent.

The GUM-ELEMI TREE.

THIS grows to be a large Tree, from whose Trunk, when the Bark is wounded, flows the Gum called the *Gum Elemi*.

The SILK-GRASS; *Lat.* Aloe Barbadiensis.

THE several saponaceous green Leaves, surrounding this Tree near the Earth, and taking their Rise without any Footstalks from its Trunk, are about two Feet and an half long, seven Inches broad near the Middle, and about

about one thick, ending in a sharp Point. Their Edges are guarded with sharp crooked Prickles. The Trunk of the Tree is green, and crouded, at several regular Distances, with numerous scaly prickly *Laminæ*, which in general much resemble those of the *American* Aloes already described; except that this Tree is much less in Bulk and Height, as well as that the Spire or upper Part of the Trunk is so weak, that it bends wavingly downwards, partly by its own Weakness, and partly by the Weight of Suckers which grow upon it, especially near the Top. These are composed of several complicated Rows of green thick Leaves: From the Centre of these rise several single Flowers standing upon a pod-like Footstalks: Each of these are made up of six blunt-pointed Petals, green on the Outside, and white within. These inclose six *Stamina* tipped with *Apices* surrounding a Pistil, which swells in an angular manner near the Middle. The above-mentioned Leaves, which encompass this Tree near the Earth, are made up of very many fine longitudinal white *Laminæ* or Filaments of Thread-like Hemp, or rather as fine as the best Flax. These are easily separable from the green pithy Substance of the Leaves, and each again divisible into innumerable Divisions as fine as Hairs. With these, twisted together, are made Lashes for Whips of a very durable Nature: And I am of Opinion, that by their Stiffness the Filaments from this Plant are mixed with the Silk in making the *Indian* Silk Handkerchiefs.

The HOP-WEED.

THIS hath many white stringy Roots. The main Stalk is quadrangular, green, and hollow, and at every two or three Inches Distance seemingly jointed; from these several Joints opposite to one another rise three or four Leaves on inch-long Footstalks, in Breadth near the Pedicles above three Inches, and in Length above four. These are high-veined, and very much corrugated, or crumpled. The Stalk, at its Extremity, bears an echinated flattish Bottom, or Bur, which, when dry, hath an agreeable fragrant Smell. The Leaves of this Plant are esteemed good to be made use of as a Gargle in sore Throats. It begins to bloom in *July* and *August*.

BRINY ROOTS.

THIS scandent Shrub takes its Name from its most useful Parts, which are its Roots. Its main Trunk seldom exceeds the Bigness of one's Arm; its Branches are weak, climbing upon any neighbouring Tree. These are thickly cloathed with winged green Leaves, sharp-pointed at their Extremities, as well as at their Footstalks, their Edges being very elegantly waved or labiated: Among these, upon short Footstalks, appear the Flowers, which are composed of five Petals, snowy-white. The Roots, when bruised, afford a very offensive Smell. This, steeped in Water, and fermented, or made into a Decoction, is very purgative, and esteemed beneficial to dropsical Persons, and very useful to prevent or cure the Scurvy, and to open Obstructions.

The

The best Method of preparing it is, to scrape a Handful of the Bark from the Root, and make a Decoction of it.

The TURK'S HEAD; *Lat.* Melocactus Americana major.

THIS is called by the Antients *Melocarduus echinatus*, and hath a great many strong fibrous Roots. Its main Body, which grows above-ground, is about a Foot long, and about two Feet in Circumference. The Inside is a stringy pulpy reticulated Substance. The Outside of this is green, and chanell'd into several deep Furrows, whose Eminencies are thickly cover'd, at certain Distances, with Tufts of Prickles, of about an Inch long; the rest of the main Stem being green and smooth, ending, as it were, in a round somewhat flattish Top; from whose Centre rises another less Body or Stem of cylindrical Form, of about eight Inches long, and three in Diameter. This is thickly covered with very sharp long fine Prickles, of a Snuff-colour. The Outside of this Part of the Stem or Stalk is of a spongy soft pappous Nature, especially near the Top, from which Place rise several many-leav'd Flowers, of a Maiden's-blush. Each of these is succeeded by a small scarlet Berry, inclosing several small flat brownish Seeds.

YAMS; *Lat.* Ricophera.

THIS very serviceable Root was formerly the peculiar Growth of *Africa*, and the most common Food of its Inhabitants. In all Probability, it derives it's Name from the Verb *Yam*, to eat; which is a Word made use of, and equally understood, by most, if not all the several Nations upon the Coast of *Guiney*, though differing otherwise in their Language, Customs, and Manners. They are planted at about three Feet asunder, in small Hillocks, such as are prepar'd to plant Hop-roots in *England*; the Ground being first hol'd about six Inches deep. The Method of doing it is, by putting about three very small Yams in each Hillock, in the Months of *May* or *June*. These sprout out soon after with the first Rain, and grow into a long trailing angular Vine, bearing a very handsome heart-like Leaf. This Root or Yam much resembles, at first Sight, by its blackish stringy Skin or Tegument, the Root of a small Tree. They are dug up of different Shape, as well as Weight; the latter from twenty to half a Pound. However, the most common Weight is about two or three Pounds: These are somewhat of the same Nature as Potatoes, mealy and white within, but generally of a closer Texture than the former. They are justly esteem'd to be very hearty nourishing Food, and are generally preferr'd to Bread by the Inhabitants of the Island. When first dug out of the Ground, they are either put in Sand, or kept in a Garret, where they are well cur'd, by admitting the dry Air to them. The Method of making use of them is, either by roasting or boiling them; if the latter, they ought to be firſt peel'd. It is thought in *England*, that they muſt be firſt ſoak'd in Water before they are uſed, to draw out their ſharp biting Taſte; but this Opinion is intirely groundleſs. Great Care ought to be had in digging them from the Ground without being bruiſed, for thoſe that are will ſoon afterward decay. The beſt Method that is yet found to preſerve ſuch, is to ſprinkle the bruiſed or broken Part over with Lime. However, after all preſervative Means, ſuch muſt be us'd firſt; the others will keep fit for Uſe for near a Twelvemonth.

The PRICKLY YAM-VINE.

THIS is a ſlender Vine, covered with ſharp-pointed Heart-like Leaves, ſet on alternately. At the Root of this Vine grows a Yam, in Taſte exactly the ſame with thoſe dug up yearly in *January*. What is moſt remarkable in this is, that it bears all the Year round; and when a Yam is dug up, a ſmall Slice of the upper Part of it, in which the Roots of the Vine are fixed, muſt be ſtill left in the Ground, which will grow downwards to the Bulk of that already cut off. This Amputation may be made for ſeveral Years, and perform'd ſeveral Times in the ſame Year, the Vine and its Leaves ſtill flouriſhing.

The WILD YAM-VINE.

THIS is a quadrangular Vine bordered or ſkirted at each Angle, with a high green thin Liſt; the Leaf is umbilicated and haſtated, of a deep-green Colour.

EDDAS; Arum maximum Ægyptiacum.

THE different Species of *Eddas* are diſtinguiſhed into the blue *Eddas*, the ſcratching *Eddas*, and the roaſting *Eddas*; the Leaves of each being haſtated, and of a blunt Arrow-head Shape. That of the blue *Edda* is very large and green on the upper Side; the under is covered over with a glaucous Mealineſs. The moſt common that are planted are the roaſting *Eddas*; theſe yield a great Increaſe, and are a very nouriſhing and healthy Food, eſpecially if ſtew'd. The ſpiral gramineous ſucculent Stalks that ſupport the Edda-bloſſoms, as well as the young tender Leaves, are uſed by Negroes by way of Sallad. A Field of *Edda*, when in Bloſſom, affords, in a calm Morning or Evening, a very fragrant odoriferous Smell. Theſe are of a round conic Shape; the outſide Skin being of a dark-brown Colour, the Inſide very white, and ſomething of the Artichoke-kind. The moſt common roaſting Kind are dug up and gathered in at one Time; but ſome of the *Eddas* may be taken from the Root of the large Sort, and if the Earth is cloſed up again, the Plant continuing to thrive, will ſtill produce more.

POTATOES.

THESE very useful Roots are distinguished in this Island from one another into at least thirteen Sorts; but as this great Variety hath but very small real Difference, I shall therefore pass by these and less necessary Distinctions, and divide them into the white and red, the long and the round Sort. Each of these differs from the *English* Potato, by being propagated by a Slip or Vine, which they produce instead of upright Stalks. Another remarkable Difference is, that the *West-India* Potatoes have all a sweetish Taste; they are here look'd upon so beneficial, that there is scarce an Estate, where there is not a considerable Quantity of Land planted with them; for these with Yams and Plantain serve instead of Bread to most of the midling, and almost intirely to the poorer Sort; tho' they are not quite destitute of a kind of Bread, made with these Roots: For the Potatoes being first grated, and the Juice pressed out, the flowery or mealy Part is mix'd with Sugar and Spice, and made into Paste, which being baked in the Oven, in the Form of a Plum-cake, its Taste is far from being disagreeable; this they call *Pone.* With the express'd Liquor of either the red or the white Potato is made what we here call *Moby,* or a Sort of cool Drink, answering to small Beer in *England.* The Method of making this, is to mix the raw express'd Juice of the Potatoes with a certain Quantity of Water; this in a seasoned Vessel will soon ferment, and in about four and twenty Hours be ready for Use; it tastes cool and sharp, and it is generally esteem'd a healthy Liquor. The Juice likewise of Potatoes, if fermented, will, by Distillation, yield good Spirit. The Vine producing each Sort is long, and trailing close to the Earth, taking Roots with its numerous Joints in wet Weather; these burrowing into the Ground bear a great Number of Potatoes: Tho' the Leaves upon these different Vines vary somewhat in Shape, yet in general they are all scollop'd, and bear bell-fashion'd monopetalous Flowers, whitish without, and of a deep Purple within, each Flower being slightly segmented about the Edges. These are succeeded by small capsular Seed-vessels, inclosing several blackish small Seeds.

The WHITE-LILY; *Lat.* Pancratium Americanum.

THE main Root of the white, as well as all other Lilies, are bulbous and round like an Onion, fastened to the Earth by several small white fibrous Strings or lesser Roots, the former being squamose, or compos'd of several Coats one over another. The Leaves are many and sharp-pointed, being about ten Inches long, and near three in breadth. From the middle of these rises a green flattish hollow Stalk, this near the Top is surrounded like the Ribs of an Umbrella, with six four-inch-long Stalks, the Flowers consisting of a double Border of five Snow-white Petals four Inches long, and about a quarter of an Inch broad, bending downwards in a very beautiful

beautiful Manner; above these, from the Centre of the same Parent-stalk, rise six small Spikes in a circular Manner; these near their Tops are green, each, as well as one other *Stamen*, which rises from the middle; these are tipp'd with falcated brownish *Apices*. The Bottom of these Spikes or Ribs are white, join'd together within half an Inch to the Footstalk with a very fine white Membrane, forming an agreeable pyramidical Flower, out of whose *Discus* rises the above-mentioned *Stamen*. The bulbous Root of this, as well as the different other Species of Lilies, here are made use of by way of Decoction, and look'd upon as a good diuretic Drench for Horses.

The Red-Lily.

AS the pale red, and the red and white Lily, differ from this only in Colour, I shall include them under the same Description. Their Roots are of the same Make and Texture as the white Lily already describ'd. The Stalk of the red Sort, *&c.* is about eighteen Inches long. The Flower is compos'd of six Leaves, somewhat sharp-pointed at their Extremities, as well as near the Stalk; their Outsides are almost intirely red, as well as the Inside, except near the Socket. There their Colour alters from a flaming Red to a greenish White; out of the *Discus* rise six purplish *Stamina* tipp'd with yellowish *Apices*.

The Wild-Lily.

THIS hath a bulbous scaly Root. The Leaves, which are thick, green and smooth, are of about ten Inches long and sharp-pointed, set on round the Stalk *squamatim*; from the middle of these the Stalk extends higher than the Leaves, ending in a russet Spike full of small Seeds.

The Wild-Tulip.

THIS hath a bulbous scaly Root, from which rises a green upright hollow Stalk of about fifteen Inches high, supporting, upon separate Pedicles, several large beautiful Flowers, compos'd of six Petals; each Petal about three Inches long, somewhat sharp-pointed at both Ends. The middle Part of every Flower-leaf is of a fine deep Red and White; on each Side these surround six white *Stamina* tipp'd with *Apices*. They are in Bloom every Evening about five a Clock, and likewise in the Morning till about Eight.

The Fit-Weed.

THIS is a small Plant, seldom rising above six Inches high. The main Stalk, as well as the Side ones, are jointed; at each Joint they are surrounded with several small Leaves, each deeply segmented, dividing the

the Extremity of the Leaf into three ſharp prickly Points: From among theſe Leaves riſes a ſmall conic ſquamoſe Spike not ill reſembling a Pine in Miniature, in whoſe different *Laminæ* are contained the Seed: A Decoction from this Plant is looked upon to be very efficacious to cure Fits ſo often incident to young Children.

Logerhead Weed.

THIS ſmall Plant hath a great many fibrous whitiſh Roots, the main Stalk, which is hollow, ſeldom riſes above ſix Inches high, this is of a purpliſh Colour; and from it riſe three Pair of Footſtalks or Pedicles oppoſite to one another, each ſuſtaining on its Extremity four ſharp-pointed narrow Leaves deeply chanelled in the middle, as well as the tranſverſe Ribs. From among the upper Leaves riſes a ſhort ſlender Spike decorated with a ſmall whitiſh red tubular Flower divided at the Top into five Parts; theſe incloſe a great many yellowiſh *Stamina*: A handful of the Leaves boiled in a Gallon of Water till reduced to half that Quantity, and that Decoction being made palatable with Molaſſes and a ſmall Quantity of Lime-juice, is very efficacious to bring away and deſtroy the Worms in Children. The common Method of uſing this Decoction is to give the Patient (if a Child of about ſeven or eight Years old) about the Quantity of two Spoonfulls cold, upon an empty Stomach at Night, by the next Morning it will have its deſired Effect; if not, let the Doſe be repeated for three Nights ſucceſſively. It muſt be uſed whilſt freſh, otherwiſe by its great narcotic and ſtupifying Qualities it becomes very dangerous.

The Pine Apple; *Lat.* Anana.

IF the general, or at leaſt if the Judgment of the moſt numerous Part of Mankind, who have taſted of this Fruit, may be relied upon, it deſerves the Preference of all other Fruits; the agreeable Variety, and the delicate quick Poignancy of its Juice is juſtly eſteemed to excel every other. The Roots of this Plant are many, ſpreading in a circular manner. From the Centre riſes a hard ſtrong Stalk ſurrounded near the Earth, and for a conſiderable Way up the Stalk with long green Leaves, whoſe Edges are finely ſerrated. Theſe Leaves are ſet on alternately; the Top of the Stalk ſuſtains the Fruit which is called the *Pine*, from ſome Reſemblance it hath on the Outſide to the Cone the Pine-tree bears. The Top of the Fruit is beautifully decorated with a *Corona* of fine green ſharp-pointed Leaves, whoſe Edges, as thoſe below on the Stalk, are finely ſerrated. When the Fruit is young, of about four Inches long, it is ſurrounded, eſpecially near the Top, with fine blue Flowers, each conſiſting of one Leaf, which is divided into three Parts, and is funnel-ſhaped. The Embryos are produced in the Tubercles: Theſe afterwards become a fleſhy Fruit full of Juice. The Seeds, which are lodged in theſe Tubercles, are very ſmall, and almoſt kidney-ſhaped. Some time before the Fruit is ripe, two, three, or four Suckers grow from the

Pl. 21 Page 281
To the Most Noble Prince Charles,
DUKE OF RICHMOND, LENOX AND AUBIGNY.
Knight of the Most Noble Order of the Garter,
and MASTER of the HORSE to his MAJESTY.
G. D. Ehret delin. et sculp.

Stalk below, yet close to the Fruit: These Shoots are taken off, and planted; and will in about fourteen Months time produce a ripe Pine. Those who cannot procure these Suckers, sometimes plant the Top or *Corona*. This, though intended by Nature chiefly as an Ornament to the Fruit, yet will grow, and in time bear a Fruit, not so soon, nor so good, as that produced by those Suckers, which Nature intended to be the Means of propagating this Fruit. The three best Sorts of Pines are the *Surinam*, the Sugar-loaf, and Queen-Pine. The Fruit ought to be eaten soon after it is cut, nor ought it to be kept upon the Stalk in the Garden, as the Custom too often is, till it be very soft on the Outside.

This is delineated in Plate XXI.

The WILD PINE.

THIS differs in Taste from the Pine already described, as much as a Crab-apple doth from the best Russet or Golden-pippin. Its Leaves are very large, in proportion to the Bulk of the Fruit, which is but small.

The LARGE WILD BARREN PINE; *Lat.* Caraguata.

THIS, in Propriety of Language, ought to be looked upon as an aquatic Plant, though suspended in the Air among the Branches of lofty Trees, to whose Boughs it is fastened by its numerous Roots, which serve not to suck, or draw from them any nutricious Juices to further its Growth, as the *Misletoe* doth from the Orange-tree, *&c.* but only to be its Supporter; provident Nature having in a very extraordinary Manner supplied this with other Means to preserve its Species; for the Leaves, which much resemble those of Pine, but only larger, surround this Plant in a circular manner, each Leaf near the Stalk terminating in an hollow Bucket, which contains about half a Pint of Water. It is by these numerous small Reservoirs of Water that the Roots, as well as every other Part of this Plant, are supplied with Nourishment without the Help of any Earth. The flourishing Condition of this, as well as the great Growth of Fig-trees upon barren Rocks, shews that Water is of greater Use to Vegetation than Earth.

The BUTTON-PINE.

THIS chiefly differs from the common Pines by its Smalness, as well as by its sour Taste: Therefore it is never cultivated. Its outward Coat is likewise redder, when ripe, than any other.

The SMALL BARREN PINE.

THE Leaves of this very much resemble a Pine. From the Middle of these, instead of a Pine-apple, grows a woody Stalk rising to about three Feet high, and divided into many Branches. These are almost intirely covered with ſmall red ſharp-pointed Berries, each guarded at the Stalk with a fine ſharp Prickle, and at the Top with two ſomewhat leſs.

The PEN-GWYN; *Lat.* Karatas.

THE Word *Pen-gwyn* is evidently a *Celtic* Word, compounded of *Pen* an Head, and *Gwyn* white; but how this Fruit came originally to have a *Celtic* Name, is foreign to my preſent Purpoſe to inquire into. The Outſide of this Plant is compoſed of ſome Scores of hard ſtiff green Leaves growing to about nine Feet high, and two Inches in Breadth, having their Sides or Edges guarded, at every Inch and an half Diſtance, with ſharp-hooked Prickles. Theſe Leaves turn very ſcoopingly inward on the upper Side, by which means they ſave and convey the Dew, and the Rain, that fall upon them, to the Roots. They grow likewiſe almoſt impenetrably thick near the Earth, ſurrounding and guarding a circular *Corona*, or Bottom, of about a Foot Diameter. From this grows a Cluſter of Fruits, each of about four Inches long, and one in Breadth, both Ends being ſharp-pointed, and the Middle of a quadrangular Form; by which means they are ſo cloſely joined, that they cannot well, until very ripe, be taken aſunder. The outward Covering of this Fruit is a ſmooth whitiſh yellow gramineous Huſk; this covers and peels off from a white pulpy Subſtance, wherein are innumerable ſmall flattiſh black Seeds. This, being the eatable Part, hath ſome ſmall Reſemblance, in its Flavour, of the Pine; and is looked upon to be cooling and wholſome. If any of theſe, when near ripe, are gnawed by Rats or other Vermin, the wounded Part will emit Drops of the moſt tranſparent Gum. This Coagulation ſhews, that its Juices are much impregnated with volatile Oil. Dr. *Towne* very juſtly recommends the Uſe of this Fruit in Fevers, provided it be uſed very moderately; for by its grateful and active Sharpneſs it is capable of penetrating through the moſt tough and tenacious Scurf, by that means uncovering the Orifices of the Salival Ducts, and enabling the Glands of the Mouth and Throat to diſcharge the Contents, which could not be done before the Impediment was removed.

The GROUND-NUT; *Lat.* Arachidna.

THIS differs very little from that called in *England* by the ſame Name. The Nut, which is the Root of the Graſs, lies three Inches deep in the Earth. This is no bigger than a black Cherry covered over with a ruſſet Skin, or tender Bark, the Inſide being as ſolid as the Kernel of a Haſel-nut, and well-taſted.

The

Pla. 22. pa.
TEMPLA QUAM DILECTA
To the Right Hon.ble Ld
Viscount COBHAM,
This Plate is humbly Inscribid.
&c.
J. Mynde sc.

The GINGER; *Lat.* ZINZIBER.

GINGER is rather of the Reed-kind, than of the *Iris*, as *Morrison* and others would have it. Its Stalk seldom exceeds eighteen Inches in Height, from whose Side grow, in an alternate Order, four or five narrow sharp-pointed gramineous Leaves, of about five Inches long, the Extremity of the Stalk ending in a soft-pointed Spire. When the Plant is dug up, its Roots are those flattish digitated Races called *Ginger*: These Races are scraped clean, and Sun-dried. It would be needless to describe the Qualities of a Root so well known: I shall therefore conclude the Description of it with observing, that the usual time of planting it is in *May* and *June*; and of digging it up, in *February* and *March*. What is most remarkable in this Plant is, that the small or Seed-Ginger, when planted, doth not decay in the Ground, as almost all other Seeds or Plants do: For though it produces the Plant, and the several Races at its Root, the first Year; yet it remains itself uncorruptible, and may the next Year be planted again as a Mother-plant, and so on the succeeding Year. However, after such a time, it becomes so far exhausted of its prolific Virtue, that it bears but a poor Crop: Therefore the most general Custom is to plant yearly with fresh Suckers, or the smallest Races of Ginger. The Land intended for this Plant must be very richly manured.

The SEA-SIDE LAUREL.

THIS beautiful Shrub grows generally near the Sea-side, cloathed with numerous Leaves. The Edges of these are remarkably indented. This is delineated in Plate XXII.

The LARGE DUCK-WEED.

THIS always grows in Ponds of standing Water, and generally covers their Surface with its broad Leaves, and much resembles in Shape and Texture those of the Water-Lily in *England*. Their upper Side is of a smooth shining yellowish Green, the under Side of a very dark Purple: Their Veins and high Ribs are tinged with Yellow; and their Footstalks are round, smooth, and of the same Colour. The Inside of this is perforated into many longitudinal Pipes, and its Length is in proportion to the Depth of the Pond; for it always grows till its Roots penetrate the Mud, or other Sediment, in the Bottom.

The Flower likewise, which is similar to those of the same Kind in *England*, always opens at Break of Day, and closes as soon as the Sun appears.

The

The SMALL DUCK-WEED.

THIS hath many long ſtringy moſſy green Roots. The Plant, ſpreading itſelf upon the Surface of the Water, is compoſed of ſeveral round-pointed Inch-long Leaves, of a greeniſh-white Colour, and regularly veined towards their Extremities. The Stalk, as well as above the firſt half of the Leaf joining to it, is bagged or ſtuffed in the Inſide of near one fifth Part of an Inch thick: This, when cut through, is full of ſmall longitudinal Veins, or open *Tubuli*. From that Part to the Extremity the Leaf aſſumes its thin proper Texture.

CAPIL-

CAPILLARY AND PARASITICAL PLANTS.

The AGNUS SCYTHICUS.

WE are now come to treat of a Plant of the parasitical Class, tho' dignified with the Name and Quality of an Animal, which (if we believe many Writers of Wonders) hath not only the Shape of a Lamb, and is woolly, but likewise feeds upon the adjacent Plants. To increase the Wonder, it is said, that, if these are removed, the Lamb soon after dies.

This, and many other surprising Qualities, are attributed to this supposed Animal, which the *Tartars* strongly affirm to exist; yet conceal the Place where it is found. However chimerical this may appear, and in reality is, yet the otherwise judicious *Kempfer* was so credulous, as to spend a great Part of his Life in Search after it; tho' a moderate Attention to the Motives of his Credibility would have spared him his laborious Search: For there is not a more suspicious, or even a surer Sign of Craft, or Cunning, and of an intended Imposition upon the Credulity of Mankind, than a sedulous Endeavour to keep any pretended Discovery of this Kind from the Inspection of the Public. What byassed human Nature to embrace low Art and Cunning, in Exchange for that true Wisdom, and best of Policy, undisguised Truth, was, and still continues to be, that groveling and sordid, yet almost universal Passion, the Love of Gain. In the following Instance we shall find, this Vice is not only the Attendant of luxurious *Asia*, or the effeminate Part of *Europe*, but that it reigns even among the rude unpolished *Tartars*, whom Luxury, one would have imagined, had not made Slaves to its unnecessary Wants. However, we find that these, like the *Ephesian* Coppersmith, are artful enough to secure their own Gain; for, as Furs are in great Esteem, and in high Value, among the Nobility of *Tartary*, and the neighbouring *Turks*, the near Resemblance of the Down growing upon this Plant, to the Wool of a very young Lamb, afforded the cunning *Tartars* room to impose upon the World, and gave Rise to that barbarous Piece of Cruelty of privately ripping up the Ewes, as soon as the Lambs had attained to the Maturity of having any Wool upon them; the Skins of which, being then delicately ſoft, theſe cunning Dealers ſubſtitute in the room of, and call them, the Skins of this ſuppoſed *Scythian* Lamb.

But, to return to my Subject; to prove that moſt of, if not all, the pretended Qualities attributed to this Lamb, how ſpeciouſly ſoever they are applied to it, as a ſtrange Mixture of a real animal Exiſtence, and vegetable Appearance, are really and truly the inſeparable Properties of this, as well as many other paraſitical Plants. However it muſt be owned, that this is the moſt ſurpriſing of any of this Claſs, and is as rare as it is curious.

As to the Proof of its ſuppoſed animal Life from its dying, as they term it, if the adjacent Vegetables are taken away; this is literally true of its Decay as a Vegetable, as all other paraſitical Plants will do, when deprived of that foſtering Nouriſhment, which they draw from the Trees they prey upon. For Inſtance, if a Branch of an Oak, which hath the Miſletoe growing upon it, decays, the latter, which before lived and flouriſhed by the Juices it received from the Oak, muſt, when this becomes dry and arid, decay likewiſe with the decaying Branch. In the ſame manner, if the Tree, or even the Branch of it, from which the *Agnus Scythicus* receives its Nouriſhment, is cut down, it muſt alſo periſh with it.

The Body of this Plant is about the Thickneſs of one's Finger, and many Feet in Length; the Inſide of a white cloſe Subſtance, ſomewhat juicy, and eaſily cut thro' with a Knife. The outſide Rind is tough, and deeply covered with a fine ſilky Down, ſomewhat reſembling Wool; and the Body of the Plant is here-and-there irregularly geniculated, having, among ſo many Bendings, ſome, which diſtantly reſemble, in Make and Bigneſs, the Legs of a young Lamb; eſpecially as it is covered over with a fine ſilky Down, or Hair, ſomewhat curled. This accidental Likeneſs of Shape and Covering it was, that gave Riſe to the Notion of its being a Lamb; and lucrative Reaſons may have help'd to keep it up. The Body of this Plant wreaths itſelf in ſeveral looſe irregular Foldings about the Body of a Tree, and with its ſeveral ſlender ligneous Roots penetrates thro' the Bark, and from thence ſucks thoſe Juices, which ought to nouriſh the Tree. From the Extremity of the Stalk riſes a Footſtalk of about twenty Inches long, ſupporting a ſingle cylindrical Leaf, very near of the ſame Length. The Back of the Leaf, on each ſide the middle Rib, is ſlightly and regularly pitted into a great Number of depreſſed Spots: In theſe are ſeen a great many very ſmall yellow Seeds, which, when ripe, are carried off by the Wind; and if they fall upon any neighbouring Tree, as ſome, among ſuch a Number, unavoidably muſt, they then, eſpecially if the Bark of it proves ſulcated, take Root, and are thus propagated.

The

The FERN-LIKE PLANT.

THIS grows to about fifteen Inches high, and is by far the moſt beautiful of the Fern-kind; its many Side-branches jutting out alternately in a very elegant Manner. I found this Plant growing in the Eſtate of Mr. *Strahan*, in *St. Andrew*'s Pariſh.

The MOUNTAIN-FERN.

THIS grows to often twelve or fifteen Feet high, very much, in its Shape and Texture, reſembling a Fern; its Root cover'd with a reddiſh Down. I found this near the Eſtate of *Benjamin Mellowes*, Eſq; to whoſe Skill, and communicative Temper, I owe the Diſcovery of many Plants.

The BLACK MAIDEN-HAIR.

I Found this beautiful Plant growing upon the Wall of *St. Lucy*'s Church.

MAIDEN-HAIR.

THIS is generally found growing on the Side of ſhady Cliffs, Sides of Wells, and other old Walls: It is thought to be a good Pectoral, as moſt capillary Plants are; and therefore it is gathered, and, with Sugar, boiled into Syrup, and made uſe of.

GRA-

GRASSES, REEDS,

AND

GRAMINEOUS PLANTS.

Of GRASSES.

IT is no small Instance of *God*'s Wisdom and Goodness, that the Face of the Earth is covered with Grass, for the Use of Cattle; and that its Verdure should, above all other Colours, be most grateful to the human Eye. The Variety likewise that is observed in their different Species, affords infinite Use and Amusement. Reeds and Rushes cover with a beautiful Green the otherwise disagreeable Surface of a miry wet Soil, and, by their numerous Roots, prevent its being washed off by Floods. It is in these, that several Kinds of Birds, such as Coots, make their Nests; and among these likewise they find their Food, and are, whilst feeding, shelter'd from the Sight of their Enemies. If we ascend up the Hills and Mountains, there the Carpet-Downs open to our View, and, with their wide Extent and Verdure, yield an innocent Delight; and if the steep Sides of Hills were not cloathed with Grass, the narrow intermediate Valleys would be scorched up with the reflecting Rays from their opposite stony Surface.

The STAR-GRASS.

THIS is the most remarkable of any of the Grass-kind. Its Roots are many and fibrous. The several Stalks join there together, as in a Bunch of Scallions. The main Stalk rises about fourteen Inches high. This is small, round, and solid; surrounded, within two Inches to the Ground, with several slender narrow gramineous Leaves. From thence upward the Stalk rises in an upright Spire, which, at the Top, spreads into seven other grassy Leaves, standing almost horizontally round a Tuft of small whitish Seed-vessels, containing a great many small black Seeds. These Leaves are remarkably stained, or, as it were, enamel'd on the Inside with a clear White for near an Inch long; the Remainder of the Leaf is perfectly Green. They are in their Bloom in *June* and *July*; but this white Part is not a transitory Embellishment, but is as permanent as the Leaf.

DUTCH

DUTCH GRASS; Gramen dactylon procumbens.

THIS hath but few Roots at its Appearance from the Ground. It is divided into many low Branches, each thickly cloath'd with narrow sharp-pointed Leaves, each main Stalk ending in two or three angular almost horizontal Spears, bearded on one Side with sharp-pointed small capsular Seed-vessels. This Grass is much coveted by Cattle of every Kind.

POND-GRASS, *or* CANKER-WEED.

THIS is a succulent jointed Grass, with sharp-pointed gramineous Leaves. The main Stalk is jointed at every four or five Inches, and as it creeps along the Ground, there sprout, from each Joint, two white strong Roots, which soon penetrate into the Earth; and by this means it is too successfully propagated, to the great Prejudice of the Planter. It is said to be of so corrosive a Nature, that a Cataplasm of this bruised Plant, steeped in Urine, will eat down any Malander, or such-like fungous Excrescences. It is further supposed, or rather believed, that, if the Juice be given to a breeding Sow, it will make her absolutely barren. This Plant grows best in wet marshy Land.

SAVANNAH-GRASS.

THIS is a long creeping Grass, full of Joints. From each Joint rises a grassy sharp-pointed Leaf, and, as it creeps along the Ground, it shoots Roots from each Joint, and so propagates. It grows to a great Length.

RICE-GRASS.

THIS Grass grows to about two Feet high. Its Roots are many and white. The main Stalk is somewhat flat, sending out from its Sides several Leaves alternately: These are grassy, and about five Inches long. The main Stalk ends in two opposite Leaves. From the Centre of these rises a two-inch-long Spear, on whose Side alternately rise several triangular Races, composed of several Seeds. This is much coveted by Cattle of every Kind.

The PLUSH-GRASS.

THIS Grass seldom grows above a Foot high, decorated near the Root with many narrow spoon-like blunt-pointed Leaves; the Spear continuing naked from these Leaves for several Inches; but decorated at the Top with several small Pannicles, set round the Stalk circularly, ending in an umbilical Form; each Division of this Pannicle

cle being bearded with a fine white Down, at whose Root lie the small capsular Seed-vessels. The green Leaves near the Ground are slightly cover'd with a very soft light-colour'd Pile.

FLAG-GRASS.

THIS hath but few Roots. The main Stalk is jointed near the Ground at every two or three Inches asunder. The grassy Side-blades are sharp-pointed, and near eight Inches long. As these grow no higher round the Stalk than about eighteen Inches, the remaining Part of the Stalk, which is generally two Feet more in Height, continues unjointed; its Extremity ending in a five-inch-long white Pannicle, whose numerous small capsular Vessels are thickly studded with very fine white Down.

SCOTCH GRASS; *Lat.* Gramen panicum.

THIS very useful Grass grows in swampy wet Places. Its Roots are fibrous and many. The main Stalk grows in Joints, at three or four Inches asunder, rising often to five Feet in Height. Its Side-leaves are many and grassy. Its quick Growth, and always thriving in such wet Places, makes it very valuable in the dry Time of the Year, when other green Fodder is scarce. Cattle of every Kind prefer this and Rice-Grass to any other. The Method of producing it is to cut it in small Pieces, leaving always a Joint to every Piece. These are stuck in, at eight Inches distant, round the Sides of Ponds, or other wet marshy Land; and will not only soon grow up themselves, but propagate others, and consequently grow prodigious thick. This may be several times cut down, and another renewed Crop be in a few Weeks reaped again.

DWRAH, *or* AMMADWRAH.

THIS is of the Rush-kind. Its Root, which hath a very strong Smell, is often steeped in Water, and mingled with the Juice of the Musk-bush. With this the *Coramantee* Negroes anoint their Skin by way of Perfume, especially when they are to go to their Merry-meetings, or public Dances.

POND-WEED.

THE Roots of this Plant are many, white, and stringy. The main Stalk is as thick as one's little Finger, strong and hairy, growing to about two Feet high. Its Leaves, which are set on alternately upon short Footstalks, are near four Inches long, and scarce one broad, sharp-pointed, and their Edges irregularly serrated. The middle Rib, as well as the Side-ones, are very regular and prominent, which makes the opposite Side of the Leaf deeply furrowed.

Dog's-Grass.

THIS derives its Name from common Obſervation, that Dogs, when ſick, often eat of this Graſs, which ſoon after clears their Stomachs by its emetic Quality. This Graſs is ſo well deſcribed by Mr. *Miller*, in his *Botanicum Officinale*, that it would ſhew more Impertinence than Judgment to give it the Reader in any other Words. " Dog's-Graſs hath " many long ſlender creeping Roots, white and jointed, ſpreading much " in the Earth, with ſmall Fibres at every Joint; from which ariſe ſeveral " tall Stalks, not ſo thick as the Stalk of Wheat, having two or three " Joints, and as many long ſomewhat broad Leaves, one at each Knot or " Joint: On the Top of each Stalk grows one long ſpiked Head, in Shape " like an Ear of Wheat, but ſomewhat flatter, conſiſting of two Rows of " chaffy Glumes." It grows in moſt Parts of the Iſland: And a Decoction of it is looked upon to be good againſt the Gravel.

The Wild-Dwrah.

THIS hath an hard bulbous Root, its graſſy Stalk, which is green and triangular, ſeldom riſing above ſeventeen Inches high. It regularly branches near the Top into a Star-faſhion, generally into three or four graſſy Leaves; from the Centre of theſe riſe two or three Spikes, cover'd at the Top with a coarſe Flag, like that of a Reed.

The Nut-Grass.

THIS is of two Sorts, the one intirely propagating its Species by its Seed, the other by its numerous Nuts or Roots. The former is leſs prejudicial to the Planter, tho', both by its very quick Growth and Increaſe, it is of great Diſſervice, by preventing, or, at leaſt, by ſucking up, the Nouriſhment of the Manure from the planted Canes, or Corn, as well as, by its thick Growth, choaking them with their numerous wreathing Roots. This was firſt brought here in a Pot of Flowers ſent from *England*, to Mr. *Lillington* in *St. Thomas*'s Pariſh: From thence it hath been more or leſs unluckily propagated throughout the whole Iſland.

Red-Flag, *or* Plush-Grass.

THIS grows to the Height of the laſt-deſcrib'd, and every way reſembles it; except that its Pannicle is longer, and of a deep-red Colour, ſurrounding the Extremity of the Stalk in a circular Manner; each capſular Seed-veſſel guarded with half an Inch long ſoft-pointed Briſtle. Neither this nor the former is eaten by Cattle, unleſs in great Want.

The Ginger-Grass.

THIS is called Ginger-Graſs, from the great Reſemblance there is between the Leaves of this, and the Flags, or ſpire-like Leaves, of Ginger.

WYTHS

WYTHS and VINES.

The BLACK BASKET-WYTH.

THIS is perhaps the strongest of all other Wyths, or of any Vegetable of equal Bulk: It is of a blackish Colour, seldom growing thicker than one's Finger. It derives its Name from its Use in making Baskets; for which Purpose its great Strength renders it very serviceable. If these grew in *Palæstine*, we may be allowed, at least, to conjecture, that they were the same with which *Samson* was bound; for we know of no other Wyths of sufficient Strength to be worthy of an Experiment of that Nature.

The COW-GUT-WYTH.

THIS is a scandent Vine, bearing Yam-like Leaves; the Flowers are of the Bell-fashion, and yellow. I never could observe, that the Flowers were succeeded by either Pods, or Berries. The Wyth itself is very strong and pliable; therefore made use of to tie the Blade of Canes, wherewith *Negroes* thatch their Houses.

The HONEY-WYTH.

I Have given this Wyth the above Name, from its sweet fragrant Smell, or rather from the great Resort of Bees to its Flower.

The PUDDING-WYTH.

THIS much resembles, in its Leaves, those of Yams: They are chiefly made use of as Bandages to tie Bundles of Straw or Rushes, with which the poorer Sort of People thatch their Houses,

The Wild YAM-VINE.

THIS hath its Name from the great Resemblance its Leaves have to Yam-vine-leaves. As it is tough and ligneous, it is made use of to make Baskets.

The WILD VINE.

THIS is of two Sorts, each bearing a monopetalous Flower, expanding horizontally; one of a pale White, the other equally chequered with a sky-coloured Blue: These are small trailing Vines, creeping over shrubby Trees, having heart-like Leaves.

The HOG-VINE; *Lat.* Convolvulus.

THIS is a creeping Vine, with a green hairy Stalk. From each Side of this Stalk rise a great many Pedicles, of above an Inch and an half long; supporting, on their respective Summits, four or five sharp-pointed Leaves, about an Inch and an half long, and an Inch broad. These spread horizontally regular round the Tops of the Footstalks. The Flowers, in general, are four in Number, yellow, and of the Bell-fashion, whose *Oræ* are much expanded: These are succeeded by as many brownish *Capsulæ*, each inclosing small Seeds.

The SWEET VINE.

THIS Vine, at some Distance from it, perfumes the Air with a very fragrant Smell; but the nearer you come to it, the less agreeable, it being then faint, if not offensive. The Stalk affords a great Quantity of milky clammy Juice. The Leaves are small and blunt-pointed, and the Flowers small and white.

The SEA-SIDE VINE; *Lat.* Convolvulus marinus.

THIS is a luxuriant trailing Vine, creeping along the sandy Banks near the Sea-side. Its Stalk is long and succulent, cloathed with middle-sized smooth shining-green Leaves. Intermixt with these appear several large bell-mouthed Flowers, of a pale White without, and a deep Purple within. As it grows chiefly upon the Sand, its continued Verdure is very agreeable, and covers it; and by so doing prevents the Reflexion of the Sun from the hot Sand, which would otherwise be very prejudicial to the Eyes of those Travellers, whose Business requires them to use such Roads.

The SILVER VINE.

THIS is a weak creeping Vine, valued chiefly for being very good to feed Horses with. It derives its Name from a shining whitish Hoariness upon the upper Side of its Leaves. Its Roots are many and fibrous. It grows chiefly in Gullies, and other such shady Places.

The WILD PURSLAIN.

THIS is of two Sorts, the Red and White, each being a creeping small-jointed Plant: They emit from each Joint many small fibrous Roots; the small upright Branches bearing narrow green Leaves of a long oval Shape. Each Sort bears a scaly conic clover-like Flower, differing only in their respective Colours, which are red and white: The white Sort, boiled to a Decoction, makes a good Gargle for sore Mouths. It grows chiefly by the Sea-side.

The POISON-WYTH.

THIS grows to a great Length, and is, tho woody, yet of a very ſpongy Nature. The Leaves are about two Inches long, and at the Stem near an Inch and a Quarter broad. The Root, when pounded, and applied by way of Cataplaſm, is of a very drawing Nature; and therefore it is often uſed to ripen Boils and Tumours. Its Flowers are ſucceeded by Berries, black when ripe.

The RIGHT-WYTH, *or the* CATS-CLAW-WYTH.

THIS is a ligneous ſtrong Vine, creeping upon the neighbouring Trees or Rocks to a very great Height. The Leaves on the upper Side are of a dark-green, ſharp-pointed, ſtanding in Pairs upon one common inch-long Footſtalk; each Pair oppoſite to one another on the leſſer Stalks. From the Boſom of the Leaves riſe numerous ſmall Claſpers, or Tendrils: Theſe not only cling round any neighbouring Supporter, as the Grape-Vine does; but, at the Extremity of each Claſper, it is fenced or provided with a forked Claw like a Fiſh-hook, but ſomewhat more lax in the Bending. This ends in a ſharp prickly Hook or Talon, by which it can not only take hold of, but pierce into, the tender Bark of Trees, as well as faſten to the Cavities of Rocks. It bears a yellow Flower, but never viſibly ſucceeded by Pods, or Berries. Its grateful Verdure, and being thickly covered with Leaves, affords an agreeable Shade all the Year.

The RIGHT-WYTH.

THIS ſcandent Wyth is an Evergreen, bearing upon its Stalks ſeveral deep-green Leaves, ſet on in a Triparture-order, iſſuing from the ſeveral Joint, which are many in Number. Theſe Wyths, when full-grown, are, near the Ground, as big as one's Arm. Their Branches have many Claſpers or Tendrils, by which they climb up any neighbouring Tree, growing ſo thick, that it makes a very dark, cool, and agreeable Shade.

The SUGAR-CANE; *Lat.* Arundo Saccharifera.

AS it would be more curious than requiſite, to examine the ſeveral controverſial Opinions, whether Canes were originally the Growth of the *Eaſt* or *Weſt Indies*; I ſhall therefore proceed to obſerve, that in the Manner of their Growth, Form of their Flags or Leaves, and Make of their * Pannicle, they reſemble the Reeds which grow in wet marſhy

* There are but few Canes, eſpecially if they grow in a deep Soil, that ſhoot out into an Arrow decorated at the Top with a Pannicle; and thoſe that do, grow generally in a ſhallow Soil; tho' the Glumes of their Pannicles contain a whitiſh Duſt, or rather Seed: Yet theſe, being ſowed, never vegetate.

The moſt natural, and perhaps the only proper, Method of producing Canes is by Suckers, or, as Experience ſhews, with the tender Tops of old Canes: Theſe being cut into Pieces of about a Foot long, and planted in Holes of about ſix Inches deep, and two Feet wide, and covered with good Manure, each Piece will produce from its Roots a great Number of Canes.

1 Grounds

Grounds in *England* or elsewhere; however, with this general Difference, that the Sugar-Canes are every Way far larger; and the Inside, instead of being hollow, is full of white Pith, containing a very sweet Liquid. The intermediate Distance between each Joint of a Cane is of different Lengths, according to the Nature of the Soil, Richness of the Manure, and seasonable Weather during their Growth; but in general from one to four Inches long, and from half an Inch to an Inch Diameter, seldom more. The Length of the whole Cane likewise depends upon the above Circumstances. It generally grows to Perfection in about fourteen Months; its then Height (the top Flag-part excluded) is from three and an half to seven Feet, a Medium between both being the most common Length, even in a very good Soil, and seasonable Years. The Body of the Cane is strong, but brittle, of a fine Straw-colour, inclinable to a Yellow. The Extremity of each, for a considerable Length, is cloathed with many long reed-like Leaves, or Blades, whose Edges are very finely and sharply serrated: And the middle longitudinal Rib in each is high and prominent.

F. *Labat*, in his History, says, that there were Canes in the Island of *Tobago*, of twenty-four Feet in Length: If he meant this in general, his Assertion is a strong Specimen of that Vanity, to say no worse, which influences many Writers to be fond of Relations of the marvelous Kind. But whoever judges of the Length of Sugar-Canes, in general, from these Instances, if there were any such, may as reasonably conclude from the Height of one *Goliah*, that the *Philistines* were in general of a gigantic Stature.

Whatsoever Difference some Soils, and very seasonable Weather, may occasion in the Growth of this Plant; yet in this all Writers agree, that it is (unhappily for the Planter) liable to one Disorder hitherto incurable, that is, the Yellow Blast.

This, among Diseases peculiar to Canes, as the Plague among those which happen to Men, too justly claims the horrible Precedence.

And as the Ingenious in this Part of the World have not as yet agreed in their Opinions about the Cause of this destructive Blast, I may without any Apology (I hope) offer my own; *i. e.* That it proceeds from Swarms of little Insects, at first invisible to the naked Eye; and as the Juice of the Cane is their proper Food, they, in Search of it, wound the tender Blades of the Cane, and consequently destroy the Vessels. Hence the Circulation being impeded, the Growth of the Plant is checked; and soon after it withers, decays, or dies, in proportion to their Degree of Ravage.

From this Supposition we may easily account for the various Phænomena, which attend the Blast, whether in its first Appearance, or its further Progress. It is difficult to distinguish the Blast in its Infancy, from the Effect of dry Weather; the Appearance in some Instances seems to be alike: However, the first seasonable Rain manifests the Difference; the uninfected Plant reaps the Benefit of it, thrives and flourishes with great Vigour; whilst

whilſt the infected, being made more ſoft and tender by the Rain, becomes eaſier to be pierced by the devouring Worms. At ſuch and other times, there are often ſeen, on the Blades of ſuch ſickly Canes, many ſmall protuberant Knobs, of a ſoft downy Subſtance, often containing in them ſmall white Maggots, which, I believe, turn afterwards into ſmall browniſh Moths, which are to be ſeen in great Multitudes among the Blades of infected Canes. It is likewiſe obſervable, that ſuch Blades will be full of browniſh decaying Spots: Theſe are ſo many Places, which have been deeper pierced by the Worms.

Multitudes of Ants are likewiſe ſeen on the blaſted Canes; theſe are invited hither to ſuck the Juice that ouſes out of the wounded Leaves, eſpecially when the Plant hath attained any Degree of * Sweetneſs. This appears by the Clamminefs, that, at ſuch times, covers the Leaves, preventing all Perſpiration. In this lacerated Condition of the Plant, the Juices want their natural free progreſſive Motion upwards; the moſt ſubtile and fineſt Part burſting through the wounded Leaves, whilſt the more groſs returns back unſecreted to the radical Veſſels. By this means they are overloaded, and, burſting, ſupply the Ants at the Roots with a nouriſhing Liquid. In this injured Condition the Roots become incapable of ſupplying the Stalk or Leaves with Nouriſhment from the Earth, if the latter ſhould ever recover.

The Blaſt is obſerved to be moſt frequent in very dry Years, there having been but little of it when ſeaſonable Rains have begun early, and continued till the Canes were ripe. In ſuch Years, a great many of theſe Vermin are perhaps drowned by the heavy Rains, as well as their Eggs made leſs prolific.

It is obſervable, that the Blaſt uſually appears ſucceſſively in the ſame Fields, and often in the very ſame Spot of Land: It is therefore very likely, that theſe are but the ſucceſſive Offspring of Parent-Eggs, from time to time depoſited there by the ſmall brown Moths above-mentioned. And when the Blaſt is found in Fields of Canes, far from infected Places, we may, in all Probability, conclude, that the Eggs were conveyed thither by the Wind. What makes this more evident is, that the Infection always ſpreads faſter to the Leeward, or with the Wind.

It is remarkable, that if Canes have been once infected with the Blaſt, although they afterwards, to all Appearance, ſeem to recover; yet the Juice of ſuch Canes will neither afford ſo much Sugar, nor ſo good of its Kind, as if obtained from Canes that were never infected. I conceive that, in this Caſe, the delicate Strainers, adapted to ſecrete the Particles, which conſtitute the Sugar, have been ſo much injured, as not to be in a Condition to perform their Offices to Perfection, although ſufficient to ſuſtain the Plant alive, and in a ſeeming Vigour.

* Perhaps the Attendance of the Ants may proceed from two Cauſes: They may be invited, as above-mentioned, when the Canes have attained ſome Degree of Maturity by the ſweet Juice, which ouſes out of the Wounds, as may be perceived by the Clammineſs of the Blades; or, if this is not the Caſe, when the Plant is very young, they may perhaps be allured to prey upon the dead and living Bodies of theſe little Animals infeſting the Canes.

2 Should

Should it be aſked, If this Blaſt is occaſioned by Worms, how comes it to paſs, that the adjoining, and often the intermixt Corn and Pulſe ſhould be free from it? it may be eaſily accounted for from ſimilar Inſtances in *England*, where the ſmall Worms, cauſing the Blight or Blaſt, which deſtroys the tender Buds of Apple-trees, never affects the Pear or Cherry-trees, tho' in the ſame Orchard; for, in all Probability, neither of theſe affords a proper Nouriſhment for them.

Having thus, till better Reaſons are offered, ſhewed the Nature of the Diſeaſe, the next Thing neceſſary will be to look for a Cure.

Hoc Opus, hic Labor eſt.

Various are the laudable Endeavours to this End, which the Inquiſitive in this and the neighbouring Iſlands have made; but, alas! made in vain: Therefore, as this Diſeaſe hath been hitherto of the Number of thoſe which are incurable, and almoſt literally as deſtructive to us, and our neighbouring Iſlands, as the Locuſts were to the *Egyptians*; a ſtudious Attempt to remove ſo great an Evil, will, I dare ſay, meet with the Approbation of every Well-wiſher to our *Weſt-India* Iſlands; eſpecially ſince what I have to offer upon this Subject is attended with the ſtrongeſt Probability of Succeſs: And as it requires very little Expence, and leſs Labour, I may with more Confidence venture to recommend it to the Public.

When the Canes appear to be firſt infected, which happens generally when they are young, take an equal Quantity of Brimſtone, Aloes, and the Bark of bitter Wood; let theſe be put in the Middle of a Bundle of wet Straw; the Whole muſt be put in a Cradle of Wire as large or larger than the Crown of an Hat, made Lattice or Net-faſhion; this is to be faſtened to a wooden Handle of convenient Length, and kept to the Windward of the infected Bunch of Cane, having firſt ſet the incloſed Combuſtibles on Fire; and holding it there till the thick Smoke hath for ſome time penetrated among all the infected Blades, and ſo on to the reſt, for a few Mornings and Evenings: This by its very Nature cannot fail of killing thoſe minute Animalcules, as well as deſtroying thoſe that are in Embryo in thoſe downy *Nidus's* already mentioned. Experience, which is the moſt convincing of Proofs, gives a ſtrong Sanction to this Method; for we find, that the Smoke of Brimſtone, in an incloſed Room full of Flour, peſtered with Wevils, will, in a few Minutes, intirely deſtroy them. If then Wevils, which have a ſtrong ſcaly Covering, and are grown to their full Strength, can be thus deſtroyed; how much more probable is it, that ſuch tender ſmall Animalcules may likewiſe, in the ſame manner, be deſtroyed? If it be ſaid, that in the former the Smoke is more confined, it muſt likewiſe be conſidered, that a far weaker Degree of this ſulphureous Smoke will deſtroy Animals of a far weaker Texture, and perhaps of but a few Days old. The Neceſſity and Uſe of the other Ingredients of the ſame Nature are too evident to be further explained.

The Cane-plant being deſcribed, and the Diſeaſes of it conſidered, and a more than probable Remedy propoſed, I ſhall proceed barely to touch on the Method of making Sugar.

The * Canes, when ripe, are ſqueezed between the iron-caſed Rollers of Wind-mills, or Cattle-mills. The Juice thus preſſed out is boiled firſt in a very large Copper or Chaldron, mixed with a very ſmall Quantity of Lime. When this is uſed in too ſmall a Proportion at firſt, a little Lime-water may be afterwards poured into the Chaldron. A ſtrong *Lixivium* of Aſhes will perform the Office of white Lime, and may be ſubſtituted in the room of it; and was originally uſed, tho' the latter is generally thought to be more efficacious. It is probable, that the Benefits ariſing from either are, in a great meaſure, owing to their alcaline Qualities. The Sugar-cane, when ripe, is of all other Plants the ſweeteſt; however, there is a latent Acid ſtill lurking in the Juice; this is apparent by its turning ſour, if ſuffered to remain unboiled any conſiderable Time after Expreſſion. The Addition therefore of *Temper*, as the Planters call it, being a certain Quantity of white Lime, is neceſſary to deſtroy, in a great meaſure, the remaining Acid, and to form a neutral Salt.

That this is one Uſe of *Temper*, is plain from the different Quantities of that which are uſed according to the different Qualities of the Cane-juice: That from unripe Canes, as more abounding with Acids, requires a larger Quantity, as doth that alſo from Canes too ripe, and tainted: For in the latter the acid Salts, that before were neutralized, ſeem to be again diſengaged, and ſet at Liberty, as may be diſcovered by its acid Taſte. And indeed many Inſtances occur in making Sugar, which demand an extraordinary Proportion of Lime; all theſe betray a Tendency to an Acidity in the Juice: But, when the Canes grow kindly ripe, the acid Particles in their Juice are few; and as the Poignancy of theſe is inconſiderable, the Juice will conſequently require a leſs Quantity of Lime. There is a further Uſe in Lime, beſides the foregoing; for it ſuits greatly in cleanſing the Liquor.

When the Quantity of Lime is duly proportioned, if the Liquor is put into a Glaſs, an immediate Separation will follow, the Impurities ſettling at the Bottom, leaving the clear Juice at the Top: But if there is a Deficiency of *Temper*, the Separation will be imperfect: If it too much abounds, there will be little or no Separation at all.

When the Lime is mixed with the Juice in the Copper or Chaldron, the *Sordes* or Impurities, being no longer intimately united with the boiling Liquor, and being forced about with the Heat of the Fire, are eaſily entangled in a viſcous Subſtance that is naturally in the Cane-juice; and then riſe with it to the Top of the Copper, forming a thick tough Scum.

* If, when Canes are ripe, the Weather ſhould prove very rainy, their Juice, if at that time expreſſed, will require a far longer Boiling, before it comes to the Conſiſtency of Sugar, than if it had been extracted in Weather moderately dry. However, this Difference in the Quality of the Juice doth not intirely proceed, as it is generally ſuppoſed, from the greater Quantity of Water at that time in the Plant, but from the greater Number of newly ſprung up Particles, occaſioned by the late Rain. Theſe, if ſoon afterwards expreſſed, having not had ſufficient time to ripen; the Make of their Particles is, as in all Acids, angular, and ſharp-pointed, and therefore diſſimilar to thoſe ripe ones: They will therefore reſiſt the Heat longer before they are broken, and brought to ſuch a Conſiſtency as to incorporate with the others that are already ripe. From ſuch a Mixture of ripe and unripe Juices, it naturally follows, that the Sugar then made will be neither of equal Conſiſtency nor Goodneſs with that made in ſeaſonable Weather, and from Canes grown kindly ripe.

This Viscidity is very apparently discovered on the leaden Beds of the Mills, as well as on the wooden Gutters, where the Juice in its Passage deposits it; and its * saponaceous Quality is no less evident in washing the Cloths that have been any ways used in cleaning the Beds of the Mills, or hath any other way been soaked in the Cane-juice.

The Clarification of the Liquor, as far as it is done in the first Copper, is perfected after the more gross Scum is taken off; the remaining Impurity, as the Liquor boils, is skimmed off from the four or five remaining Coppers or Taches, into which the Liquor is successively poured; each of these being gradually less, as they are to contain a Quantity of Liquor still wasting as it boils.

In conveying this to the fourth Copper, it is in its Passage strained thro' a thick Woolen Cloth, where it leaves all the Remainder of its Impurities, that had escaped the Scummer.

After this a light white Scum is taken off; and, when this ceases to arise in any considerable Quantity, and the Liquor, by long boiling, becomes more of a Syrup than a thin Liquid, it is then poured into the first Tache, and from this to a lesser, till it is conveyed to the last. When it hath here attained the due Consistence necessary to become Sugar; it may be asserted in general, that no more than a seventh Part of the Whole remains; which Diminution is occasioned by the Impurities being scummed off, and the watery Particles evaporated.

From this last Stage, whilst of the Consistency of a thick granulated Syrup, it is conveyed into a large Brass Cooler, where it begins, as it cools, to shoot into Crystals, which are the genuine and essential Salts of the Plant. These are forwarded and helped to shoot, by gently stirring the whole Mass; by which means the Air is admitted to every Part, and the Particles of Sugar disengage themselves from the clammy Substance of the Melasses.

If the Syrup be continued longer on the Fire, than is necessary to bring it to a proper Thickness, the Particles of Sugar cannot grain, or crystallize, when afterwards in the Cooler, for want of a sufficient intermediate Fluid; the whole Mass in such a Case being too well united, to suffer the Melasses to separate from it.

On the other hand, if the Syrup hath not undergone a sufficient Evaporation, the Grains or Salts will be larger indeed, but close to each other; Hence several of them being too much separated from their neighbouring Particles, they become too weak to resist singly, and are therefore drained away in the intermediate Fluid, the Melasses. Upon this Principle we may account for the Make of Sugar-candy, whose large Crystals are

* This saponaceous Quality in the Cane-juice is capable of resolving viscid Concretions: It is to this, chiefly, that we may attribute the surprising quick Recovery of those sickly Negroes, who are permitted to drink freely of this Cane-juice when intirely ripe. It is likewise so nourishing, that Slaves have subsisted upon this alone for a whole Week.

Repeated Draughts of it are very efficacious, to remove the Effect of the poisonous Cassado-juice.

From this Juice likewise, when mixed with Water, and fermented, is made a Drink, called the Sugar-Drink. This, tho' it appears muddy, yet is very wholsome and diuretic.

obtained

obtained from a Syrup too thin to ſhoot into Salts capable of uniting cloſe together.

The proper Time to remove it from the Cooler to the Pots or Mould, is when it hath grained or cryſtallized; the better the Sugar is, the ſooner this is completed: Hence that juſt, but ill-expreſſed Notion, that good Sugar may be potted ſooner than bad. The Pots or Moulds made uſe of are earthen, and of a pyramidal Form, containing from eight to thirteen Gallons.

About twenty-four Hours after the Sugar is potted, the ſmall round Hole in the Bottom of each Pot is unſtopped, and the Pots put upon earthen Jars containing about four Gallons, into which Veſſels the Melaſſes drain from the Sugar, the latter becoming fit for Exportation in about a Month's time, and ſometimes ſooner. The Sugar in this Degree of Perfection is called *Muſcovado*, which is a Term too well known to want any further Explanation. What is called here clayed Sugar, is brought to that Degree of Whiteneſs, by making a Batter of the ſofteſt fineſt white Clay mixed with Water: And after the upper Part of the *Muſcovado* Sugar in the above-mentioned earthen Pots is dug up, and cloſely laid on again in a level Manner, or rather ſomewhat ſhelving towards the Middle, a ſufficient Layer of this Batter is poured upon the Top of the Sugar in the Pot. The Water from this by Degrees gently ouſes from the Clay, thro' the Sugar; and when all the Moiſture from the Clay is abſorbed by it, which is generally done in about a Month's time, another Layer of freſh Putty is laid on, the former old one being firſt taken away. In about five Weeks after the latter is put on, this becomes dry, and is taken off; and the Water iſſuing from it meeting with leſs groſs Viſcidities than the former, waſhes the Particles of Sugar clean, and carries away with it thoſe leſs feculent Impurities. This completes the Work, as far as it is manufactured here; tho' this is brought to a far greater Degree of Whiteneſs and Perfection in *England*.

Out of the above-mentioned Skimmings, when mixed with a certain Quantity of Water and Melaſſes, and fermented, is extracted that ſpirituous Liquor called Rum. And from the great Quantity of Oil in the Cane-juice, which is conſiderably tranſmitted to the Rum, proceeds the Excellency of this Spirit, when compared with Brandy: The latter, wanting this Oilineſs, ſtimulates and lacerates the Coats of the Stomach; whereas the former, if firſt meliorated by Age, and made into weak Punch, and drank moderately, by its Oilineſs preſerves the Bowels.

Moſt of our Planters are yearly great Sufferers (eſpecially when they firſt begin to diſtil) for want of proper Knowlege how to raiſe and continue a regular Fermentation in the *Mulſa* intended for Diſtillation: Yet I imagine, that their want of Succeſs may not only be accounted for, but likewiſe remedied.

In order to do this with Certainty, we muſt obſerve, that no Fermentation can be raiſed under thirty-ſix Degrees of Heat, or kept up after ninety;

ninety; a leſſer than the former will not be ſufficiently warm to raiſe an Ebullition, and a greater than the latter diſſipates the ſpirituous Particles too much. Therefore if Experiments were made with a Thermometer in every Diſtil-houſe, to fix the certain Degree of Heat, that a well-proportioned *Mulſa* would ferment in, it would be eaſy, by the Help of this Inſtrument, always afterwards to aſcertain this neceſſary Degree of Heat, let the Change of Weather be ever ſo ſudden or conſiderable. For if the Heat proved ſo great as to exceed that Degree, in which ſuch a well-proportioned *Mulſa* was uſed in Time paſt to ferment beſt, then the Windows towards the Eaſt ought to be opened ſo as to admit ſuch a Quantity of cold Air as would reduce the Heat to a proper Standard.

On the contrary, if the Spirit in the Thermometer ſinks below the neceſſary Degree of Heat, then theſe Windows ought to be intirely or partially ſhut up, in order to procure a ſufficient Degree of Heat. By this means the Diſtiller may come to a Certainty, and proceed by Rule, and not by Chance.

If after ſuch Rules, and neceſſary Cautions, the *Mulſa* doth not ferment, if this happens in the Beginning of the Crop; ſuch a Failure ought to be attributed to the then, comparatively ſpeaking, ſour and unripe Juices of the Canes: For the Juices of theſe, as well as moſt, if not all other unripe Fruits, witneſs that of the Grape, ſeldom or never ferment well; becauſe their Particles, in that unripe State, are not ſufficiently meliorated by the Heat of the Sun. In ſuch a Caſe, I am apt to believe, that a greater Proportion of Sweetening, than when the Canes are ripe, ſhould be added to the *Mulſa*.

On the other hand, a diſproportionate Quantity of Sweets, as they are oily, will prove too inactive, and will incline the Liquor more to a Rancidity, than Fermentation: Therefore a greater Quantity of Water, and thin returned Liquor, which hath a great deal of Acidity in it, ſhould be added to the Skimmings of Canes that are full-ripe, and conſequently very ſweet.

As to thoſe who keep their fermenting Veſſels in the open Air, or ill-covered under Sheds, their bad Succeſs may be evidently accounted for, by the Inequality of the Heat and Cold they are expoſed to.

I cannot conclude the Deſcription of this very uſeful Plant, without taking Notice of a moſt ſurpriſing Inſtance of the Effect of ſome Effluvia, or Vapours that aroſe from the Mudgeon or Dregs of the Liquor returned from the Still, and which for ſome time had been reſerved in a Ciſtern.

In the Month of *April* 1743. *Abel Alleyne*, Eſq; the then Manager at the Eſtate of the Honourable and Reverend Society for propagating the Goſpel in foreign Parts, ordered one of the Ciſterns, which the returned Liquor was kept in, to be cleanſed: The Quantity of this thick Sediment in it was not above ſeven Inches deep. The firſt Negro Slave who attempted to clean it, was no ſooner at the Bottom, than dead; the second and third met with the same Fate instantly. A white Person, who was a Workman on the Estate, being near at hand, determined, if possible, to bring them up, imagining they were only in a swooning Fit. To this Purpose he went down to the Bottom of the Cistern, which was about nine Feet deep, and found the Negroes dead: He went down a second time with a Rope, in order to sling it round them, and to bring them up; but he had no sooner reached the Bottom, but a sulphureous suffocating warm Blast took away his Senses, and he was taken up for dead; however, being blooded, though he was for a long time afterwards very sickly, yet he at last recovered. The best Method of dissipating these noxious Vapours is to admit into them a free Circulation of the Air, as well as to pour in, by Gutters, a considerable Quantity of Water. This Plant is delineated in Plate XXIII. Fig. 1.

The DUMB-CANE.

THIS Plant grows to four Feet high, having, at the Top, two green shining Leaves, about nine Inches long; between these rises up a small Spire: The Body of the Plant hath some Resemblance of a Sugar-Cane. From hence, and from the Effects it hath upon Persons who ignorantly taste it, it derives the Name of *Dumb-Cane*; for as soon as any of the Juice is swallowed, the Tongue, Fauces, and Oesophagus, immediately swell so prodigiously, that the Person afflicted cannot speak. It so affects the salival Glands, that it causes an immediate Salivation. If what is thus discharged, hath the same Quality as that occasioned by Mercury, and if the narcotic Quality could be corrected, it might, perhaps, be of great Use in Physic; for it hath been experienced, that several poisonous Plants, whose Juices were properly corrected, and exactly dos'd, have been successful Remedies.

A Physician, who accompanied the Duke of *Albemarle* formerly to *Jamaica*, says, that the Juice of the Dumb-Cane, mixed with a certain Portion of fresh Fat, is a sovereign Remedy in Dropsies, externally used, by rubbing the Part affected with this Ointment.

The REED.

THIS is a slender long grassy Stalk, creeping upon the adjoining Wood, jointed at every six Inches: From these Joints issue several smaller Side-branches, cloathed alternately with long sharp-pointed gramineous Leaves. The main Stalk is hollow; the Side-ones solid. The Flowers are succeeded by several small grisly-grey Colour Grains, like *Guiney* Corn.

OF

Plate. 23. Page
Fig. 2
Fig. 3
Fig. 1
To the Honourable
Coll: Yorke
this Plate is humbly inscrib'd
&c.

OF

GRAMINEOUS PLANTS.

MR. *Ray*, in treating of the Wiſdom of *God* in the Creation, juſtly obſerves, that it is no ſmall Inſtance of his Goodneſs, that Wheat, and we may add Rye, the moſt common Corn uſed for Food, ſhould be the Growth of moſt Parts of *Europe* and *Aſia*. It is likewiſe no leſs worthy our Notice to obſerve, that where the Soil under the Torrid Zones is too hot to produce ſuch, the ſame divine Wiſdom hath appointed other Kinds of Corn to grow and ripen there in great Plenty. Thus in *Africa*, and the *Weſt-Indies*, the Want of Wheat is ſupplied by *Indian* and *Guiney* Corn. And in ſome Places, where the exceſſive Heat of the Climate renders Labour, and the Cultivation of Corn, painſul, there bountiful Providence, with an unſparing Hand, ordains Food without Labour, by cauſing Plantain and Banana Trees to grow in great Plenty; whoſe Fruit is, by many Perſons, preferred to any Kind of Bread whatſoever. It was under the Shade of theſe that many harmleſs Nations of *Indians* lived ſecure, until Luxury taught their more artful Neighbours unneceſſary Wants: And ſucceſsful Tyranny called that a glorious Conqueſt, which was the Effect of Fraud and Oppreſſion.

The INDIAN CORN; *Lat.* Mays.

THE Roots of this Plant are many: Its Stalk, which is jointed at uncertain Diſtances, and within pithy, is ſtrong, and of a Reed-like Subſtance: It hath likewiſe ſeveral long Reed-like Leaves. The Extremity of the Stalk is decorated with a Tuft of waving chaffy Glumes, which are called the Male-flowers.

From the Side of the Stalk, generally near the Summit, appear the Ears, which are ſometimes three in Number: Theſe ſtand upon ſhort Footſtalks, and are inwrapped in ſometimes ten or eleven green huſky Leaves, the Whole having the Appearance of a long Cone, from whoſe Summit appears a Taſſel of long ſilky Filaments, each having its Riſe from one of the Grains: Theſe are looked upon to be the Female Flowers. The cloſe-folded Leaves prevent the incloſed Grains from the Injury of Weather whilſt young, and from Vermin and Birds when ripe; and anſwer every Intention of a Pod in leguminous Plants. The Ear, which is generally from five to eight Inches long, contains often three hundred Grains, which cloſely, and in perpendicular Rows, regularly ſurround a ſtrong chaffy Huſk, whoſe Inſide is pithy. This is delineated in Plate XXIII. Fig. 2.

JOB'S TEARS; *Lat.* Lacrymæ Jobi.

THE Stalk and Pannicle of this Plant much resemble that of the *Guiney* Corn; and their Seeds, which are inclosed in small *Capsulæ*, are about the Bigness of an *English* Pea, and of different Colours. These are strung upon Silk, and used instead of Bracelets by some of the poorer Sort, but especially by the *Negroes*.

GUINEY CORN; *Lat.* Milium Indicum arundinaceum.

THE different Sorts of this Corn are generally distinguished into that of the Loose-ear *Guiney* Wheat, the White, and the Red, the Loaf, or the Clove-corn. The Stalk of each is a Reed or Cane-like Substance, jointed at about nine or ten Inches asunder, more or less: Its Roots are many and long, closely matted together, and to be seen, some Parts of them, above-ground: The Stalk often grows to eight Feet high, sometimes higher, according to the Richness of the Soil it is planted in. It bears long Reed like Leaves, gently waving into several Bendings, ending in a sharp Point. The Top of the Stalk terminates in a spicated Tuft, composed of numberless Fibres, each supporting many small Grains of about the Bigness of a Coriander-seed. I have reckoned on one large Ear near four thousand Grains. There is likewise sometimes one, if not two other small Ears upon the Stalks, somewhat below the large one, which terminates the Stalks; however, where these Side-ears are, the main one is generally less. This Grain made into Bread, or otherwise used, is justly esteemed very wholsome. It is with this that the Slaves are generally fed, each being allowed from a Pint to a Quart apiece every Day. This is delineated in Plate XXIII. Fig. 3.

OF

BOOK IX.

THE ſeveral foregoing Books being taken up in conſidering the Nature of our Land-Animals, Vegetables and Minerals, I ſhall proceed to make ſome Remarks on the Shores and its Inhabitants; firſt curſorily obſerving, that thoſe are (a few Bays excepted) intirely invironed with high Cliffs, from three to ſixty Feet high; which, under Providence, are in a great meaſure, our Bulwarks againſt our Enemies, as well as fix'd Boundaries to the raging of the Sea in the Time of Hurricanes.

As far as the Sea in moderate Weather waſhes them, they are almoſt one continued Rock from thence upwards; they are in ſome Places (eſpecially at the North End of the Pariſh of St. *Lucy*'s) ſomewhat ſhattered and divided, by Veins of gritty Marl, Earth and loomy Chalk; likewiſe all the Stones found on the Shores (except a few Flints on a Place called the *Green-Shoal)* are of the Aſterites-Kind; and the Sand (except in a few deep Bays) ſeems, in a great meaſure, to be only Fragments of Star-ſtones and Corals.

The ſeveral Cavities in the Cliffs facing the Sea, are proper Dens for Racoons, and ſuch like wild Beaſts. They are likewiſe a Place of Safety for ſeveral Sea-birds to breed in, eſpecially, at a Place called the *Bird-rock*; where are to be ſeen, at moſt Times of the Year, a great many of their Neſts and Eggs: The young ones are ſharp-bill'd, web-footed, and very fat, but taſte fiſhy. The old ones are ſeldom or ever ſeen in the Day-time; for they are obliged to range to ſo great a Diſtance from the Shore for Food, that they have been ſeen Scores of Miles from Land; yet they direct their Courſe, in the darkeſt Night, with a ſurprizing Exactneſs, to their reſpective Neſts.

The Sea Weeds, which grow in Plenty near the Shores of this Iſland, are no leſs beautiful to the Eye than they are uſeful; affording Food for ſeveral Kinds of Fiſh, ſuch as the Tortoiſe, Chubbs, Mullets, Blue-fiſh, and the Moſs-grooper, *&c.*

And as in Northern Climates they prove a Shelter from the Inclemency of the cold Weather, to ſuch Fiſh as are obliged to ſeek their Food in ſhallow Water; ſo likewiſe, in hot Countries, they equally ſhelter them from the intenſe Heat of the Sun. Nor are the large Beds of Sea-Weeds, which are ſeen floating in the Ocean, without their peculiar Uſes; among others, in theſe ſeveral Kinds of the leſſer and moſt helpleſs Species of Fiſh, depoſit their Spawn; and here likewiſe they are protected whilſt young from their Enemies.

The Variety of theſe Weeds, eſpecially the ſoft *Fuci*, growing on the Rocks and Stones near the Shore, are not only curious in their Make, but they are likewiſe, in all Probability, (were we to know all the Intentions of Providence in creating them,) as many, as uſeful, and valuable, as the terreſtial Vegetables we are better acquainted with.

Some of them, like Groves in Miniature, extend their numerous Branches to a conſiderable Diſtance: Others, like weak creeping Vines, are of ſo ſmall, ſoft and delicate a Texture, that they may be intended for Food, to the ſmalleſt Animalcules. So that upon the whole, from the little that we do know of ſubmarine Productions, we may juſtly conclude, that, as the Earth is full of God's Goodneſs, ſo is the great and wide Sea.

The Black-fringed Sea-Weed.

THIS is very finely fringed, and very buſhy, eſpecially near the Top. Its ſeveral Subdiviſions are neatly bearded with infinite Numbers of ſhort, hairy, ſoft pointed Briſtles. Its moſt ſubſtantial lower Branches are thickly ſtudded with ſeveral ſmall Berries.

The Fringe Sea-Weed.

THE Colour of this is generally White, and exactly reſembles the Fringe or Border of Cuſhions: It is found upon almoſt every ſandy Bay.

The White, Narrow-leafed Sea-Weed.

THE Roots of this are cloſely matted together. Its ſeveral, thick-ſet, capillarious, fringy Branches, are divided into lateral, ſmaller ones; and theſe again, elegantly ſubdivided into others, ſtill leſs; each generally ending in a triparture Diviſion, and ſometimes, tho' rarely, in a white, leafy, broad Point.

The LONG BROWN SEA-WEED.

THIS Moſs is the longeſt of all others, being often above twelve Inches in Length; the main Branches are ſeldom ſubdivided into leſſer ones; each being thickly cloathed with ſmall oval Leaves, intermixed with browniſh Berries. This chiefly grows upon Rocks, at ſome Diſtance from the Shore.

The WHITE BROAD-LEAF'D SEA-WEED.

THE Leaves of this ſomewhat reſemble thoſe of a ſmall curled Lettuce, but far more membranouſly thin; and the Leaves more ſinuated and involved in the Foldings of each other.

The PALE-GREEN SEA-WEED.

THIS grows into a Cluſter of thin, flat, pale-green Leaves, divided ſideways into many Branches, each terminating in a blunt, forked Point.

The CORAL SEA-WEED.

THIS extends itſelf from the Rocks, in a long, naked Stem, of a brown yellowiſh Colour, in Subſtance near as thick a as Raven's Quill; each, at its Extremity, divided into a Group of many blunt-pointed Diviſions.

The BROWN HEAD-LEAF'D SEA-WEED.

THIS grows in a thick Bunch; the Leaves, as they grow, widen; and, towards the Top, divide into five or ſix deep round-pointed Sections.

The FINE, WHITE SEA-WEED.

THE Branches of this are very ſmall, and all over hairy: I have found this Sort growing upon the Sea-Cliffs, near a Place called the *Spout*, in St. *Lucy*'s Pariſh: There is likewiſe a black Sort, which differs from this only in Colour, as well as a dark green ſimilar to it. This, as it grows, becomes a ſtronger Plant, and expands into Branches, reſembling, in Miniature, a Deer's Horn.

The

The NEW-ENGLAND, *or* RIBBAND SEA-WEED.

THIS resembles a Half-inch-wide green Ribband, and is about two Foot long; it is generally to be found in deep Water.

The RUSSET NARROW-LEAF'D SEA-WEED.

THIS grows in Bunches, upon one common Foot-stalk; its Branches, which are about two Inches long, are flat, and laterally divided into many others, ending generally in a blunt, forked Point.

The FEATHER SEA-WEED.

THE Roots of this are long, and creeping, having several upright, single Stalks; these are about four Inches high, each exactly resembling a small blunt-pointed Pen-feather, whose Plumes are wet, and separated: These are in general, of a dark Colour.

The SEA-JEWS-EARS.

THESE are thin and pellucid; of a semioval Form, growing upon loose Stones and Rocks.

The ORANGE-COLOUR SEA-WEED.

THIS seldom grows above four Inches high, is of a fine bright Orange-colour, and very much resembles a small, thick-branch'd Tree, in Miniature.

The BLACK BRANCHING SEA-WEED.

THIS divides into many upright streight Branches, each forked at the Top.

The KNOTED SEA-GRASS.

THIS is by far less pliable than the last described, and about as slender as a strong Horse-hair, the whole being seemingly knoted, the intermediate Space being very small.

The PALE-YELLOW SEA-GRASS.

THIS is of a fine, soft, silky Texture, growing upon old Logs of Wood under Water, and hanging wavingly in it.

The

The SHORT PURPLISH SEA-MOSS.

THIS is ſeldom above an Inch long, and generally grows upon the Extremity of a pendent Rock or Stone.

The PALE PURPLE SEA-WEED.

THIS very beautiful Weed grows like a Tree in Miniature; it is very full of Branches, and the Whole tranſparent.

The CARBUNCLE.

THIS is a ſtony or ſhelly Excreſcence, growing hollow upon Rocks, and ſometimes even upon the Back of a Tortoiſe-ſhell: When upon the latter, they are generally of a Limpet-ſhape, the former, of many irregular Shapes: Each Sort is inhabited by a Shell-fiſh peculiarly.

The TUFTED PENNATED SEA-WEED.

THE Roots of this are ſmall and long; the main Stem is generally ſingle, each Side elegantly adorned with ſmall oval Leaves, growing oppoſite to one another, and, when dried, of a tranſparent yellow Colour; theſe grow in ſeveral Pairs, and then diſcontinue; ſo that the Stalk ſhall appear naked for about a Quarter of an Inch in Length; and then another Tuft riſes; and ſo on alternately to the Top. This Plant ſeldom grows above four Inches high.

The PENNATED SEA-WEED.

THIS ſeldom grows above three Inches high, and exactly reſembles a very ſmall Pen-feather. It generally grows upon old Logs of Wood in the Sea, and moſt commonly not far from Shore.

The GREEN BROAD-LEAF'D SEA-WEED.

THIS differs not materially, in its Make, from the White-lettuce Sea Weed, already deſcribed. It is obſerved, that Tortoiſes feed more upon this, than any other Kind of Sea-Weed.

The RED-LEAF'D SEA-WEED.

THESE grow in many oblong Leaves, about three Quarters of an Inch long, and of a light-purpliſh Colour; generally between high and low Water-mark.

The SEA-GRASS.

THIS is of a green Colour, hanging in many very fine long silky Shreds; growing generally upon the Rocks, between high and low Water-mark.

The THICK PLUSHY SEA-MOSS.

THIS is generally of a green Colour, of a soft plushy Texture; growing upon Stones and Rocks in the Sea.

The GRAPE-SEA-WEED.

THIS is of a dark Amber-colour. Its several Branches are so many hollow Tubes, the Extremity of each generally ending in a small, round Berry, about as big as a Grain of Pepper.

The GREEN SILKY SEA-MOSS.

THIS grows generally between high and low Water-mark; and very much resembles, in its Texture and Appearance, a confused scattered Cluster of raw Silk.

OF
EXANGUIOUS ANIMALS.

THESE are divided into Terrestrial, Aquatic, and Amphibious; and, according to *Aristotle*, they may be farther classed, by the following Order, into three Kinds.

(1) First, Those that are composed of a soft fleshy Texture; such as Snails.

(2) Secondly, Those that are more firm; as all the crustaceous Kinds, such as Crabs and Lobsters.

And (3), Thirdly, Those that inhabit the firmest Shells; such as Oysters, as well as all other Shell-fish.

These, and such bloodless Animals, are called Exanguious.

The MARBLED SNAIL.

I Found these in Plenty upon the Bank of a Pond, whose fresh Water had a Mixture of Salt-water, sometimes thrown into it at high Tides.

The young Snails hang in great Numbers upon the Grass, in the watery Edges of the Pond.

(1) Μαλακία ἐστὶν ὅσα ἄναιμα ὄντα ἐκτὸς ἔχει τὸ σαρκωδες, ἐντὸς δὲ τὸ στερεὸν, καὶ τὰ ἐναίμα τῶν ζώων, οἷον τὸ τῶν Σηπίων γένος.

(2) Μαλακόσαρκα ἐστὶν ὅσα τὸ μὲν στερεὸν ἐκτὸς ἔχουσι, ἐντὸς δὲ τὸ μαλακὸν καὶ σαρκῶδες, τὸ δὲ στερεὸν αὐτῶν οὐ θραυστὸν, ἀλλὰ φλαστὸν, οἷον τὸ τῶν καράβων γένος, καὶ τὸ τῶν καρκίνων.

(3) Ὀστρακόδερμα ἐστὶ, ὧν ἐντὸς μὲν τὸ σαρκῶδες, ἐκτὸς δὶ τὸ στερεὸν, θραυστὸν, καὶ κάτακτον, ἀλλ' οὐ φλαστον, τοιοῦτον δὲ τὸ τῶν κοχλίων καὶ τῶν ὀστρέων γένος.

The SEA-LEECH.

THE common People call this the Sea, or the Black-pudding: However, I ſhall call it, the Sea-leech; it hath the Name of Black-pudding from its great Reſemblance to it.

Its ordinary Bulk is about five Inches long, and two in Breadth; yet I have known it to extend itſelf above eight Inches in Length.

They are generally found clinging to the lower Sides of Rocks or Stones, between high and low Water-mark.

They have no Appearance of Eyes, Legs, or Feet; at leaſt, to the naked Eye.

Their Motion is very ſlow, and of the muſcular Kind.

When the Tide of Flood begins to waſh the Holes they are in, they extend from what we muſt now call the Head-Part, in a Bunch, ten or twelve Snouts like the Horns of a Snail, all coming from one common Root, like a Bunch of Coral; the Top of each being far wider than the Shank.

I am confident, that theſe (if I may be allowed to compare ſmall Things with great) are deſigned, as Trunks are to Elephants, to be the Inſtruments of receiving their Nouriſhment thro' them.

Having examined the Inſide of this, it is one continued Circumvolution of white Guts, not bigger than the baſe String of a Fiddle.

The Water that it emits, when ſqueez'd, is of ſo corroſive a Nature, that it ſmarts very much, and raiſes Bliſters upon that Part of the Skin, upon which it falls.

The NAKED SEA-SNAIL.

THIS is generally of about an Inch and an half long; but, as it is capable of greatly contracting or expanding itſelf, it appears of various Bulk, according to its different Motions; ſlender, when it extends itſelf, and thick almoſt to a Roundneſs, when it is contracted.

It is found under the Rocks and Stones near the Shore.

The SEA-SCORPION.

WHAT we call here the Scorpion, is by *Petiver* called *Stella marina Scolopendroides.*

Its five Rays might perhaps properly cauſe it to be called the *Stella Marina.*

The

The SEA-FORTY-LEG.

THIS is about two Inches and a quarter long; the Back jointed; and divided into ſeveral *Annuli*, thick-ſet, with ſtiff ſhort white Briſtles.

Theſe, when they are all erect, exactly reſemble Pluſh.

The Head is guarded with a *Forceps*, reſembling that of a Land Forty Legs; and is found under Stones and Rocks near the Shore.

The SEA-SUCKER.

THIS almoſt ſhapeleſs Animal is of a brown fungous tough Jelly-like Subſtance, ſticking to the Rock always under Water: At a Diſtance, it hath the Appearance of a Sponge.

It is generally of an irregular roundiſh Shape, of about an Inch and an half in Diameter.

In the Centre of it appears a round Hole, about a quarter of an Inch broad: This I take to be its Mouth; for when this is touched, it cloſes its Sides together.

I SHALL finiſh this Claſs of Exanguious Animals, by deſcribing one of a ſurpriſing Make, fixed to the Rocks and Stones, without any ſhelly Covering.

Its Length, is about an Inch, and three Quarters; and is an Inch in Diameter; the Outſide granated, very much reſembling the Roe of a Fiſh, interſperſed with Veins of a Snuff-colour, and pale-red, mixed with very pale green Liſts: When it is feeding, both Ends are flat, the one cleaving to the Rock, the other expoſed to the Flux and Reflux of the Waves: In the Centre of the latter is its Mouth: In this, when it opens, or gapes to receive what Nouriſhment the Sea, by the daſhing of its Waves, throws in its Way, are ſeen innumerable ſmall blunt ſpiral Feelers, of about a quarter of an Inch long; interwoven within one another: Theſe are all in Motion, as well as the Mouth open, when the Waves daſh againſt the Rocks they are upon, as if they were, at that Inſtant, in Search of Food. In the Intermiſſion of the Waves, the Mouth cloſes; and theſe Feelers, which are very many, and much reſembling, in Shape, the ſmall *Veſiculæ*, which conſtitutes the pulpy Part of Oranges or Lemons) are contracted, and with a ſtrong, but ſlow muſcular Motion cloſe together, and are, as it were, ſucked back into the Body of the Animal: Yet they immediately dart out upon the Return of the Waves; but when the Whole is intirely left dry, by the Abſence of the Water, this flat circular End, in which the Mouth and Feelers appeared, and which, juſt before, when feeding, were of an Inch Diameter, is in an Inſtant contracted, like the Mouth of a Purſe drawn together; and becomes of a blunt conic Form, the Baſe being the Part fixed or clinging to the Rock.

Its Appearance, in this its quieſcent State, is repreſented in *Plate* X. *Fig.* 4.

There is likewiſe a leſs Sort, of the ſame Species with the above deſcribed, differing chiefly in Colour, which of the ſmaller Kind is black.

OF

OF

CRUSTACEOUS ANIMALS.

As those Kinds of Fish, whose Safety depends upon their Swiftness in Swimming, have not only their Tails and Fins properly adapted for that Purpose, but their Eyes likewise (witness the Dolphin) never jutting out above the Surface of the Head, lest they should retard their Motion thro' so thick a Medium; so, on the other hand, as the Safety of Crabs and Lobsters, &c. depends not upon their Celerity, and as these inhabit the Cavities of submarine Rocks and Cliffs, where their Enemies often lurk under the same Shelter; to prevent therefore a sudden Surprize from any of these, Providence hath not only guarded them with a strong shelly Covering, and often Prickles; but hath so commodiously formed the Eyes of these Creatures, that by their great Prominency, and jutting out, sometimes above half an Inch, they can, without moving their Bodies, take at once a circular View all around them.

The *West-India* Lobsters, or what are so called, being in Reality the same Make in this Part of *America*, differ in general, from the *English* Kind, by their Want of Claws, and being guarded with many Prickles.

These, as well as all other Kind of Shell-fish, as *Milton* elegantly expresses it,

> *Within their pearly Shells at Ease attend*
> *Moist Nourishment, or under Rocks their Food*
> *In jointed Armour watch.*-------

The

The Dominico-Lobster.

THIS Lobster seldom weighs more than about two Pounds.

The Red Lobster.

THIS is the largest Lobster that we have in this Sea; for it often weighs twenty Pounds; its Colour as above; but the larger it is in Bulk, the less delicious it is.

The Green Lobster.

THIS derives its Name partly from its Colour; and generally weighs, when full-grown, between three and four Pounds.

Its two largest Horns, or Feelers, are about eighteen Inches long: From between these come two lesser ones, forked or divided near their Extremities.

The Eyes are guarded by two sharp-pointed crooked Horns.

The Extremities of the Feet are hairy; and the Shell upon the Back, especially about the Roots of the great Feelers, is very thickly studded with sharp Prickles.

The Queen *or the* Bonne-Grace Lobster.

THE largest of this Sort never weigh above two Pounds.

This is justly esteemed one of the best of the Lobster Kind.

The Lazy Crab.

THIS is a very large beautiful Crab.

The Back is generally full of small Knobs of a pale-scarlet Colour; guarded here and there, but especially about the Edges of the Back-Shell, with short sharp Prickles.

It hath four strong Legs on a Side; these are covered over with a short brownish Hair or Pile, and are likewise, in the Male Crab, defended with Prickles, the last Joint of each Leg ending in a sharp Point.

Its two great Claws, from the Setting-on to the Body to their Extremities, are often ten Inches long.

The very Tips of the two meeting Claws, with which it holds its Prey, remarkably differ from all other Crabs, by their great Breadth, as well as by their several regular Indentings, which, when they close together, fall as regularly into their Sockets as the opposite Sides of a Pair of Nippers.

A Claw of one of these uncommon Crabs may be seen in the Collection of Mr. *Peter Collinson.*

The Whole is delineated in Plate XXV. Fig. 1.

The

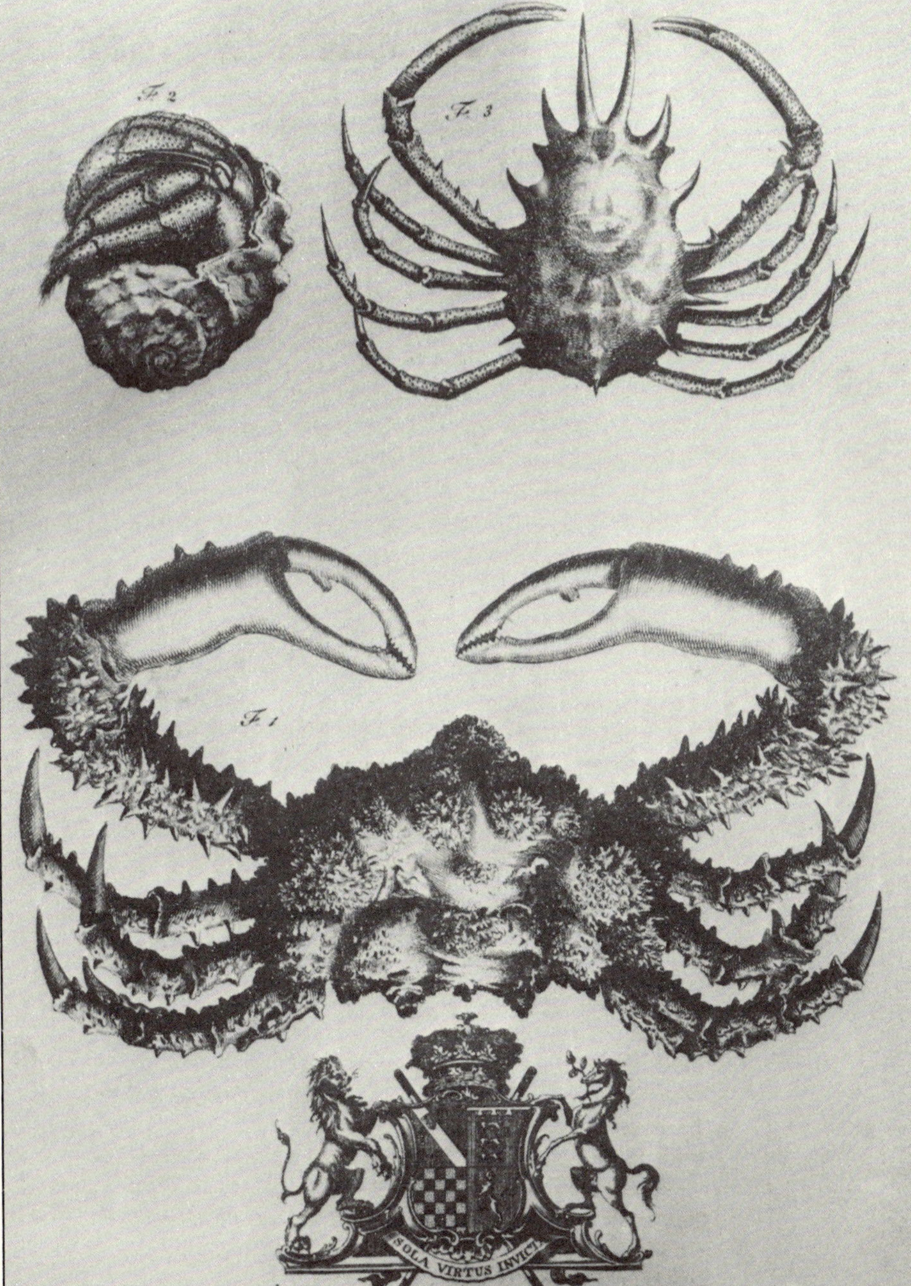

To His Grace the DUKE of NORFOLK,
This Plate is humbly Inscrib'd &c.

J. Mynde sc.

The HORSMAN-CRAB *or* BEN-TROTTERS.

THIS ſmall whitiſh Crab, which is generally to be ſound between high and low Water-Mark, and is called *Horſeman* from its great Speed in ſecuring itſelf, either by running into its Hole, or the Sea, is far leſs than the white Land-Crab: Its Number of Feet are equal.

The CLUB-MEN.

THESE are very ſmall Crabs, near as ſmall as thoſe called the She-Biters.

They are to be ſeen near the Edges of ſalt-water Marſhes; they burrow in theſe ſandy and clayey Places.

Their Claws are very large, in proportion to their Bodies: From hence it is ſaid that they derive their Name.

The SHE-BITERS.

THESE are very ſmall Crabs, inhabiting the Shoals.

Their Backs are not wider than an *Engliſh* Shilling.

Their Claws, in proportion, are very large.

The LONG-LEGGED VERY SMALL WHITE CRAB.

THE Body of this is not larger than an *Engliſh* Silver Groat.

Its very ſlender Legs are at leaſt a Foot and an half long, and reſemble knotted Thread.

This Species of Crabs are ſo ſeldom to be met with, that I never ſaw but one, and that dead, which was in the Collection of my learned Friend Dr. *Warren*, deceaſed.

The SCUTTLE-CRAB.

THIS is a ſmall browniſh Crab, marbled on the Back with darkiſh Lines.

It feeds generally upon the Moſs that grows upon the Rocks, between high and low Water-Mark.

Their Armour, when caſt, which they caſt once a Year, looks very beautiful.

This Crab will creep up and down the moſt ſteep perpendicular, or even a projected Rock.

The RED SEA-CRAB.

THIS differs very little, if any thing, from the same Species in *England*, except in the Colour, which is of a fine Scarlet; and therefore abundantly brighter than that of the *English* Crab, which is of a dirty Red.

They grow here to a great Size; but are not, however, so large as those of the same Kind in *England*, but better tasted.

The SIR-EAGER CRAB.

THE Back-Shell of most Crabs is somewhat roundish; but of this 'tis near four Inches long, and not full two broad, the opposite Parts in Length ending in a sharp strong Prickle.

From each of these Points to the other the Shape of the Shell comes nearest to that of a Semicircle.

This is regularly and deeply indented like the Teeth of a Saw, each ending in a sharp Point.

The Colour of the upper Shell is blackish, clouded or stained with some scattered Spots of Pale-white.

The two Claws are long, but slender; the upper Joint, as well as the Claw itself, guarded with three or four Prickles or Teeth.

The upper Part of the last Joint is irregularly squared, and of a brownish Colour; the Part underneath somewhat rounder, and of a bluish Colour.

Their Legs are equal in Number to those of the Crabs already described; and they are chiefly caught in salt-water Ponds near the Sea.

They are justly esteemed good Eating.

The LARGE WHITE LAND-CRAB.

THE Claws of this are very long and large, in proportion to the rest of the Body.

These always burrow in a sandy Soil, near the Sea; and come chiefly out at Nights, to feed upon the green Grass, and tender Vines, growing upon such Places.

They often likewise feed upon Manchaneel Apples, as well as upon the Leaves or Berries of Poison-trees.

At such times they are dangerous to be eaten, unless very great Care be taken to wash the Fat, as well as the other Meat on the Inside, with a sufficient Quantity of Lime-Juice and Water.

The MULATTO CRAB.

THIS differs from the white, already described, chiefly in the Colour of its Shell, which is yellowish, and its Residence, which is often up in the Country, above two or three Miles distant from the Sea.

They generally burrow about the Edges of fresh-water Ponds, the Bottom of their Holes being always so deep as to reach the Water.

These feed in the Night upon the green tender Grass about the Sides of the Ponds, and are in general esteemed as good Eating as the white Crabs; but prove equally as poisonous, if eaten soon after they have fed upon the Berries of Poison-trees; unless, as before, the Fat and the inside Meat is first washed in Lime-Juice; which either corrects or washes off the venomous Quality.

The BLACK LAND-CRAB.

THIS small Crab, though I have never seen it, yet, as I am credibly informed, is generally an Inhabitant of old Timber-Houses, and such Ruins; and it is very fat and delicious, but rare and uncommon.

The RED LAND-CRAB.

THIS is but a small Crab; the Back seldom above two Inches long; its Colour of a blackish Red.

It hath two Claws, and four Legs, on a Side.

These Crabs, after a heavy Shower of Rain in the Months of *March*, *April*, and *May*, are to be seen in great Numbers, loaden with Spawn, going down to the Sea-Side to deposit it in the Sand, near the Wash of the Water; and soon afterwards to return into the Country, but generally not very far from the Sea; though I have often seen them at above two Miles Distance from the salt Water.

As these are very fat, they are esteemed good Eating.

The common Method of dressing them among the Negroes is, to roast them.

This Species of Crabs were perhaps very plentiful in *Italy* in the Time of *Virgil*; for, in his fourth *Georgic*, he forbids roasting of red Crabs near an Apiary, the Smell of them being disagreeable to the Bees.

The SOLDIER-CRAB.

THIS is amphibious.

Sometimes it is caught upon the Rocks at a considerable Distance from the Land; but they are most commonly seen upon the Shore.

It

It is thought to have derived its Name from its frequent Change of its Quarters; for its firſt Appearance is generally in a very ſmall Periwinkle-ſhell; as it grows too big for this, it looks out for another empty Shell, agreeable to its preſent Bulk; ſoon after, it takes up its Abode in a large Wilk-ſhell.

It hath two Claws, reſembling thoſe of a Crab.

The Tail-Part affords a great Quantity of Oil, which is looked upon to be of great Service to lubricate ſwoln or ſtiff Joints.

Theſe are often found cleaving to Rocks in the Sea, but oftener in Granaries on Shore, where they deſtroy much Corn.

The Tail, or the hinder Part, is covered with a thin Skin.

This is delineated in Plate XXV. Fig. 2.

The HORNED CRAB.

THIS Crab is of a middle Size.

The Shell upon the Back, in ſome Places, is much elevated, and again indented.

Its Head is guarded with two Pair of Horns, both very ſtrong, and ſharp-pointed; the one Pair about an Inch and an half long, the other not above half an Inch.

Each Side of the Trunk or Body is likewiſe armed with four ſharp Prickles, reſembling a Cock's Spur.

The Body is ſupported by four Legs on a Side, beſides two long feeding Claws, each ending in a *Forceps.*

The whole Crab is covered with browniſh pluſhy Hairs.

As to the reſt of its Make, it differs from other Crabs, in that its Shell is longer in proportion than moſt other Crabs, the *Sir-Eager* excepted, and more ſharp-pointed than any towards the Head.

This is delineated in Plate XXV. Fig. 3.

The SEA COCK ROCH.

THIS is of the Crab-Kind, and about an Inch long.

Its upper Shell, which is of a browniſh White, freckled with Black, much reſembles, in its Make, that of a Tortoiſe.

The Head is provided with two hairy Feelers, like the Land-Cock-Roch.

It hath likewiſe near the Mouth two long Legs, or Providers, one conſiderably longer than the other.

The Body is likewiſe ſupported by ſix other ſhorter hairy ſtrong Legs.

The Tail-Part ends in a ſharp-pointed Flap, which covers Part of the Belly, and in the Females is the Repoſitory of the Spawn.

Theſe live in Holes in the Sand, in the Waſh of the Water.

They are ſometimes eaten; and have a Crab-like Taſte.

SHELLS

SHELLS, *and* SHELL-FISH.

THESE are alſo of the exanguious Tribe; and differ from the cruſtaceous Kind, by having their Organs leſs perfect; for, as the admirable Mr. *Locke* obſerves, an Oyſter or Muſſel hath not as many, nor as quick Senſes as a Man, or ſeveral other Animals: Nor, if it had, would it, in that State, and Incapacity of transferring itſelf from one Place to another, be bettered by them. What Good would Sight and Hearing do to a Creature that cannot move itſelf to or from an Object wherein at a Diſtance it perceived Good or Evil? And would not Quickneſs of Senſation be an Inconvenience to an Animal, that muſt be ſtill where Chance hath once placed it, and there receive the Afflux of colder or warmer, clean or foul Water, as it happens to come to it?

It may be expected, that I ſhould here conſider theſe beautiful Productions in a phyſical Light, and as they are a Curioſity belonging to the Cabinets of the *Virtuoſi*: Yet I ſhall beg leave to ſpeak of them, as they contribute to the Amuſement of the Ladies. And this I am the rather induced to do, as I have heard ſeveral of the Fair Sex, who are fond of Shell-work, frequently ridiculed, as waſting their Time in a trifling and uſeleſs Manner; and this not without a very conſiderable Expence. That this Employment is by no means trifling or uſeleſs, is eaſily proved, as it is abſolutely grounded on the noble and elegant Art of Deſigning, and as its End is equally that of Imitation. What the Painter performs by an Aſſemblage of various Colours properly diſtributed, is here produced by a Diſpoſition of Shells, as their different Tinctures, Shapes, or Sizes, happen to direct the Fancy: Add to this, that the Repreſentation is by far more ſtriking in Shell-work, than on the Canvas.

I have ſeen a Roſe in Shells, glowing with as exquiſite a Red, as that which it received from the Spring. And hence it appears, that Shell-work partakes of the Nature of Painting, if not of Statuary; imitating not only by Colours, but a full Relievo. In ſhort, nothing can be more properly termed a new Creation, than a well-executed Syſtem of Shell-work.

What can be more delightful to the Imagination, than a Grotto completely embelliſhed with this Kind of Imagery? With what truly romantic Ideas muſt it inſpire one, to ſit in a Room furniſhed with the Riches of the moſt diſtant Shores and Oceans! And theſe not confuſedly thrown together, but judiciouſly diſpoſed into regular Repreſentations of natural Objects! We are ſtruck with Aſtoniſhment! Nor can we determine, whether we ſhould moſt admire the Sun, which gave to each particular Shell its delicate Hue, or the Hand of the fair Artiſt, which ranged them with ſo much Skill.

It will be needleſs perhaps to add in this Place, that one of Mr. *Addiſon*'s Sources of the Pleaſures of Imagination may be found in a Work of this Sort, in its higheſt Perfection; namely, *The Beautiful.*

Thus I have obviated the Objection, that this Amuſement (or rather Art) is uſeleſs and trifling. And here it will be proper to take Notice, that the Genius of Women is much better fitted for this Employment, than that of Men. For as Shell-work conſiſts in forming pleaſant Pictures, and agreeable Repreſentations, and as theſe are effected by the means of Shape and Colour artificially put together, a certain Facility and Fancifulneſs is required, which our Sex is an intire Stranger to.

But, allowing this Diverſion to be trifling and uſeleſs, the moſt ill-natured Caviller cannot deny, but that it is innocent. How many Ladies do we ſee faſhionably murdering their Time in Gaming! A Diverſion attended with ſome of the moſt abject Vices, and ſhocking Conſequences: Vices, which one would think inconſiſtent with the Softneſs of the Sex, and Conſequences, which, I am ſure, that Softneſs can but ill ſuſtain. A Practice this, which ruins their Honour, their Families, and Fortunes; and (what one would think the Ladies would take more to Heart)---their Faces. Whereas the Joys of this our Purſuit are pure, and intirely founded on a contemplative Turn of Mind; eſpecially, if we ſuppoſe, that one of our modern *Calypſo*'s, after having thus adorned her Grot, would no doubt chuſe to reap the Fruit of her Labours, by making it a Place to cultivate her Mind in by Muſing.

Some Diverſions muſt indiſputably be indulged the Fair Sex; and this, I think, is not only as becoming, but as proper, as any. We cannot ſuppoſe, that our *Cynthia*'s and *Flavia*'s can leap a five-barr'd Gate, or walk half a Day with a Gun in queſt of a Wood-cock; Sports, which are better ſuited to the Strength and Roughneſs of the oppoſite Sex. But it may be objected, that there are other Diverſions equally adapted to the Delicacy of Women; for Inſtance, the Cultivation of Flowers. In the firſt Place, I ſhall remark, that the Study of the Floriſt is not ſo pleaſing; whoſe chief Pride conſiſts in a Variety and Beauty of Colours. Now I appeal to the moſt experienced Naturaliſt, Whether the moſt choice Collection of Tulips, can vie in beautiful Colours with the Pamana, or Sun-ſhell, the Ear-ſhell, and the *Murex*, which affords the genuine *Tyrian* Purple? And now I am particularly inſiſting on the

Beauty

Beauty of Shells, I cannot but obſerve, that no antient Author, beſides *Lucretius*, has taken notice of them in that View. His Lines are theſe:

> *Concharumq; genus, parili ratione, videmus*
> Pingere *telluris gremium, qua mollibus undis*
> *Littoris incurvi bibulam lavit æquor arenam.* L. II. v. 374.

> The Race of Shells, with ever-varying Birth,
> So *paint* the Boſom of the bounteous Earth;
> Where the calm Sea the concave Margin laves,
> And bathes the thirſty Beach with gentle Waves.

In the next Place, the Study of Flowers is more expenſive. I confeſs indeed, that a Collection of Shells cannot be prepared without ſome conſiderable Charge. But do not we daily ſee large Sums thrown away on a ſingle Tulip-root, the Succeſs of whoſe future Beauty is intirely precarious? And are not the Hopes and Fears of its Maſter determined by Showers and Sun-ſhine? Beſides, after much Pains and Solicitude, when the Flower is arrived at its utmoſt Luſtre, how ſoon does it fade! Whereas the Strength and Beauty of the Shell is fixed, and will ſtand through many Ages. Indeed, the only Argument that the Study of Flowers has to recommend itſelf to the Ladies, is, that it will admirably teach them how frail and fugitive is Beauty!

I ſhall only add, that I hope the following Book, from what has been ſaid, will be conſidered as not only written to gratify the Curioſity of the inquiſitive *Philoſopher*, but to improve the Imagination of the Female Artificer.

Of the MUREX.

BUT of all the Shell-fiſh, which are picked up on the Shores of this Iſland, the Purple-Fiſh is the moſt curious and valuable, and deſerves a diſtinct and a more accurate Deſcription.

We make no Scruple to aſſert, that this Purple-Fiſh is the ſo much celebrated *Murex* of the Antients; ſince it ſo well agrees with the Account which *Pliny* and other *Naturaliſts* have tranſmitted to us of the latter; as we ſhall endeavour to evince in the following Remarks.

And that we may proceed with as much Clearneſs and Perſpicuity as we can; it may not be amiſs to enumerate the ſeveral Appellations with which this particular Species or Sort of the Purple-Fiſh was diſtinguiſhed from the reſt, and to account for each as well as we may.

I. It is too notorious, that claſſical Authors, and more eſpecially the Poets, have *indiſcriminately* uſed the Words *Purpura*, and *Murex*; and have given the general Name, *Purple*, to the ſeveral particular *Tinctures* of theſe different Shell-Fiſh, as if the *Dye* of the one did not *differ* from the *Its ſeveral Names.*

the *Dye* of the other; and the *Purpura* and *Murex* were one and the *same* Fish. But let us, if it be possible, be more exact, and distinguish them as well as we can; which we shall attempt to do, by giving the Reader as clear and precise a Description of the *Murex* [for to that more especially shall we confine our Inquiries] as we are able to collect from antient Authors.

Κῆρυξ, *and* Buccinum. 1. The *Murex* is often called by *Greek* Authors, κηρυξ, and *Buccinum* by the *Latins*; both which Appellations are evidently bestowed upon it, because it belongs to the Tribe of those Shell-fish which are stiled *Buccinatores*; or, as *Horace* names them, *sonantes Conchæ*, sounding Conchs; whose Shells, being twisted, hollow, and fit for the Retention of the Air, resemble, in an imperfect Degree, that Shell, which the Antients used in giving their Alarms and Summons, both at Sea, and on Land; and which some of the Moderns at this Day carry and use in their Ships.

Murex. 2. And as these Names are attributed to the *Murex* from the *Form* or *Make* of its Shell; so we are induced to believe, that the Appellation *Murex* owes its Rise to the *Roughness* of it. For its Shell is studded with *Prickles*, ranged in regular Rows, as is the *Purpura* also, and many other Shells of this Class: So that the Name *Murex* is by no means *proper* to the Purple Conch we are speaking of; but belongs to a large Tribe, whose particular Families are variously distinguished. It is confessed, that *Pliny* (*a*) [if we understand him, and the Copy be clear of Errors] asserts, "that the *Buccinum* hath no *Prickles*, as hath the *Purpura*; and "that in the Shells of both there is a Number of circular Rows, or Ribs, "corresponding to the Number of their Years." But we presume this laborious Naturalist either means, that the Prickles of the *Buccinum* are not so *prominent* or *large* as those of the *Purpura*; or herein has forgotten himself. For if the *Buccinum* and *Murex* are the *same* Fish, and allowed by him (*b*) to be so; it follows, that the *Buccinum* must have the *Aculeos*, as well as the *Purpura*, since the Name *Murex* was evidently bestowed upon it, because its Shell is *pointed* or *prickled* (*c*). This is the proper Sense of the Word *Murex*; which is therefore applied to *Calthorps*, or *Chausse-trappes*, warlike Engines, which are armed with Iron Teeth, or Spikes; and its Derivative *Muricatus*, to every thing that is sharpened and pointed.

From these two Appellations, taken together, we may form a just Description of the *Murex*, and define it to be "*a Conch, studded on its*

(*a*) *Alterum* [*viz. Purpura*] *clavatum est ad turbinem usq; aculeis in orbem septenis fere; qui non sunt Buccino; sed utrisq; orbes totidem, quot habeant annos.* Nat. Hist. L. ix. c. 36.

(*b*) See Notes, *ibid.* And *Vossius* in his *Etymologicon* shews, that the *Murex* answers to *Aristotle*'s Κῆρυξ [i. e. the *Buccinum*], and is supposed by some to be derived from it. *In Voce* Murex.

(*c*) So *Isidorus* says, "*Murex* est Cochlea Maris, *dicta ab Acumine & Asperitate.*" Lib. ii. c. 6. Originum. Here *Isidorus*'s *Acumen* signifies just the same Thing, as *Pliny*'s *Aculei*.

"*outer*

" *outer Shell with sharp and regular Points*; *and twisted or wreathed in the* " *Form and Likeness of the Blowing-shell.*" In which *last* Particular it is distinguished from the other Purple-Fish, which is usually stiled the *Purpura.*

3. I need not add, that the *Murex* is often called *Ostrum* also: For from this Name we shall receive no further Information in our present Inquiry; unless it be, that as this Word means really no more than the *Shell* (*d*); so we may from hence conclude, that the Antients bestowed this Appellation upon it, by way of *Eminence*; to testify, as it were, the just Sense they had of its superior Excellence to all *other* Shells. Ostrum.

II. Thus far then as to the *Names* attributed by the Antients to the *Murex*; which have enabled us to discover the *Form* and *Fashion* of its Shell; which the curious Reader will find to correspond with the annexed Draught [Plate XXIV. Fig. 2.], which is taken from a Shell now lying before me. The *Murex* is a small Fish; the best and largest Sort of them, if we believe *Athenæus* (*e*), not exceeding one Pound in Weight. Those we met with on the Shore of this Island (as far we could discover) are much less, falling very short, even the greatest of them, of a Pound; and most of them being little more than half as heavy. The Dimensions of the Shell before us, which is one of the *common* Size of those, that are here found, will justify this Account: For on the *flat* Side, or *Mouth*, it measures no more than three Inches and an half in Length, and two in Breadth; and on the *Curve*, or outward Shell, four Inches and an half in Length; and over the thickest Part, three Inches and an half in Breadth: And this Measure rather *exceeds*, than *falls short* of the Truth. *Its Size.*

The Colour of the outward Shell is a dark Grey, sometimes faintly tinged with a Yellow.

(*d*) *Ostrum* is derived from the *Greek* ὄστρεον, a Name, which the *Grecians* bestow on all Sorts of Shell-Fish. It was also stiled *Conchylium*, for the same Reason, *i. e. the Shell-Fish*, from the same *Greek* Origin, κογχύλιον. But neither this nor that Appellation was peculiar to the *Murex*, properly so called; but attributed also to the *Purpura.*

(*e*) Γίνονται δ᾽ ἔνιαι τῶν μεγάλων καὶ μναιαῖαι· L. iii. C. 11. " There are some of the " *great* ones, and those of a Pound Weight." He calls them indeed Πορφύραι, the Purple-Fish; but I make no Doubt this Word must be taken for the *several* Species; for the *Murex*, as well as for the *Purpura*. But if this be not granted, it will follow, that if the largest *Purpuræ* do not weigh more than a Pound, the largest *Murex* will not weigh so much: For *Pliny* assures us, that the *Purpura* is *larger* than the *Murex*, Buccinum *minor* Concha, *L.* ix. *C.* 36. And though *Rondeletius* here corrects the present Reading, and changes *minor* into *major*; yet this Liberty with the Text is not to be allowed. And *Athenæus*'s Authority in another Place, *L.* iii. *C.* 11. is against it, who says, that the *smaller Purpuræ* are bred on the *Shores* and *Sea-coast* [the usual Haunts of the *Murex*, as we shall see hereafter]; and that their *Flos* is *red*; whereas *that* of the *larger Purpuræ* is mostly black, and sometimes, ἐρυθρὸν μικρὸν, a *little red*, *i. e.* of the *Violet* Colour. These two Circumstances convince us, that by the smaller *Purpura*, *Athenæus* means the *Murex*.

The Inhabitant of ſo ſmall a Shell cannot reaſonably be expected to abound with a large Quantity of ſo valuable a Liquor; and indeed an hundred of them are ſcarcely ſufficient to wet and diſcolour a Cambrick or Lawn Handkerchief of an ordinary Size.

Where found. III. As the *Shell* of the *Murex* differs from that of the *Purpura*; ſo the *Places*, which each more particularly delights to frequent and inhabit, are alſo different. *Pliny*, in recounting the various Kinds or Sorts of the *Purpura*, diſtinguiſhes them very circumſtantially (whether juſtly or no, let the Reader determine) by the Places of their Abode; ſtiling ſome *Pelagiæ (f)*, that love and haunt the *deep* Water; and are thoſe of which he ſpeaks, when he ſays, that ſome are caught with ſmall Nets, caſt into the Deep: Others again he calls *Littorales (g)*, ſuch as frequent the Shore; others *Lutenſes*, which delight in *Mud* and Slime; and others *Algenſes*, which ſhroud or hide themſelves in the *Sea-Weed*. But the *Buccinum*, or, what is the ſame, the *Murex*, ſticks and faſtens, he ſays, only on *Rocks*, and there or thereabouts is gathered or picked up *(i)*. And in this Situation I firſt diſcovered it. For, walking one Day on the Shore, on the North Side of the Iſland, and ſeeing a Shell-Fiſh, fixed to the Rock before me, I ordered a Slave to fetch it. The Lad, advancing towards it with too much Hurry and Impetuoſity, graſped it with ſo much Roughneſs, that his Hand was immediately ſtained with a florid and moſt delightful Crimſon.

This Accident, the Reader will ſuppoſe, awakened my Curioſity in no leſs a Degree, than one of much the like Nature did that of the *Egyptian Hercules*; whom the Inhabitants of old *Tyre* acknowlege to have been the firſt Diſcoverer of the *Murex*, and Inventor of the Purple-Dye. "He," as *Julius Pollux* relates the Story, "amuſing himſelf on "the Sea-coaſt, and perceiving the Hair of his Dog's Lips to be tinged "with a fine Red, was excited to examine more nicely into the Cauſe; "and diſcovered that his Dog had found, and been eating, the *Purpura*, or "Purple-Fiſh *(k)*." Or rather thus; as the Story is related by *Polydore*

(f) The Scholiaſt on *Apollonius*'s *Argonauts*, commenting on the Word πορφύρεσκεν, *L.* i. *v.* 461. explains it, κατὰ βάθος ἐνεθυμεῖτο, "He thought upon it *deeply* or ſeriouſly in "his Mind;" and adds this Reaſon, Πορφύρα γὰρ εἶδος ἰχθύος ἐν βάθει τῆς θαλάσσης εὑρισκόμενον; the *Purple* is a Kind of Fiſh found in the Bottom of the Sea. See alſo *Schol. in L.* ii. *v.* 548.

(g) I am induced to believe theſe *Purpuræ littorales* are the *Murex*; and that what he ſays of them, means no more, than what he tells us of the *Buccinum* or *Murex*, when he ſays, it is gathered or taken about or near the Rocks. And then the *Purpuræ littorales* of *Pliny* are the ſame as the *ſmaller Purpuræ* of *Athenæus*, which are bred on the *Shores* and *Sea-coaſt*.

(i) Buccinum non niſi petris adhæret; circaque ſcopulos legitur. L. ix. C. 36.

(k) No one will doubt, I believe, that this was the *Murex*; and yet we ſee the Grammarian calls it the *Purpura*: An Inſtance of what hath been before obſerved, that theſe Names are attributed *indiſcriminately* by Authors to the ſame Fiſh. And *Polydore Vergil*, in the Citation that follows, is guilty of the ſame Error.

Vergil,

Vergil (*l*), who hath given it a more sprightly and gallant Turn. "*Hercules* was in Love with a young Lady, a Native of that City, whose "Name was *Tyros*. A Dog that used always to attend her, creeping one "Day among the Rocks, found a *Purple-Fish*; and, having broken the "Shell, and eaten the Meat, returned to his Mistress with his Chaps tinged "with a *purple* or *scarlet* Dye. The Lady, delighted with this uncommon "and beautiful Tincture, declared to *Hercules*, when he was paying her the "next Visit, she would no longer receive his Addresses, till he had procured "her a Gown of a Colour as bright and splendid, as what he saw on the "Dog's Lips. This full and positive Injunction forced the Hero to go "in quest of the Shell-Fish; which being found, he not long after "brought the Lady the Present she had required, and became the Inventor of the Purple-Dye." The *Tyrians*, it is evident, gave so much Credit to the most particular Circumstance belonging to this Tradition, viz. *the Manner whereby the Purple-Fish was first discovered*, as to preserve the Remembrance of it long after on their Coins; some of which may be now seen, with the *Dog* and *Purple-Fish*, in the Cabinets of the Curious.

But to return: I shall offer my Remarks on the several Particulars of these Stories hereafter, in the Process of this Dissertation, and as they occur: And in this Place only take notice, that though these Authors differ very much in *other* Circumstances; yet they agree in *this*, that the Dog, both that belonging to *Hercules*, and that of the young Lady, found the Shell among the *Rocks*, which *Pliny* says is the *usual* Habitation of the *Murex*, and the *Place* where I first discovered the *Barbados* Shell; which we presume is a *third* Circumstance in favour of our Shell, towards proving it to be the *Murex* of the Antients.

IV. The *Flos* or *Tincture* of the *Barbados Conch* stained, as we before observed, the Hand of the Slave, with a *florid* and *bright* Red. Now *Aristotle* (*m*), in describing the κήρυκες, *i. e.* the *Murices* (for of them only he must be supposed to speak), uses almost the very same Words, saying, "The Juice or Liquor, being either squeezed or emitted, dyes the Hand "with a florid Colour." And *Pliny* (*n*), after him, assures us, that the Tincture of the *Buccinum*, mixed with that of the *Purpura Pelagia*, gives the latter *austeritatem illam, nitoremque illum, qui quæritur, cocci, the Fulness and Brightness of Scarlet.* Lastly, we observe, that the Shell which *Tyros*'s Dog picked up, emitted a Liquor, which tinged his Chaps with a *Scarlet-Dye* (*o*). This *Colour*, then, we suppose to be, among *Of the Flos, or Tincture.*

(*l*) *De Rerum Invent.* Lib. iii. C. 6.

(*m*) The Words are inserted into *Athenæus*, L. iii. C. 11. from *Aristotle*, by his *Latin* Translator; and are these, "*Expressus ille succus manum tingit florido colore.*"

(*n*) Nat. Hist. *Lib.* ix. *C.* 38.

(*o*) "*Sua labia* puniceo *colore infecit.*" P. Vergil. ibid.

other

other Marks, a *distinguishing* Note and Sign of the *Murex*, properly so called. For *Pliny* (*p*) assures us, there are only two Kinds or Genus's of the *Purple-Fish*; and that, from the different Mixture of the respective Juices of each, the various Sorts of Purple were made.

The *Colour* therefore of the *Flos* or Tincture is another favourable Circumstance, which strengthens our Comparison, and renders it probable, that the *Barbados Conch*, we are now treating of, is the *Murex* of the Antients. And here we desire the Reader to recollect *another* Particular in the forementioned Story, which must not be omitted: It is this; The Lady required, that *Hercules* should present her with a Vest or Garment of as beautiful and lively a Colour, as that, which he then saw on the Dog's Lips (*q*).

This Purple-Juice is reposited in a small transparent Bag, or rather Vein, on the Back of the Fish, not far from the Head. And when it is taken with a Pencil or otherwise from the Vein, the Shell being broken, it is of the Consistence of a thin Cream, of a yellow-greenish Colour; smelling raw, and very disagreeably; and the Wool, Cotton, or Linen, that hath been soaked in this Liquor a *few Hours*, afterwards assumes a deep scarlet Hue. But what shall we say is the Reason, why this same Liquor, when transfused or ejected through the Mouth of the living Fish, dyes the Hand of him who plucks it from the Rock, *immediately?* Is it because, when the Juice is percolated or strained through the Vein by the *living* Fish, it is mixed with some other Liquor, that *accelerates* this Alteration of Colours? Or because, when the Fish is *dead*, the *Flos* is itself *debilitated*; because less vigorous and able to exert its extraordinary Virtue? Whatever be the Cause, the Fact is certain: For the Liquor, when emitted by the Fish, *immediately* stains the Hand; but when taken out of the Vein of the dead Fish, it is *some time* before the Wool receives the Tincture.

But though we are not able to account for this sudden *Transmutation* of Colour; yet the *Use* of it arising to the Fish itself we think is evident; and affords us a fresh Instance of the Wisdom and Contrivance of the great Creator; who hath hereby furnished the *Murex* with the Means of

(*p*) *Concharum, ad purpuras & conchylia, eadem quidem est materia; sed distat temperamento. Duo sunt genera.* Buccinum *minor Concha —— alterum* Purpura *vocatur.* Lib. ix. C. 36.

(*q*) *Canis illius labris* splendidiorem, says *P. Vergil. ibid.*

(*r*) Buccinum *per se damnatur, quia fucum* remittit. Plinii L. ix. C. 38.

(*s*) *Athenæus*, from *Aristotle*, says, ἀνὰ μέσον τῦ μήκωνος ϗ τῦ τραχήλυ ἄνθος ἔχυσιν. L. iii. C. 11. Μήκων, or *Papaver*, is τὸ ἐνἶὸς τῦ ἰχθύος, *internum, seu intestinum piscis*, as he acquaints us from *Epænetus, ibid.* *Pliny* says, the *Flos* lies *in mediis faucibus*; which his Annotator corrects, and says, *inter hæc* [viz. *cervicem & papaver*] *supra ventrem flos situs est.* L. ix. C. 36.

securing

fecuring itfelf from Danger; which, as the Ink-Fifh, is directed, at the Approach of its Enemy, to emit this purple Liquor, which, being *nau-feous* and *offenfive*, annoys and beats off its Affailant.

The LARGE CONCH.

Buccinum maximum, labro maxime patente purpureo, clavicula muricata. Lift. Hift. Conchyl.

THIS kind of Fifh is of two Sorts, diftinguifhed by the Thicknefs or Thinnefs of their Shell.

That with the thinneft Shell is generally the largeft, and the other the moft ponderous.

The Outfide is of a brownifh White, ftudded at uncertain Diftances with blunt-knobbed Protuberances.

The Infide is finely polifhed; and its Colour, near the Extremity, of a pale Red; farther in, of a deep Maiden's-blufh.

The Head of the Fifh is guarded with a black horny Beak, or Tongue.

This, being extended out of the Shell, and fixed in the Sand, by a ftrong mufcular Motion, drags the Fifh with its cumberous Weight of Shell after it.

Ariftotle and *Pliny* are of Opinion, that the Ufe of this Beak, in fome of the *Conch* Kind, is to pierce thro' other leffer Shell-Fifh, which they prey upon.

Though this may be one of their Ways of Feeding, yet they are feldom deftitute of Food at lefs Trouble; for there grows upon the Outfide of their Shells a fine whitifh Mofs. This, in all Probability, is no lefs agreeable to them, than the green broad-leaved Mofs is to the *Tortoife*; for the whole Fifh (except the Tail, which remains in the Shell) is feen to come out of it to feed upon this Mofs, which it licks very clean from the Shell; fo that they always carry fome Part of their Food along with them.

They are likewife to be met with feeding, after great Floods, in the Sea, oppofite to deep Gullies, or Rivers; for as thefe Torrents generally carry a great Quantity of Land-Fruit, Leaves, and fuch Fæces, into the Sea; the *Conchs* are often feen to feed upon them.

They are evidently Male and Female. The *Penis* of the Male is above two Inches long.

The horny Beak above-mentioned is near two Inches in Length, and three Quarters of an Inch broad, fharp-pointed, and fomewhat crooked.

This is fortified with a ftrong middle Rib, and faftened to a tough cartilaginous Neck, as thick as one's Thumb.

The upper Part of this, by feveral crofs Indentings, is made as rough as a Rafp: It is with this Part that it licks, or rather fcrapes off, the Mofs that grows upon its own Shell.

About two Inches lower appear three cartilaginous Protuberances of a blunt conic Form.

Two of these are about an Inch and an half long, on whose Extremities appear the Eyes, surrounded by two bluish Circles.

The Third, which stands between these, is near as thick as a large Swan's Quill, and two Inches long; the Extremity of it ending in a Mouth, which, when the Fish doth not feed, is strongly contracted.

Somewhat below these, in the Male, appears the *Penis.*

When the Tail-part, which is gritty, and somewhat sandy, is taken away, the rest of the Fish, being fry'd, eats like Tripe, but shorter, sweeter, and more luscious.

The empty Shell, especially those that are thin, is made use of, instead of Sounding-Horns or Bells, to call out the Slaves to their Work in the Morning; for the Tip-end of the Shell being broken off, till the Cavity appears of about the Bigness of a Mouth-Piece to a French-Horn, they blow into it, and it sounds so loud, that it may be heard in a calm Morning, above a Mile off.

It seems that these were made use of as early as the Time of the *Romans*; for *Persius* says,

Buccina jam priscos cogebat ad arma Quirites.

Neither was the Use of the Fish unknown to the *Romans*; which appears by the following Lines.

Lubrica nascentes implent conchylia lunæ.
Sed non omne mare est generosæ fertile testæ.
Murice Baiano melior Lucrina peloris.

Hor.

------ *viles pellent obstantia conchæ.*

Hor.

Manilius likewise very well describes their burying themselves in the Sand; which they do during the Winter Months,

Sic submersa fretis concharum, & carcere clausa.

These are found in great Plenty at the Bottom of the Sea, in about five or six Fathom Water, in most of our Bays; but chiefly to the Leeward of the Island, especially in the Summer-Months.

If the Day is windy, the Divers cannot well see them from their Boats.

To remedy this rippling of the Water, they sprinkle a Spoonful of Oil upon the Surface of it.

This

This for a ſhort time unites the broken Surface of the Water; by which means the Divers can clearly perceive any Object at the Bottom: When the Conchs are thus ſeen, they dive, and bring them up.

The COW-HEEL CONCH, *or* HELMET-SHELL.

Buccinum roſtratum grande, raris lineis circumdatum, læve, non niſi ima parte cujuſque orbis ſtriata. Liſt. Hiſt. Conchyl. Tab. 911.

THIS is often five Inches long; the Outſide of a dirty White, and clouded at uncertain Diſtances with browniſh Spots: The firſt Circumvolution takes up the greateſt Part of the Shell: The Lip is not expanded, as in the large grey Conch; but is thickly welted, turning outwards: The incloſed Fiſh is eatable, as in the former.

The SMALL BROAD-LIPPED CONCH.

Buccinum breviroſtrum labroſum craſſum nodoſum, columella late plana. Liſt. Hiſt. Conchyl. Tab. 989.

THIS Shell is about three Inches long; its Mouth very wide, and its Lips much expanded: The firſt Circumvolution of this takes up the greateſt Part of the Shell.

The SMALL CONCH.

Buccinum minimum oblongum læve, e cinereo & fuſco variegatum, rictu anguſto.

THIS is about an Inch long: The Aperture for the Mouth is long, and narrow; the whole Shell ſmooth and ſhining, more or leſs ſpeckled with greyiſh-white and browniſh Spots.

The LARGE BROWN CONCHA VENERIS.

Concha Veneris major fuſca, cui maculæ fuſcæ albis circulis circumdatæ. Liſt. Hiſt. Conchyl. Tab. 698.

THIS is often near four Inches long, and about half as broad over the Middle, growing bluntly tapering towards both Ends: The Colour of this is of a ſhining Brown, interſperſed with whitiſh and black Spots: The Slit or Mouth is on each Side dentated.

The

The BROWN BLACK-SPOTTED SMALL CONCH.

Concha Veneris parva ſubfuſca lævis nigris maculis donata.

THIS is a very ſmall well-poliſhed Shell, not much exceeding a Quarter of an Inch. Its Back is high-raiſed, and variegated with blackiſh Spots.

The SMALL DARK-COLOURED CONCH.

Concha Veneris ſubfuſca lævis, elato dorſo biſaſciata. Liſt. Hiſt. Conchyl. Tab. 670.

THIS is from an Inch to two Inches long, of a dark-reddiſh Brown: In every thing elſe it reſembles the laſt-deſcribed.

The SMALL WHITE CONCHA VENERIS.

Concha Veneris alba.

THIS is a very ſmall white *Conch*, not much above a Quarter of an Inch long: Theſe often go here by the Name of *Rice-ſhells*, and at a Diſtance very much reſemble that Grain.

The MUSIC-SHELL.

This is what Doctor Liſter *calls* Buccinum muſicum grave, faciatum ex lineis quibuſdam interſectis, et maculatum, clavicula leniter muricata.

THIS Shell is about two Inches long, and near an Inch broad towards the Clavicle: Its Mouth is large: The Circumvolutions are ſcarce perceivable, ſeveral longitudinal Riſings ſupplying ſeemingly their Places: The Shell is thick and ponderous, and of a Fleſh-colour, regularly marked with blackiſh ſtrait Lines and ſeeming Notes, as if drawn for Muſic: Theſe are as regularly croſſed with other leſſer Lines: It is from theſe that it derives the Name of *Muſic-ſhell*.

The

The BEEF-SHELL.

Patella oblonga, articulata articulis striatis, extus subfuscis, intus e viridi cœruleis.

THIS is from one to two Inches long: The Shell, which is of a blackish-grey Colour, is divided into eight Joints laid over one another: By the Help of these the inclosed Fish can either bend inwardly, or expand and streighten its Shell at Pleasure. The Edges of the Shell are covered over with a greenish strong bearded thread-like Substance: The Fish, which is of a pale-reddish Colour, is very firm eating, short, and well-tasted. These Beeves are likewise distinguished into the Day and Night-Beeves.

The CORNUA AMMONIS.

THESE are properly called *Tubuli vermium albidi, vel e rubro fusci*; and are generally of a dull-white Colour, incurvated in many Revolutions like a Ram's Horn, but far closer together; the small gyral Extremity ending in a Point, the other being sealed or glazed over. There are some to be met with by far less curvated than the above-described. They are often found upon a Place called *Long Bay*, in the Parishes of *St. Andrew*, and *St. Joseph*.

The SEA-EGG.

THESE are distinguished into three Sorts; the black, the grey, and the shooting *Sea-Egg*. The two former are chiefly the eatable Kind: Their Appearance very much resembles an Hedge-hog in a defensive Posture: It is of a globular Form, depressed at both Ends; its Mouth being a small round Hole in the Middle of one of the depressed Surfaces, having five Teeth, the Points of all meeting in the Centre: The whole Shell is very thickly studded round with sharp-pointed Prickles of about an Inch long. These serve as offensive and defensive Weapons, the latter to guard it against the white Gavally, a Fish, which, instead of Teeth, hath strong bony Mandibles, and preys upon Shell-Fish. These Prickles serve also instead of Feet; for, by moving these, it can roll on in an orbicular manner to find out its Prey, or to avoid Danger. These Prickles are jointed to the Shells by a Socket, which exactly fits a protuberant Knob arising from the Surface of the Shell. From the Socket of these Prickles arise likewise many small cartilaginous Sinews, which terminate in the innumerable small Holes, which in a very beautiful Manner perforate the Shell. It is by the Help of these that the Prickles are capable of Motion. When it moves, which it does but ſlowly, it ſeems as if a Grove of Spears was in Motion. The Inſide of the Shell is lined with about five Lobes of a granulated yellow Subſtance, reſembling the Roe of a Fiſh: Theſe Lobes are in Length about three Inches, and in Breadth near an Inch: However, their Bulk depends much upon the Time of their being taken; for theſe Lobes are larger, and even better taſted, in the (*a*) Full than in the Wane of the Moon; but if not quickly eaten, or put into ſtrong Vinegar to harden, they very ſoon diſſolve into a rich reddiſh Liquid: A full-grown Shell is about fourteen Inches in Circumference. The Difference between the black, the grey, and the whitiſh Sort, is not worth mentioning: However, the black ſhooting *Sea-Egg* differs from all the reſt by the extraordinary Length of its Prickles, and its great Force in darting them to annoy its Enemy; which it does with that Violence, that I have known them to ſtrike or dart them thro' the thick fleſhy Part of the Toe-nail of a Fiſherman. The *Sea-Eggs* were known to the *Romans*; which appears from the Words of *Horace*.

---------*Miſeno oriuntur echinie*

Horret capillis, ut marinus, aſperis,
Echinus. Epod. V. 27.

It appears from a Paſſage in *Lucilius*, that the Antients were not well acquainted with the proper Seaſon for taking this Fiſh, as being larger and more delicate in the Full of the Moon.

-----*Luna alit oſtrea, et implet echinos.*

Theſe *Sea-Eggs* are to be found in the Bottom of fine ſandy Bays, in about ſeven or eight Feet Water. This is delineated in Plate XXVI. Fig. 1. Fig 2.

The PLATE-FISH, *or the* FLAT SEA-EGG.

THE *Plate-Fiſh* hath a flat Shell ſomewhat diſhing inwards on one Side: It is generally from two to three Inches Diameter. The upper and the under Shell are ſo thin, and ſo cloſe together, that the incloſed Fiſh is very ſmall. The ſeveral thouſand little Holes in theſe Shells, by which, I ſuppoſe, Nouriſhment is conveyed to the Fiſh, are no leſs remarkable for their Number, than their Regularity and Beauty. This

(*a*) This evidently confutes Mr. *Rohault*, who ſays, that the Moon hath no Influence in repleniſhing at the Full, or leſſening at the Wane, the Meat that is incloſed in theſe and ſuch-like Shell-Fiſh.

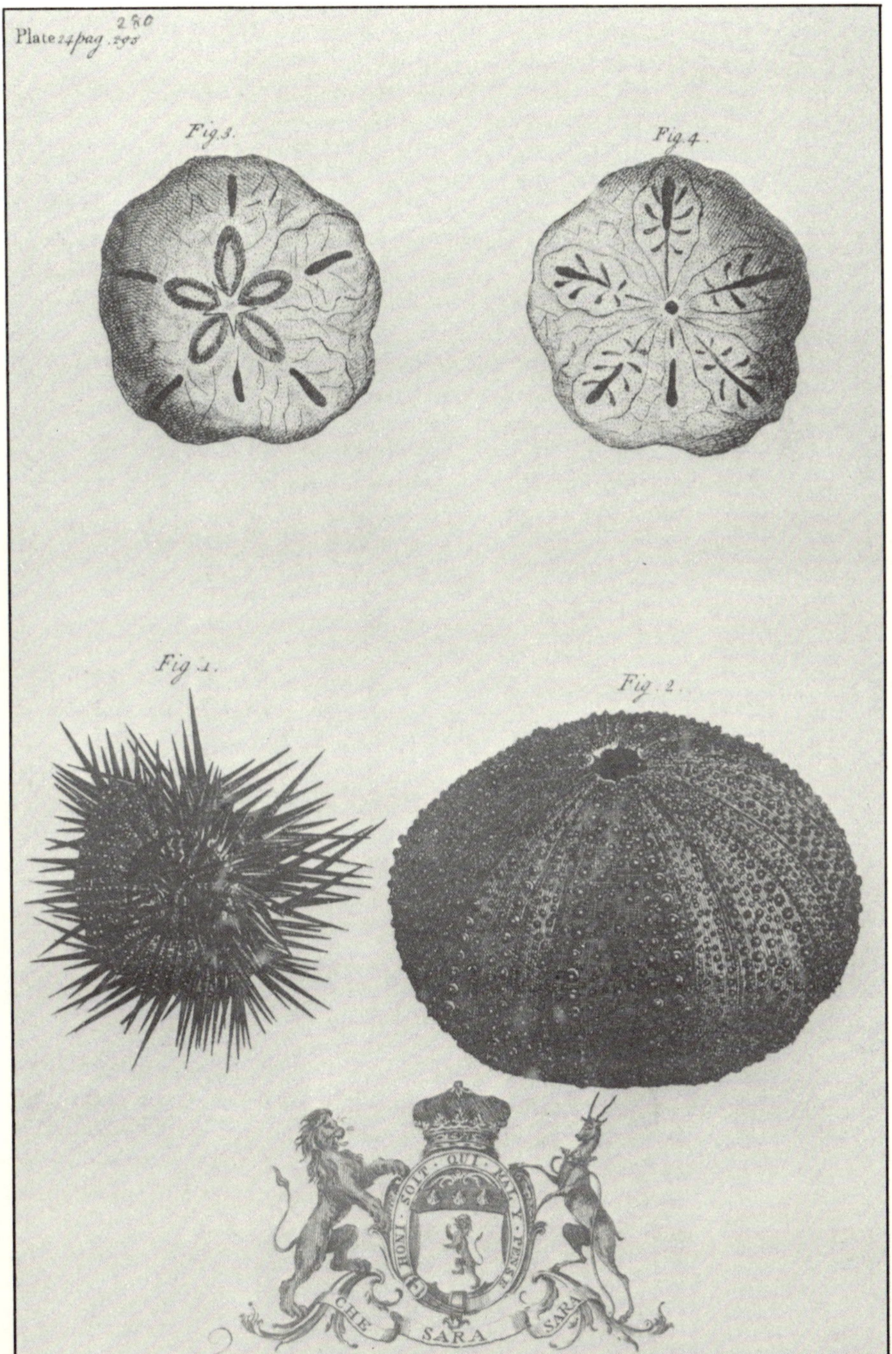
Plate 24 pag. 295
280
Fig. 3.
Fig. 4.
Fig. 1.
Fig. 2.
HONI SOIT QUI MAL Y PENSE
CHE SARA SARA
To His Grace The DUKE of BEDFORD
This Plate is humbly Inscrib'd.
&c.
J. Mynde sc.

Shell-Fish is generally found slightly covered with Sand, which I take to be their Security from being perceived by Crabs, Conchs, or other Animals, which prey upon them: Their Surfaces are covered with thick setaceous *Villi*; and the Middle of the Shell is by small Punctures divided into the Appearance of five narrow rosaceous Leaves, as well as four oblong Holes of near half an Inch long, and a quarter wide. Perhaps Nature intended these, that the Fish might fasten itself by them to the Points of Stones or Rocks in the Bottom, lest the Current should drive them away from their Feeding-place, or Place of Safety. Plate XXVIII. Fig. 3. Fig. 4.

The MUSSEL-SHELL.

THIS is not here above an Inch and an half long at most; and the inclosed Fish is proportionably smaller, a dozen of them not weighing an *English* one: These I have found in Beds to the Leeward of the Island in the Parish of *St. Lucy*.

The LARGE WHITE COCKLE.

THESE are found near *Needham*'s *Fort*. The Shell is a great deal larger than that of the *English* Cockle, and flatter.

The RIBBED LIMPET.

THESE are called here the *Nipple-Shell.* Some are of a white Colour, others brownish, and some spotted: There are likewise of these Species some that are smooth.

The THIMBLE-LIMPET.

THIS is smaller, and more compressed, than the last-described.

The ROCK-OYSTER.

THIS is a rocky rather than a shelly Substance. It grows to the Side of mossy Shoals and Rocks at different Distances from the Shore. When opened, they have the Resemblance of a Shell on the Inside, and contain a fleshy Substance, in Taste very much resembling an Oyster, differing very little, if any thing, from the *Mangrove-Oysters* so ſo commonly growing to the Roots of thoſe Trees, in ſeveral of our *Weſt-India* Iſlands.

The RED SPECKLED WILK-SHELL.

THESE Shells are beautifully ſtreaked with ſmall broken Veins of Red upon a white Ground: They are to be found in Plenty clinging to the Rocks on the Leeward-ſide of the Iſland, eſpecially in *St. Lucy*'s Pariſh.

The BLUE *and* WHITE-MARBLED WILK-SHELL.

THIS is of the Make and Bigneſs of the laſt-mentioned, and differs only in this, that the white Ground of this is finely ſpeckled with blue broken Streaks: Theſe are likewiſe found chiefly in the Leeward-part of the Iſland.

The BLACKISH WILK-SHELLS.

THESE are of the Bigneſs of the laſt deſcribed; and I take them to be the ſame Species, but that the Shells of theſe are turned black with Age.

PUNCHES.

THESE long ſlender Worms are very ſeldom ſeen. Their Habitations are innumerable Cluſters of wreathed vermicular open-mouthed Tubes, cemented together into irregular Lumps, and ſometimes into large Beds many Feet in Circumference: Theſe Tubes differ in Colour, being ſome of an Amber-colour, ſome black, and others of a dirty White. The Aperture, by which the Worm enters, is round and open, of the Bigneſs of a Raven's Quill. The Edges of theſe Tubes are ſo ſharp, that they will, if trod upon, cut a circular Inciſion in the Fleſh, leaving the middle Space of their Diameters untouched: This Puncture exactly reſembles the Impreſſion made upon Leather by an Inſtrument, which the Shoemakers call a Punch. Theſe are generally found between high and low Water-Mark.

The

The PRICKLY HAM-MUSSEL.

THIS Shell is often eight Inches long, ſmooth within, and rough without; being ſtudded without jutting ſeeming Tubes. This is ſeldom found with Fiſh in them on our Shores.

The LARGE THREAD-GIRDED COCKLE.

THIS is a large flat Cockle, generally found in the Sand near *Needham's Fort.*

GRATER SCALLOP.

THESE are here but ſmall, and very ſeldom found on our Shores.

CURL-GIRDED NEEDLE.

THIS is very ſharp-pointed, of about two Inches long and wreathed. From the Uſe made of it, it is often called here *the Tobacco-Stopper.* I have found theſe upon the green Shoal near *Six Men's* Fort in *St. Peter's* Pariſh. The Fiſh inhabiting this Shell, when it moves, carries it upright: When Numbers of theſe are together, and in Motion, they reſemble a Grove of Spears.

TRITON'S TRUMPET.

THIS is the largeſt, as well as the moſt beautiful of the Turbinate Kind, and very ſeldom to be met with upon our Shores: Theſe are beautifully ſtained with black and white Spots, and the Shell is often nine Inches long.

The Smooth Grey Casket.

A Small Species of theſe are often found on our Shores, but ſeldom any of the large Sort: What are here found are generally of a yellowiſh dirty-white Colour.

The Spire-Shell.

THESE are generally ſmall, and very ſmooth, of a reddiſh Brown, ſpotted with Black. There are others of a Pale White, ſpotted with Black; both Sorts generally well poliſhed. Theſe are very common upon almoſt all our Shores.

The Welted Casket.

THIS is generally of a brown Colour, and furrowed length-ways: It is of a far ſtronger Conſiſtence than the laſt-deſcribed.

The Grey Casket.

THIS is generally of a grey Colour; and ſometimes they are found of a browniſh Grey. Their Texture is generally very thin.

The Whirligig.

THIS is generally of a bluiſh Colour, and is often found upon the ſeveral Shores of the Iſland.

The Knobbed Trumpet.

THIS Shell is of a dirty White, as well as ſome of a brown Colour; and found in many of our Bays, but not in great Plenty.

The Smooth-lipped Casket.

THIS Shell is often found on the Shore at *Maycock's Bay* in the Pariſh of *St. Lucy*.

Agate-

AGATE-STAMPER.

THIS Shell is from half an Inch to two Inches in Length, generally of a reddish Brown spotted with Black: They are likewise found of a pale-white as well as bluish Colour, and very common upon the Shores of this Island.

The ROYAL STAIRCASE.

THIS is a small white spiral Shell, belted with a sharp rising Ridge from the Top to the Bottom in the Form of a wreathed Staircase.

The SPOTTED CYLINDER-SHELL.

THIS is about two Inches long, of a dirty Colour, white speckled with sometimes black, sometimes brown Spots.

The BROWN CYLINDER-SHELL.

THIS is thin, and well polished; seldom longer than two Inches, and of a reddish-brown Colour.

The MOLE COWZY.

THIS is very seldom found on the Shores of this Island: What I saw was three Inches long, of a brown Colour, and faintly spotted with Black.

The TRIANGULAR STRIATED BUCCINUM.

THIS is a middle-sized Shell, thickly labiated, and pretty sharp-pointed; the several *Striæ*, on the Back-part especially, running transversly and deep.

The

The Top-Shell.

THIS is a large Shell, ſpotted moſt commonly with Blue and White, ſometimes with Red and White: When this rough outſide Coat is taken off, the inner Part conſiſts of fine Mother of Pearl. The Mouth-piece, or Cover of the Mouth, is of an horny Subſtance, of a browniſh-black Colour, incircled with ſeveral Ringlets of a Gold-colour. Theſe are generally found in deep Water, and in great Plenty in the Sea, oppoſite to *St. Lucy*'s Pariſh. The laſt Stage of an *Hermit-Crab* or *Soldier* is, to inhabit one of theſe empty Shells.

THE

THE Deſcription of our ſeveral exanguious Tribes of Animals being completed; our next Inquiry will naturally fall upon our Submarine Plants. I ſhall divide theſe into three Claſſes: The Firſt and moſt perfect is the large black Sea-Rod, which often grows to the Height of a ſmall Tree; this, as well as each of this Species, is ſmooth, black, and ſhining, of a very tough Subſtance, ſomewhat between Horn and Wood, ſmelling ſtrongly of the latter when burnt. The Second is very buſhy, ſeldom exceeding four Feet in Height; its very numerous Branches ariſe almoſt together from very near the Root, and are in general of an equal Height and Bigneſs: This differs from the naked black Sea-Rod, by having their Surface covered with a calcareous Cruſt (of about the Thickneſs of an *Engliſh* Shilling), and here and there marked with Aſterisks; the innner Texture of the Plant being of the ſame Nature as the above deſcribed. The Third and laſt are all Kinds of Coralline Bodies.

The LARGE BLACK SEA-ROD.

THIS often grows to about fifteen Feet high, and always in a conſiderable Depth of Water; its Branches are few, but ſtrong and tapering; the Outſide appearing almoſt as ſmooth, black, and ſhining, as Ebony; and if firſt boiled in freſh Water, and afterwards often oiled, they will be very pliable and tough, and make very beautiful Switches.

The LESSER SEA-ROD.

THIS is a ſlender ſmall black Rod, having a great many very ſmall weak undivided Side-branches. This Species grows between high and low Water-Mark in the Clefts of Rocks, ſeldom riſing above five Foot high.

The SEA-FEATHERS.

THESE ſeldom grow, on the Shores of this Iſland, above two Feet and an half high; they likewiſe ſeldom or never differ in their Make or Texture, being each compoſed of a ligneous horny Subſtance, and beautifully reticulated. Some are of a darkiſh Colour, and others of a light Grey: I obſerved ſome of the latter to have their Edges bordered with a faint Yellow, ſlightly tinged with Green.

The INCRUSTED SEA-RODS.

THESE are always in Bunches, ſeldom growing above four Feet high; their many Branches ſpring from one common Stalk, very near the Root: Theſe are almoſt all of a Size, in Length and Thickneſs; and the Outſide covered with a calcareous Cruſt, of about the Thickneſs of an *Engliſh* Shilling. The Surface of this is pitted with many very ſmall Star-like Holes. This is delineated in Plate XXVII. Fig. 1.

The SMALL INCRUSTED SEA-ROD.

THIS ſeldom grows above a Foot and an half high; and its Surface, inſtead of being, as the laſt-deſcribed, covered all over with a thin pliable calcareous Cruſt, is here-and-there annulated with irregular knotty Bandages, of a white coralline Subſtance, reſembling the ſmall Star-Coral. This Kind of Sea-Rod branches very little, and its Branches are but very ſlender.

OF

Plate. 27. Page
Fig. 1
To the Right Hon.ble the
Earl of Sandwich,
First Lord of the Admiralty
This Plate is humbly inscrib'd
Fig. 2
Ehret del.

OF

CORALS.

THOUGH these, in general, are Vegetables; yet we are not to look upon them among the lowest Class of this Kind, because they bear, at least here, neither Leaves, Flowers, nor Fruit, having likewise their Consistences so brittle, that they are neither malleable, nor any ways pliable: However, their innumerable Shoots and Branches are not void of Beauty, nor useless in medicinal Preparations; the white Sort being, when pulverized, esteemed good to free the Stomach from acid sour Juices; and the red Coral is not less efficacious in stopping Fluxes: From the Growth of these, we may likewise observe, that Providence is not tied down in its Operations to mechanical Rules: For among Vegetables, which flourish upon the the Surface of the Earth, all our Art and Contrivance are in vain, unless we can procure them either natural or artificial Heat, and a Communication of Air; whereas these grow to a great Length in above forty Fathom Water, where the Heat of the Sun cannot penetrate.

I shall begin the Description of Corals with that called the *Harts-horn Coral.*

The SEA-GINGER, *or the* PALMED HARTS-HORN.

THIS is a digitated Sea-Coral, and is called Ginger from its very hot Quality; for if a Piece, newly broken, be apply'd to the Tongue, it tastes excessively hot. It is found in great Plenty upon the Shores in the Island of *Antigua*, as well as among many other Places upon the *Pelican shoal* in this Island.

This is delineated in Plate XXVII. Fig. 2.

The

The HART'S-HORN CORAL.

THIS takes its Name from the great Resemblance its wide-extended Branches have to an Hart's-horn, both in Colour and Shape: They grow upon Rocks, at different Depths under Water: The Branches are of a brownish White, but always at the Top sharp-pointed, and tipped with a clear White: They are of so close a Texture, that, when struck by any Piece of hard Wood or Iron, they afford a metallic vibrating Sound: This Sort grows to be often above five Feet in Length. I have likewise seen a lesser Sort, resembling the palmed Hart's-horn.

The PAN SHOAL, *or* HONEYCOMB STONES.

THESE are to be found under Water at different Depths, from two to twenty Feet: Their Shape generally resembles a Honeycomb, and that not only as they are broad with a similar Thinness, but likewise as they are full of Holes: Several of these, especially near the thickest Edge, are cemented, or, as it were, waxed over, exactly resembling that Part of an Honeycomb which is filled with Honey. From the Stone when broken, drops a liquid Substance of the Consistence of thin Oil. These in few Years grow to a considerable Bigness, and are to be found upon almost all the Shores, more particularly at a Place in *St. Lucy's* Parish, called *Fryer's Well.*

The WHITE CORAL.

THIS Coral is found upon the Shores of most Bays in the Island: The Stars are composed of so many thin Partitions; the intermediate Space, between each Line, hollow; and the solid waved Interstices of the Coral-smooth, hard and well-polished: These intervening Ribs rise higher than the Star-like Resemblances.

This is delineated in Plate XVII. Fig. 7.

The MILLIPORA PERFORATA.

THIS is all over perforated into innumerable small Holes, generally of an equal Size: I am of Opinion, that these Holes were formerly filled with star-like Substances; but, as the Rays which compose these, are from their softer Texture, as well as their hollow Interstices, less able to bear the violent Tossings of the Waves against the sharp-pointed Rocks, that these Stars, in time, were so intirely broken, nothing now remains, but the Holes which they once occupied, among the stony Ribs.

The

The LARGE WHITE FINE-RAYED STAR-CORAL.

THIS is very often of a considerable Bulk, and its Surface thickly crouded with very small Stars, whose Rays are scarce perceptible to the naked Eye.

The KNOTTY STAR-BEAMED CORAL.

THIS is to be found on all our Shores, and very much resembles that of the same Class found in the *Mediterranean* by Dr. *Shaw*, with this Difference only, that the Extremities of this are of a blunt conic Shape.

The TUBULAR CORAL.

THIS spreads into several thick short Branches studded all over with innumerable small open-mouthed Tubes, the Whole generally of a brownish Colour without, and white within.

The COMMON BRAIN-STONE CORAL.

Fungus coralloides encephaloides, gyris in medio sulcatis, lamellatis, serratis. Boerh. Ind. Plant. p. 1.
Lapis fungites cerebriformis. Ray's Hist. App. p. 1850.

THIS, as do many others of the same Species, derives its Name from its Resemblance to the human Brain, the waving white Ribs rising higher than the intermediate Spaces: These Stones are to be seen very common on all our Sea-shores, as well as on the Land, especially about *Black Rock*, in *St. James*'s Parish, and generally near the Sea.

The LARGE-RAYED BRAIN-STONE.

Fungus coralloides, laminis magis undulatis. Boerh. Ind. Plant. p. 2.

THIS hath its several sinuated Risings and Depressions intirely rayed, whereas these Risings in all other Kinds of *Brain-stones* are so many solid Veins dividing the porous Part into many sinuated Interstices.

What is called here the fine ribbed *Brain-stone* differs from the above by having many hard sinuated Veins running through it, and rising higher than its porous Interstices.

The Small Star-Coral.

THIS differs from the last described, by having its Stars even with the Superficies of the intervening Ribs.

The Very small Star-Coral.

Fungus coralloides, stellis minoribus, rupibus corallinis accrescens. Boerh. Ind. Plant. p. 2.

THE Surface of this is almost intirely covered with the Appearance of very small Indentings like Stars.

The Large Star-Coral.

THE Surface of this is indented with the Appearance of Stars, far larger than those already described.

The Wire-Sponge.

THIS is of the Sponge-kind, though it hath no Elasticity: Its Substance is far more reticulated than any other Sponge: It grows chiefly upon the broken Stumps of Sea-Rods: It generally grows of a cylindrical Form, and hollow in the Middle: It is of the Colour of a common Sponge, and its Parts are as stiff as if they were composed of very fine Wire.

The Soft Sponge.

THIS Class cleave to the Rocks, or rather grow from them, as *Jews-Ears* do from Trees.

The Beamed Coral.

Madrepora ramosissima fusca, halcyoniis et ostreis accretis insignis. Boerh. Ind. Plant. p. 5.

THE Specimen I saw here has no Oysters or *Halcyonia* on it; but it hath a fine brown *Tubulus Marinus.*

I Shall conclude this Book with the Description of that surprising Creature the *Animal Flower*, first observing that the remarkable Place called the *Spout*, in *St. Lucy*'s Parish, is a large Hollow under a rocky Cliff impending over the Sea: As the Waves continually rush with a violent Force into this hollow confined Part, it throws upwards, through an Hole in the Surface of the Rock, a considerable Body of Water, from fifteen to fifty Feet high, containing several Hogsheads. It may be seen at several Miles Distance.

THE

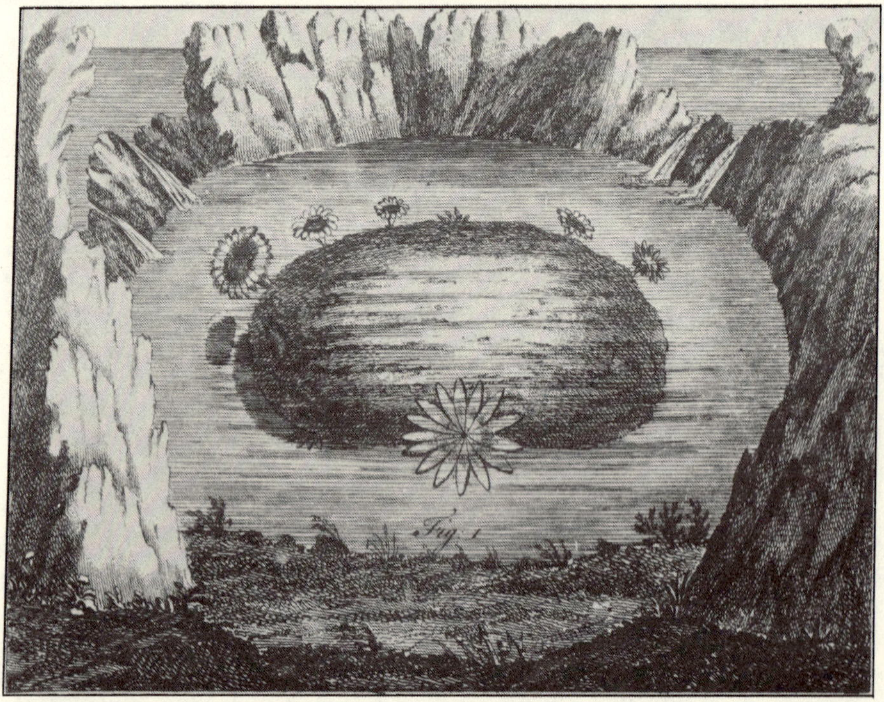
Fig. 1

Fig. 2.
NEC CUPIAS
NEC METUAS
To the Right Honourable
Philip Lord Hardwick
LORD HIGH CHANCELLOR OF GREAT BRITAIN;
This Plate is humbly inscrib'd.
&c.
J. Mynde sc.

THE

ANIMAL FLOWER.

AS in Man, the moſt perfect Part of the ſublunary Creation, there are apparently ſeen ſeveral different Degrees of Perfection of Body and Mind ; and in Animals the Sagacity of ſome is evidently ſuperior to that of others ; ſo likewiſe in this ſeemingly confuſed Species of animal Life, and vegetable Appearance, the Chain gradually deſcends with a ſurpriſing Mixture and Connexion.

Whoever hath Leiſure and Abilities to purſue a general Inquiry of this Nature, will ſoon find, that this progreſſive Series runs through the whole Creation ---- From the moſt exalted Genius to the almoſt ſenſeleſs Idiot ---- From the moſt ſagacious ſenſible Creature to the almoſt inſenſible Muſſel ---- From the towering Cedar to the Hyſſop ſpringing from the Wall, or the humble Moſs.

Such is that univerſal Harmony and Connexion, that runs through the numberleſs Ranks and Orders of Beings, till we come at laſt to inanimate Matter.

This ſurpriſing Creature, that I am to treat of, hath, for a long time been the Object of my own ſilent Admiration ; and it would even now be thought chimerical to mention, much more deſcribe, the Qualities of ſo ſtrange a Phænomenon, if the *Polypus* of late Years had not afforded a ſurpriſing Inſtance of Almighty Power.

The Cave that contains this Animal, is near the Bottom of an high rocky Cliff facing the Sea, in the North Part of the Iſland, in the Pariſh of *St. Lucy :* The Deſcent to it is very ſteep and dangerous, being in ſome

Places almost perpendicular; and what adds an Horror to this dreadful Situation, is, that the Waves from below almost incessantly break upon the Cliff, and sometimes reach its highest Summit.

As soon as you are freed from this complicated Apprehension of Danger (in your Way down) you enter a Cave spacious enough to contain five hundred People. The Roof of this is in some Places imbossed with conglaciated Incrustations intermixed with small Tubes, through whose Extremities a small Quantity of the most limpid Water drops.

From this you enter another Cave, small in Comparison of the former. The Bottom of this is a natural Bason of Water of about sixteen Feet long, and twelve in Breadth. This, at low Water, is about eleven Feet perpendicular Height from the Sea, which, when the Wind is high on that Point, dashes into it; so that the Water in it is intirely salt, except a small Mixture of fresh, which ouses and drops through the Roof of the Cave.

In the Middle of this Bason there is a fixt Stone, or Rock (as I shall call it), which is always under Water.

Round its Sides, at different Depths (seldom exceeding eighteen Inches) are seen at all Times of the Year several seemingly fine radiated Flowers of a pale Yellow, or a bright Straw-colour slightly tinged with Green.

These have in Appearance a circular Border of thick-set (1) Petals, about the Size of, and much resembling, those of a single Garden Marigold, except that the Whole of this seeming Flower is narrower at the *Discus*, or Setting on of the Leaves, than any Flower of that Kind.

I have often attempted to pluck one of these from the Rock to which they are always fixt; but could never effect it. For as soon as my Fingers came within two or three Inches of it, it would immediately contract, and close together its yellow Border, and shrink back into the Hole in the Rock; but, if left undisturbed for the Space of about four Minutes, it would come gradually in Sight, expanding, though at first very cautiously, its seeming Leaves, till at last it appeared in its former Bloom: However, it would again recoil with a surprising Quickness, when my Hand came within a small Distance of it.

Having tried the same Experiment by attempting to touch it with my Cane, and a small slender Rod, the Effect was the same.

These were strong Appearances of Animal Life; yet, as its Shape, and want of local Motion, classed it among Vegetables, I was for some time in Suspense, and imagined it might be an aquatic Sensitive Plant: And though its Contraction to avoid the Touch was quicker than any Plant of that Kind; yet, as its seeming Leaves might be, and in reality

(1.) Petals are the fine coloured Leaves, which compose in a Marigold, and such-like Flowers, the yellow circular Border.

They are called Petals to distinguish them from the green Leaves of the Plant.

were,

were, of a far thinner and more delicate (1) Texture than thoſe of any Plant; and as Water is eight hundred times heavier than Air, the ſudden Weight of ſo thick a Medium, by its Undulation cauſed by the Preſſure of my Hand or Stick, might very well account for its ſudden Contraction.

This was my Opinion, till a ſubſequent Viſit cleared my Doubts; for I plainly ſaw four dark-coloured Reſemblances of Threads ſomething like the Legs of a Spider, riſing out of the Centre of what I have termed a Flower. Their quick ſpontaneous Motion from one Side to the other of this circular yellow Border of ſeeming Leaves (which in reality were ſo many Arms or Feelers), and their cloſing together in Imitation of a *Forceps*, as if they had hemmed in their Prey (which the yellow Border likewiſe ſoon ſurrounded and cloſed to ſecure), fully convinced me, that it was a living Creature.

Its Body at a Diſtance appears to be about as big as a Raven's Quill, and of a blackiſh Colour; the one End ſticking to the Rock, the other extending a very ſmall Diſtance from it; and incircled round with a yellow Border, as above deſcribed.

Thus what in its firſt Appearance ſeems to be of the vegetative Kind, by its Motion, and quick Senſe of Self-preſervation, proves an Animal.

Now, ſince the ſame Wiſdom and Goodneſs, which give Being to Creatures, often preſerve them in that Exiſtence by Ways and Means as wonderful as their Creation was before; this leads me to offer a probable Conjecture, why God's amazing Providence (which doth nothing in vain) endued the Arms or Feelers of this Animal with a fine yellow Colour, and hath ordained it to differ in this Particular from the ſeveral Tribes of fungous Animals, that are always found cleaving to the Rocks in the Sea.

As theſe latter may be fed with Spawn, or ſome Animalcules, which the Flux or Reflux of the Waves may throw in their Way, there was no Need of any uncommon Means to intice their Prey (if Animals) within their Reach; whereas the Water in the Cave is, for the moſt part, void of any Motion that can convey Food for theſe Animals. Therefore there was a Neceſſity of ſome extraordinary Temptation to allure their Prey within their Power, to ſeize it; otherwiſe they might ſtarve in the midſt of Plenty.

To this End, that Divine Goodneſs, which filleth every thing living with Plenteouſneſs, hath finely deviſed this providential Stratagem (if I may

(1) Tho' I could not by any means contrive to take or pluck from the Rock one of theſe Animals intire; yet I once cut off (with a Knife which I had for a long time held out of Sight near the Mouth of an Hole, out of which one of theſe Animals appeared) two of theſe ſeeming Leaves. Theſe, when out of the Water, retained their Shape and Colour; but, being compoſed of a Membrane-like Subſtance, ſurpriſingly thin, it ſoon ſhriveled up, and decayed.

be allowed the Expreſſion), and given theſe Animals that fine tranſparent Colour, to be a Means to provide for them their daily Food: For as bright Rays of Light (or ſomething ſimilar in its Effect) are very inviting to ſeveral Animals, eſpecially thoſe of the aquatic Kind, the beautiful Colour of this circular Border may ſerve as a Decoy for very young Fiſh, or other Animalcules, to divert themſelves (as Flies about the Flame of a Candle) in ſwimming about the Verge of this ſeemingly harmleſs Flower, until they come within the Circle; then theſe bright Leaves in Appearance prove, in Reality, ſo many Arms or Feelers, that with a quick Motion cloſe together, and ſurround the Prey; which, being thus ſecured, is conveyed to the Mouth, as above-mentioned.

There are likewiſe on the uppermoſt Part of the Rock, in the above-deſcribed Baſin, innumerable Cluſters of (what are here called) Water-Bottles, very much reſembling ſcattered Cluſters of unripe Grapes; the Outſide conſiſting of a bluiſh ſkinny Tegument, like that of a Grape; the Inſide full of Water ſomewhat turbid.

Among theſe alſo are a great Number of Animal Flowers of the ſame Species with the yellow large ones. Theſe now to be deſcribed are likewiſe fixt to the Rock, not in Holes, as the above-mentioned, but ſticking to the Surface among theſe Water-Bottles, and generally not above nine Inches under Water.

The Leaves, or rather Feelers, of theſe are of a greyiſh-purple Colour variegated with black Spots. Their Motion likewiſe to avoid the Touch is not ſo quick.

Having plucked one of theſe from the Rock, I perceived the Body, which was about an Inch long, to have, whilſt between my Finger and Thumb, a ſenſible vermicular Motion. The Feelers likewiſe, which decorated one End of it, when expoſed to the Air, ſhrunk up, and remained as lifeleſs: But as ſoon as the Whole was dipped in their proper Element the Water, they would immediately, as it were, aſſume a new Life, and appear again in their full Vigour.

Soon after the Diſcovery of theſe ſurprſing Animals, a great Number of People came to view them: But as this was attended with ſome ſmall Inconveniency to a Perſon, thro' whoſe Land they were obliged to paſs; he therefore, to get rid of the Company, reſolved to deſtroy the Object of their Curioſity: In order to do ſo effectually, he took a Piece of Iron prepared for that Purpoſe; and then carefully bored and drilled every Part of the Holes where theſe ſeeming Flowers were bred; but, to his great Surprize, they in a few Weeks appeared again iſſuing from the ſame Holes.

Let us here, for awhile, ſtop, and ſee whether our much boaſted Reaſon can find out how even a latent Principle of Life can be preſerved, after the whole organic Body is torn in Pieces.

When we ſee this Animal, in a ſhort time after, reſuſcitate, and appear in its former Proportion, Beauty, and Life, can we, after ſuch an ocular

lar Demonſtration of ſo aſtoniſhing a Change in a Creature deſtined for this Life only, and removed (in all Appearance) but a few Degrees from the vegetable Creation, any longer entertain Doubts about the Poſſibility of another Doctrine of a far greater Conſequence? And as every paſt Age hath been, ſo undoubtedly every future will be, bleſſed with ſome ſurpriſing new Diſcovery of God's unſearchable Power and Wiſdom.

Our own hath produced a wonderful Inſtance of this; for what Sceptic, ſome Years ago, would have believed the Poſſibility of ſo extraordinary a Production as the *Polypus*? Who would not have ſaid, with the unbelieving *Jews* in the Wilderneſs, Can God do this? And yet we find, that this ſurpriſing Generation is now a known Matter of Fact.

That the above-mentioned Conjecture about the Uſe and Efficacy of its Colour is not groundleſs, may be made ſtill more evident, by many analogous ſtriking Inſtances.

For thoſe ingenious Gentlemen, Mr. *Turberville Needham*, and Mr. *Trembley*, obſerve, that Polyps, and aquatic Inſects, kept in Glaſs Veſſels, by excluding the Light from every Part, except one little Opening, after ſome time all aſſembled at this Opening; and yet theſe Polypes have not, perceptibly to the ſtrongeſt Magnifier, any Organ that in the leaſt reſembles Eyes.

If Light is therefore ſo attractive to theſe Animals which are viſible, why may it not be likewiſe ſo to other Animalcules to us imperceptible? And may we not further ſuppoſe, that the Appearance of the former towards the Light may be in Search of theſe Animalcules, their deſtined Prey?

But in what manner the Rays of Light affect theſe Animals, whether by its Motion acting upon their whole exceedingly delicate nervous Syſtem, which, like the *Retina* of the human Eye, is in every Part ſenſitive, is, I believe, inexplicable.

Where Sight is apparently wanting, as in Polyps, a Delicacy of Touch may, for ought we know, and indeed in all Probability doth, take up the gradual Chain, and, in a ſurpriſing manner, ſupply its Place.

Such is the inſenſible Gradation, which is progreſſively continued by imperceptible Degrees thro' the whole Creation, from animate to inanimate, rational to irrational, that we know not where preciſely to determine their reſpective Boundaries. In like manner, Light and Darkneſs, Motion and Reſt, we ſpeak of as Things very different and oppoſite: Yet no one will preſume to ſay, what is the preciſe and abſolute Boundary between langid Motion, and abſolute Reſt; or determine the Period where the laſt dying Sound expires in dead Silence.

And perhaps this gradual Chain and Connexion terminates not with ſublunary Things; but may be progreſſively continued far above the Ken of the moſt exalted Genius, or even the Comprehenſion of perhaps celeſtial Beings, till all created Perfection is loſt in him, who is Perfection itſelf.

This

This Animal, and the Cave containing it, are delineated in Plate XXIV. Fig. I.

The SMALL BLUISH ANIMAL FLOWER.

THESE grow in Clusters upon the Rocks between High and Low-water Mark. The Edges of each are composed of a circular Border of small fistular thread-like brown Petals (if I may so call them) surrounding a fungous Substance, of about the Breadth of an *English* Silver Two-pence, and of a bluish-green Colour. This Species is by far less quick in avoiding approaching Danger than those already described; consequently their Organs of Sensation are less perfect; for they will suffer themselves to be touched, before their guardian Petals or Arms close together to defend or preserve the Whole. I have observed a larger Sort of the same Species, having their brown Petals or Arms longer than the above-described, as well as of an irregular unequal Length: These likewise gradually lessen in their sensitive Perfection, and are generally found at some Distance under Water, whereas the former, in Neap Tides, are often for a short time exposed to the Air, and that seemingly without any Prejudice.

The

THE

NATURAL HISTORY

OF THE

Iſland of *BARBADOS*.

BOOK X.

Of the SEA *and its* INHABITANTS.

AS the Earth is full of God's Riches, ſo is the great and wide Sea, wherein are Things creeping innumerable, both ſmall and great. We are no ſooner advanced to the Shore, at leaſt a few Yards farther upon the ſmooth Surface of the Water, on the Weſt Side of this Iſland, but we are, eſpecially in calm Mornings and Evenings, pleaſed with the Sight of ſeveral Groves of Coral, Sea-Feathers, and Sea-Rods; the former grow in thick Cluſters, yielding an imbrowned Shade, and remain as ſturdy Oaks unmoved; the latter, with their numerous, pliant Branches, wavingly bend with the undulating, ſlow Motion of the Water. The interſperſed, vacant, ſandy Spaces reſemble ſo many bright Lawns which pleaſe with a Kind of regular Confuſion.

The Whole is made more grateful ſtill by the Sight of numerous Tribes of Fiſhes, which either feed upon, or wantonly ſkim the Surface of the deep Water, or lie groveling at the Bottom.

There the bulky *Conch*, as it moves, is ſeen to furrow the glittering Sand; and the Warlike *Sea-Egg* ſlowly waves its thouſand Spears.

A Scene of this Nature cannot be better repreſented than in the Words of *Milton:*

These
With their Fins, and ſhining Scales
Glide under the green Waves in Sculls that oft
Bank the Mid-ſea: Part ſingle, or with Mate
Graze the Sea-weed, their Paſture, or through Groves
Of Coral ſtray, or ſporting with quick Glance
Shew to the Sun their waved Coats dropp'd with Gold.

As the Inquiries we have made in the foregoing Books have ſufficiently convinced us, that Nature hath provided terreſtrial Animals, ſome with Strength, others with Swiftneſs, and even the weakeſt and feebleſt of the exanguious Tribes (ſeemingly the moſt helpleſs and imperfect) with ſome Skill or Contrivance to preſerve its Species; ſo we ſhall be no leſs agreeably ſurprized to find that the ſame Divine Power is not leſs conſpicuous in the Make of the Inhabitants of the wide-extended Ocean, from the huge *Leviathan* to the ſmalleſt Fry. Each Individual is not only adapted, by its Blood and Juices, to the Element they are in; not only taught to propagate their Species, but alſo to preſerve them from their Enemies; for the Lamb flies not the Wolf with greater Dread than the *Flying-Fiſh* doth the *Dolphin*, and that in like Manner its Enemy the devouring *Shark*.

It is remarkable in the *Flying-Fiſh*, that theſe, whilſt ſmall and unfit for Uſe, are never ſeen near the Shores; and the few that accidentally appear, avoid the Bait or Net with a ſurpriſing Shyneſs; but when full grown, and in Perfection, they croud to a convenient Diſtance from the Shores, and by their Tameneſs court, as it were, the deſtroying Hand of Man; for they will neither fly out of his Way, as they do from the *Dolphin*, nor will they dive into the Deep; but ſuffer themſelves to be taken up with the Hand, or with a ſmall Hoop-net, which is the common, and moſt expeditious Way of taking them.

Many have been the fruitleſs Inquiries to find out the certain Periods and Cauſes of the various and oppoſite Sea Currents, ſo remarkable near this Iſland. As I have obſerved theſe to be higher and ſtronger at the Full and Change of the Moon, as well as more rapid, and more inclined to the Northward, this induced me to believe that, tho' the primary Cauſe of the Current, as well as the higher Flow of the Water, at ſuch Times, is owing to the Influence of the Moon; yet the Byaſs of this Current, more to the Northward than to any other Part,

Part, is occasioned by the Nearness of the South-West Continent of *America* to this Island, and the Remoteness of the Continent on the North-East, or opposite Part from it.

For, when the Waters at these Spring-Tides are elevated and depressed, by Turns, by the Influence of the Moon, they will naturally recede towards that Point where they meet with the least Resistance; therefore, tho' the Course of the Current may, and often doth, run towards the South-West; yet, as soon as it is resisted by the Continent of *America*, on that Point, which is not quite three hundred Miles distant from this Island, it recoils back with redoubled Force towards the opposite Point, which is the North-East; and, as it meets there with no Resistance, 'till it comes to the Continent of *Africa*, which is about two thousand Miles distant from this Island, it is no Wonder that the Current, especially in the Decrease of the Moon, is more inclined that Way; and, when it varies from these Courses, it may be attributed chiefly to the Change of the Wind from the true Trade Easterly Winds to the South-West; or perhaps to the great and violent Land-Floods from the many and large Rivers from the neighbouring South-West Continent of *America*, which may sometimes byass these Currents.

It hath been observed by many Writers, that the Sea-water is salter in these hot Climates than in other Northern Seas. This Difference the ingenious Mr. *Rohault* says, arises from hence, *viz.* " That the Sun's " Heat, being more intense near the Equator than in those Seas which " are at a greater Distance from it, a much greater Quantity of fresh " Water must continually ascend up in Vapours, than elsewhere; " which do not often descend again in Rain, 'till they are carried to a " great Distance from thence: so that there being a less Quantity of " that which temperates the Salt to be found in those Seas which are " between the two Tropics, than in those Seas which are in the frigid " and temperate Zones; it is no Wonder if their Waters are salter:" And he likewise adds, " That the Ocean is of much larger Extent be-" tween the Tropics than any where else, and yet there are fewer Ri-" vers that discharge themselves into it."

The Cause of the Water's greater Degree of Saltness in these Parts, being thus explained, the Wisdom of such Allotments of Providence will soon appear, when we farther consider, that as the calm Latitudes are within the Tropics, therefore if the Sea was not much impregnated with Salt, its Surface in those Parts, called the calm Latitudes, would in all likelihood stagnate, and send up so many impure and stinking Vapours, that it would infect not only those but the adjacent Climates with Distempers, perhaps very little less pernicious than the Plague itself.

Of all the Fish caught in our adjoining Sea, I shall take notice of such only, as have been either insufficiently described, or whose Descriptions have been totally neglected by other Authors. Among the former is the following.

The

The TOAD-FISH.

THIS Fish, which is the Dread and Terror of Fishermen, is no less ugly in its Shape, than mischievous by its Prickles; its Length in general, is from Eight to Eleven Inches; the Colour of the Skin, is of a brownish-red, intermix'd with blackish Spots; the Head is very large, in Proportion to the rest of the Body, and almost intirely cover'd with sharp Prickles; the Eyes are pretty large, incircled with a white Iris; its Back is remarkably arm'd with a Row of very sharp Prickles; these, when the Fish is neither in an offensive nor defensive Posture, are almost couchant; and, to prevent their Points being by any Accident blunted, they are sheath'd in a strong muscular Membrane, or Web, dividing each Prickle. When the Fish is disturb'd, this Skin or Membrane is immediately contracted, and lies in Folds close to the Back, so that almost the whole Length of these Spears are then unsheath'd, and better fitted to make deeper Wounds, than if the Web had been immoveable, as it is in almost every other Fish, guarded for common Defence with Prickles. When these Spears are erect, if the unwary Fisherman treads upon it, (for it will not attempt to get out of his Way,) gives a strong Flutter with its Fins, and strikes or darts them into his Foot: the Pain that immediately ensues, is not to be described, being so very violent for the Time it lasts, (which is generally 'till the same time of the next returning Tide) that the Agonies they endure are inexpressible: And as few, (if any) that have been once wounded, will be so hardy as to venture a second Time to the Sea, without being first guarded; perhaps the Proverb, *Ictus Piscator sapit*, might have its Rise from hence; especially if these Fish are in Plenty upon the *Italian* Shores. It is observ'd that the Part affected near the Wound turns livid; but, if the Liver of the Fish be immediately applied to it, it gives Ease in some small Degree. This I take to be owing to the Oylinesss of the Liver, which blunts the Spicula of the poisonous Particles; for I am very confident, that this severe Pain doth not proceed from, or is occasion'd by the Wounds, or Punctures made in the Feet, however sensible that muscular and nervous Part may be, but from some liquid Poison injected into the Wounds, thro' these Prickles.

Of FLYING-FISH *in general.*

OF this Class there are no less than Five Sorts, in this Part of the *West Indies*; tho' there hath been I believe, as yet, but one of them taken Notice of, or described by the Writers of Natural History; which, by way of Eminency and Distinction, is called the *Flying-Fish*: the rest, no less deserving of that Name, are the *Flying-Gar-Fish*, the *Sea-Bat*, the *Guinea-Men*, and the *Ballahws*.

The

The FLYING-FISH.

THIS Fish is preposterously described by most Travellers, who often take more Pleasure in setting Men a staring at Representations in themselves incredible, than in candidly and honestly relating such Matters of Fact, as come within the Sphere of their Knowledge.

Thus, when this Fish is made the surprizing Topic of Wonder, as the Generality of Mankind know no other Method of flying but with Wings, they are immediately induced to believe, that this Fish hath, according to the real Acceptation of the Word, Wings to fly with; whereas what are so called, are only Fins common in their Make with the Fins of several other Fish, differing only in Bigness, which in Proportion to the Size of this Fish, are indeed far larger than any other, the *Flying-Gar-Fish* excepted. This, that we are now describing, is not, at its utmost Growth, more than Nine Inches long, flattish on the Back, and growing broader towards the Head; the Back and Head are of a dark Blue, and whitish under the Belly; they have two large Fins, situated *post Bronchia*; these in a full-grown Fish are six Inches long; and each in Breadth at their greatest Expansion, about three Inches. When pursued by a *Dolphin*, as their Swiftness in swimming is far inferior to that Fish, to avoid him they give a Spring out of the Water, and immediately (if I may be allowed the Expression) unfurl or expand their Fins, not with a tremulous, but a steady, quick, gliding Motion; which they do with great Swiftness; but, if the Day be hot, or the Weather sultry, their Flight is short; for, as soon as the Heat of the Sun dries up the Moisture upon their Fins, which made them pliable, they must dip into the Water to receive a fresh Supply. By the best Guess that I could make, I never perceived any of them to fly above seventy or eighty Yards, tho' oftener but forty or fifty at a Time. Their Increase is prodigiously great; otherwise the whole Species must have long ago been destroyed; for they are a Prey to Men, Fish, and Birds; having no certain Tenure of Life, either in Air or Water: for, when pursued by *Dolphins*, or other voracious Fish, if, to avoid these, they seek Refuge in the Air, a Bird called the *Cobler*, among a great many others, darts with the Swiftness of an Eagle to destroy them. I cannot help observing here, that there is something very peculiar in the Make of the Wings of this Bird; for, as it seeks its Prey at a great Distance from Land, and being obliged to be long upon the Wing, often at the Distance of a hundred Miles off, to be able therefore to endure so great a Flight, the Wings are not only large in Proportion to the Body, the better to enable them to bear its Weight; but the Tip-end also of the Pinion, instead of having one, the common Make of most Birds Wings, it hath two strong Bones; by which means the Vibration of the Wings is stronger and more steady.

The FLYING-GAR-FISH.

THIS is a far larger Fiſh than that properly called the *Flying-Fiſh:* it hath a long Duck-like Noſe; and its Fins are ſo large, that it can raiſe itſelf out of the Water, but its Flight is ſhort.

The FLYING-SEA-BAT.

THIS is of different Size, from ſix to fifteen Inches long; tapering from the Head to the Tail; the Mouth is cloſe to the lowermoſt Part of the Head; the lower Jaws are expanded, and terminate in two long Prickles, in Length an Inch; the Eyes are large; between theſe the Head ſinks in, or is much indented; this, as well as the Neck, is cover'd over with a hard Scarf, or thick Coat of Mail, which terminate in two long ſharp-pointed Lobes in each ſide of the Back; the reſt of the Body of the Fiſh, which is of a dark Grey above, and whitiſh under the Belly, is cover'd over with hard ſmall Scales; from behind the Gills on the Belly-ſide riſe two Fins; theſe are in Length about two thirds of the Length of the Fiſh; with theſe it flies; for, when expanded, they are from the Extremity of the one, to the Extremity of the other Fin, full as broad as the Fiſh is long; theſe Fins are narrow and ſtrong when cloſe to the Body; but towards their Extremities very membranaceous, thin, and of a black Colour; the Back is ſupplied with two ſmall Fins; the Belly likewiſe under the Wings, is guarded with two Inch-long, ſoft pointed Horns, the whole Fiſh ending in a middle-ſized forked Tail; theſe when caught with a Hook, and brought to the Surface, will fly a conſiderable Way, unleſs check'd by pulling the Line: they, as well as the two-following, are very ſeldom to be met with.

The GUINEA-MAN.

THIS is conſiderably larger than the *Flying-Fiſh*, and hath a ſomewhat Duck-like Noſe, and Fins ſo large, that he can raiſe himſelf out of the Water, and fly a ſmall Diſtance, tho not ſo far as the *Flying-Fiſh.*

The BALLAHW.

THIS is a ſmall Fiſh, ſeldom above ſeven or eight Inches long; the under Part of the Noſe is a great deal longer than the upper: this likewiſe by its large Fins, flies a ſmall Diſtance; but not near ſo far as the *Flying-Fiſh.*

The

The INK-FISH.

THIS is of the cartilaginous Kind, and hath been by ſeveral Authors, ſo well, and ſo often deſcribed, that I have nothing to add, unleſs that the Curious hitherto have in their Accounts look'd upon this to be the only one that emits a black Liquid; whereas the two following are provided with ſuch, which they ſpirt out to diſcolour the Water when in Danger of being caught. By this Means they eſcape the Ken of their Enemies; for, as they are not provided with defenſive or offenſive Weapons, this Liquid, which is as black as Ink, is as neceſſary for their Preſervation, as Swiftneſs is to a *Dolphin*, or their large Fins to any of the *Flying-Fiſh* Kind. The blackeſt Liquid lies in the Tail of this Fiſh.

The CAT OF NINE TAILS.

THIS cartilaginous Fiſh, notwithſtanding its Name, hath but five inſtead of nine Tails; which in general are not above ten Inches long, tapering to the Point; and, tho' they are but ſlender, yet their Strength and elaſtick Power is almoſt inconceivable. The largeſt of theſe ſeldom weighs above a Pound, yet it is found to be a difficult Taſk for a very able Man to uncling one of them from the Rocks to which they are fix'd. As cartilaginous Fiſhes in general have no Air Bladders, it is no Wonder that none of this Claſs are found in deep Waters, their Reſidence being chiefly among Shoals and Rocks between one and ſix Foot deep. They feed chiefly upon *Wilks*, young *Conchs*, and *Crabs*; the latter it breaks to Pieces, and the former it ſucks out of their very Shells.

The SCUTTLE FISH.

THIS hath eight Tails, and often weighs near two Pounds; every Part of this Fiſh is full of ſtrong elaſtick Muſcles, eſpecially its numerous Tails, which are often fifteen Inches long, tapering to the Point. This, as well as the *Ink-Fiſh*, is provided with a black Liquid, to diſcolour the Water, whilſt it makes its Eſcape from its Enemy, with this ſmall Difference, that the Liquid is lodged in the Head of this, whereas that of the *Ink-Fiſh* lies in its Tail.

The TRIANGULAR FISH.

THIS often grows to be near fifteen Inches long, weighing ſix, ſeven or eight Pounds; the Mouth is ſmall, and pouched and tinged with a blackiſh Hue for near an Inch round; the Teeth are many,

many, long and blunt-pointed; the Body of the whole Fiſh is a triangular Trunk, the Belly-part making the Baſe of the ſmalleſt Angle; the Skin is black, hard and rough; being raiſed into innumerable ſmall Angles repreſenting Fret-Work; the Mouth lies level with the Baſe of the Angle which conſtitutes the Belly-Part; its Eyes are large, and the upper Part of the Head much indented between them; the Back is very crooked and ſharp-edged, being almoſt of the Conſiſtence of a Horn; it hath two ſmall Fins *poſt Bronchia* one upon the Back near the Tail, and one nearly oppoſite to it under the Belly; each Side of the loweſt Angle being the Belly-Part is, near the Tail, guarded with a ſhort horny Prickle; the whole Fiſh ending in a forked Tail.

This is delineated in Plate XXVIII.

The HORNED CONEY FISH.

THIS chiefly differs from that already deſcribed, by having its Head guarded by a Pair of ſharp-pointed Horns of, ſometimes, near an Inch long, bending ſomewhat downwards. There is another remarkable Difference, which is peculiar to this Fiſh alone; for if eaten, eſpecially the Liver and the Head, the Perſon feeding on it will, in a few Minutes after, be as drunk as if he had drank ſtrong Liquor to Exceſs: For his Reaſon and his Limbs will fail him, until reſtored to both by a ſound Sleep.

The SOAP FISH.

THIS is a ſmall Fiſh, not above ſix Inches long; it is of ſo ſoapy a Nature, that when caught and ſtrongly agitated in Water, it will cauſe almoſt as ſtrong a Lather as an equal Quantity of the beſt Soap. I am of Opinion that Nature intended it this Quality (ſince it is incapable of every other Defence) to be as much a Means of avoiding its Enemy, as the Swiftneſs of ſwimming is to ſome, or their Prickles to others: But as the extraordinary Slippineſs of this Fiſh can be no Defence againſt an Antagoniſt provided with Teeth, therefore I beg Leave to conjecture that its natural Enemy is among the cartilaginous Claſs of Fiſh (eſpecially as it is alway found feeding near the Shore;) and none more likely of this Kind than the *Cat of Nine Tails*, and *Scuttle-Fiſh*; for theſe, by their numerous Tails which have ſo ſtrong a muſcular Force, graſp and hug their Prey to Death, unleſs by its Slippineſs it can diſentangle itſelf from them: The Strength of the above-mention'd Tails is otherwiſe ſo great, that by fixing their Heads, and, perhaps, ſome of their Tails to a Rock in the Bottom under Water, and the remaining round the Arm of the Perſon diving for them, there hath been one Inſtance of the Diver not being able either

to

To the R^t Honble
Henry Pelham Esqr
First Lord Commiss^r of y^e Treasury
This Plate is humbly inscrib'd

G. Bickham de. sc.

to pluck up the Fiſh clinging to the Rock, nor to extricate himſelf from it, and conſequently periſhing in the Attempt.

The OLD MAN.

THIS Fiſh is about eleven Inches long from the Head to the Tail, and about ſeven in Breadth; the Head and Mouth are ſmall; the Teeth many and ſharp; the whole Skin is black and rough: It hath two very remarkable Prickles upon the Back, not far from the Tail; theſe, when the Fiſh feeds, are couchant cloſe to the Back; but, when annoy'd, are erect and prove offenſive or defenſive Weapons; but as they have no membranaceous Web to ſupport them by its Elaſticity, as moſt other Fiſh (which Nature hath endued with Prickles) have, to ſupply this ſeeming Want, this Fiſh can, by the Help of very ſtrong Muſcles, join'd to the Root of theſe Prickles, raiſe them in an inſtant upright, ready to guard himſelf or annoy his Enemy; and, even when the Fiſh is dead, if both theſe Prickles (the uppermoſt being above an Inch and a Quarter long) are lifted up into an upright Poſition, the largeſt falls into a Socket in the Back-Bone: where it will remain unmoveable, and will ſooner break than bend. But as a ſmall Touch upon the Trigger of a Gun's Lock will bring down the Cock from its full Bent, ſo the leaſt Motion or Touch upon the lower Prickle will, in an inſtant, bring down the upper cloſe to the Back.

The OLD WIFE.

THIS differs very little from the laſt deſcribed; each having their two Fins *poſt Bronchia*; one on the Back, and one under the Belly, of equal Size and Poſition, as well as the Make of their forked Tail alike: However there is this material Difference, that as the former hath its Prickles upon the Back near the Tail, this hath them near the Head, with this additional Circumſtance, that, inſtead of two, it hath three Prickles, but equally intended to be its Safe-guard.

The JACK.

THIS Fiſh is about ſix Inches long, and near two broad from the Back to the lower Part of the Belly; the Eyes very large in Proportion to the Bigneſs of the Fiſh, and is ſurrounded with a yellow Iris; it hath two narrow, ſharp-pointed Fins *poſt Bronchia*; [one ſtreight, one under the Belly, and another upon the Back as far as the Navel;] likewiſe a Pair of ſmall ones under the upper Part of the Belly: Oppoſite to theſe on the Back are likewiſe two ſmall, prickly Fins; the Back from Head to Tail is blackiſh, and the Belly of a Silver-white: Theſe are often caught in very great Numbers in Nets.

They are in ſome Seaſons of the Year, eſpecially, when caught in *Chriſt-Church* Pariſh or thereabouts, very poiſonous; at ſuch Times there are generally in their Gills two ſmall red Lumps; and when they are ſuſpected to be poiſonous, they try the Experiment upon a Duck, by giving her one of theſe to ſwallow; and if at that Seaſon it is poiſonous, the Duck dies in about two Hours.

The DOLPHIN.

THE ridiculous Repreſentation of this Fiſh by moſt Painters and Engravers (even after its true Shapes being ſo well known) is a ſtrong Inſtance that vulgar Errors are but with great Difficulty got over. It is ſo far from having a large Head and Gills, a crooked, thorny Back, &c. as generally repreſented, that it is one of the ſtreightest Fiſh that ſwims. This is delineated in Plate XXIX.

The SEA TORTOISE.

MOST, if not all Authors, who have treated of this ſcarce-to-be-called amphibious Animal, have inadvertently rank'd it among the teſtaceous Kind; tho' it is apparently otherwiſe by its numerous ſtrong Ribs and Back-Bone: The ſeveral *Lamina*, which adorn and fortify its Back, are ſo far from being of the Shell-Kind, that they are rather an elaſtick, tranſparent, horny Subſtance. A late ingenious and very uſeful Author hath been much impoſed upon, by a pretended Deſcription of this Animal; eſpecially where he aſſerts that the Under-Shell is what alone is made Uſe of; and that to ſeparate this from the upper, Fire is made underneath; and that as ſoon as it is heated, the Shell becomes eaſily ſeparable, and ſo taken off with the Point of a Knife, and that with ſo little Prejudice, that the *Tortoiſe*, being afterwards permitted to go into the Sea, not only lives, but hath its Armour or Shell ſoon renewed. Thus that induſtrious and learned Gentleman, whilſt laudably deſirous to give Mankind as true a Deſcription of this, as he does of almoſt every Thing he treats of, was much impoſed upon by his Informer. This Misfortune will ſometimes happen, when our Informations come from Perſons, who are fond of aggrandizing, or leſſening what is, or is not, agreeable to their own favourite Opinions, or groundleſs Imaginations. The Deſcription of the *Tortoiſe* juſt now mentioned, is ſo far from being true, that the lower Shell, or what is ſo called, is of no Manner of Uſe, the upper alone being of Service; nor can this be come at 'till the *Tortoiſe* is not only dead, but all the Fleſh likewiſe taken from the Ribs and Back; it is not till then that a Fire is made under the inſide of the Upper-Shell (as I ſhall in Conformity to the received Opinion call it; though it is manifeſt it is not a Shell from its elaſtick Quality.) And when

Ephraim Chambers, vide Chambers's Dictionary.

Plate 29. Page 308

To the Right Rev.[d] Father in God, –

Zachary. Lord Bishop of Bangor.

This PLATE is humbly inscrib'd &c.

G. Bickham de. sc.

when this Heat hath penetrated thro' the Ribs, and a cruſty hard Subſtance which covers them, that the ſeveral foliaceous Lamina, generally about ſixteen in Number, begin to exfoliate on the outſide: thoſe about the Shoulders are the thickeſt and beſt. The Back of a Tortoiſe is convex, and the Belly-part flat; it is cover'd all over, except near the Neck, the Anus, and under the Fins, with what we have called a ſhelly Subſtance; the upper and the under Parts being both of them cloſely join'd together, by this ſtrong Armour. The Penis of a Male Tortoiſe of any conſiderable Bigneſs, is full twelve Inches long; and its Teſticles are as delicious as theſe of Lambs: in Cooting-time, the Male and Female remain in Coition, eight or nine Days, ſome ſay more; at the Expiration of which Time the Male is ſo reduced, that he is ſcarce able to ſwim; his Callapee (being the Belly-part) becomes from a hard horny Subſtance, as ſoft almoſt as a Jelly. The Penis dryed and given in Powder, is looked upon by ſome to be a ſovereign Remedy againſt the Gravel.

Of the three different Sorts that frequent, or are bred near theſe *Weſt-India* Iſlands, the *Hawk's Bill* alone affords what is commonly call'd the Tortoiſe-ſhell. The two other Species, *(viz.)* the Green and the Yellow, or Mulatto *Tortoiſe*, have each of them ſuch Shells, divided into as many regular Lamina, but they are ſo very thin as not to be fit for uſe. A *Tortoiſe* hath four Fins, with which it paddles whilſt in the Water, not much different from the Strokes of Oars; and it is likewiſe by the Help of theſe that the Female glides along the Sand, when ſhe comes on Shore to lay her Eggs. The common Method of taking them, is to pitch Nets with very large Meſhes, in the Bays where they frequent, to feed upon the green, broad-leaf'd Moſs that grows at different Depths in the bottom of the Sea; when taken entangled in theſe Nets alive, they may be brought aſhore, and kept ſome Weeks alive without any Suſtenance; for ſeveral Days after they are taken, they ſigh heavily; if they die in the Net, they ſtink in leſs than an Hour's time; but if killed, which is done by cutting the Throat, (to give vent to the Blood which is always as cold as Water) the Fleſh will keep not only uncorrupted, but tho' cut in Pieces, the fore Quarter and Callapee will continue to have a ſtrong, lively, muſcular, convulſive Motion, for fourteen, or even eighteen Hours; for if at that Time it is pricked with a Pin or Fork, it will move and contract itſelf viſibly. Some Part of the Fleſh cuts reddiſh, reſembling coarſe Beef; another Part as white as a Chicken; the Fat about the Fins and the Guts, is ſomewhat Yellowiſh, but the far greater Part, cloſe to the upper and under Shell, is as green as a Leek; they are caught of different Sizes; the largeſt that hath been taken in this Iſland, within my Remembrance, did not exceed four hundred Weight. They are often ignorantly repreſented to have three Hearts; this Miſtake ariſes from their having two large Auricles, one on each Side of it; the Blood in this Animal is very groſs, cold and viſcid; and as the Heart hath but

one

one Ventricle, its Circulation is performed after the same Manner as in a Frog; in which, not above one third passes through the Lungs; it is likewise owing to the Coldness of the Blood, its slow Motion, and thick Armour with which the *Tortoise* is surrounded, that it is at a very little Expence of Spirits by transpiration; and consequently hath less Need of a fresh Recruit: The Liver is large, and of a dark Green, and the great Gut or Colon, on the outside full of seeming Prickles; however their Points are soft and pliable, the Flesh when baked or stewed, is a most delicious and nourishing Diet; the young Ones are often caught with a Hook and Line: the properest Bait for this Purpose is a Sea-Bladder; and they are likewise sometimes drawn ashore in Nets. There is another Method of taking the larger Sort, esepecially the Females, by watching their coming ashore in the Night, upon the dry, sandy Bays, in the Months of *June*, *July*, *August*, and *September*, in which laying Seasons, after they have crawled above High-Water-Mark, they dig with their Fins (which are strong, nervous, and fleshy) a Hole of about two Feet deep, in the loose Sand, in which the Female lays sometimes an hundred or more Eggs; the outward Tegument of these is rather skinny than shelly, its Shape is round, of about an Inch and a Quarter Diameter; the inside of the Egg is yellow, and to the Taste somewhat gritty. After these Eggs are thus deposited in the Sand, the Tortoise fills up the Hole in so nice a Manner, that it will be scarce perceivable that the Sand had been disturb'd; and the Eggs, by the Heat of the Sun, will, in nine Weeks, be hatched, and the young Tortoises immediately crawl into the Sea.

Before we conclude this Description, it will not be impertinent to observe, that the Ancients knew the Value of the Tortoise-Flesh in Physick, as well as of its Shell for making the Lyre. The former appears from *Nicander*, in his Poem call'd *Alexipharma*; and *Horace* speaking of the Lyre made of the Tortoise-shell, says,

O testudinis aureæ
Dulcem quæ strepitum Pieri temperas
O mutis quoque piscibus
Donatura cycni, si libeat, sonum.

By the Word *piscibus*, it is evident, that the Lyre mentioned by *Horace* was made with the Shell of the Sea-Tortoise, which maketh Use of Fins to swim with as a Fish; whereas the Land-Tortoise (tho' the Shell of this hath been often us'd in *Greece* to the same Purpose) is arm'd with Claws, and is at most but barely an amphibious Animal; whereas the other, applied here by *Horace*, is almost intirely an Aquatic.

The SEA-HORSE.

AS I never saw this Animal 'till it was shrivel'd and dry'd up, I will not pretend to describe it; any farther, than that by its Trunk it seems to be a Fish which sucks in its Nourishment.

Of FRESH WATER FISH.

OF this Kind we have very few intirely peculiar to the fresh Water; that is, which breed in it; the greatest Part being generally, when very young, the Inhabitants of the Sea; when the great Current of our Land-Floods breaks thro' the Banks of our fresh Water Ponds, causing by this Means a Communication for some Time with the Sea, several Sorts of young Fish swim from the salt into the fresh Water; and from these Ponds they are caught, and carried into others several Miles up in the Country.

A CAFFUM.

THIS is a large, scaly Fish, often weighing fifteen Pounds; and measuring about three Feet in length. The Head is small in Proportion to the rest of the Body; the under Jaw is longer than the upper; its Eyes are large, incircled with a broad, golden Iris. The Back is of a dark blue, and the Sides and Belly of a shining Silver-Colour; it hath one Fin upon the Back, two *post Bronchia*, two under the Belly, and one near the Tail, which ends forked. This Fish is generally very poor whilst in the Sea, but soon grows very fat in Land-Ponds, where they are fed with the Guts of Fowls, small Fish, or upon Pond-Bugs, their most natural Food in this Situation.

The MULLET.

THESE are of two Sorts, the Sea and the fresh Water *Mullet*; the former is of a light-grey upon the Back and Sides; the latter of an iron-grey, and the Scales very remarkably divided in regular Rows upon their Sides and Backs.

The CARMOW-MULLET.

THIS differs from the former by the Greatness of its Bulk, especially its Head, which is remarkably large and flat; these often weigh above two Pounds, being very fat and delicious Eating.

The MUD-FISH.

THERE are three Sorts of *Mud-Fish*; the first and smallest somewhat resembling *Smelts* in its Make; the second is the *Mangrove Mud-Fish*; this is thicker and larger, and hath often large Roes in it; the third and largest is the *Logger-Head-Mud-Fish*; this hath a large Head and Eyes, the former somewhat flattish; and are often a Foot long; every Species of these are justly esteemed to be very delicate Eating.

The two first Fins appear *post Bronchia*, with two smaller in a direct Line under the Gullet; it hath likewise one continued membranaceous Fin from the Navel to near the Tail, with another opposite of equal Length upon the Back. The whole Fish is covered with many small Scales, and ends with a round pointed Tail.

The STAR-FISH.

THIS is here likewise call'd the *Five-Finger'd-Fish*, and but very seldom taken.

The SHARK.

I SHALL take my Leave of the Deep and its Inhabitants with a Description, or rather cursory Observation, of the *Shark*, a Fish well known both in the Northern and Southern Seas for its ravenous Nature; preying upon most Animals that come in its Way. The young *Sharks*, upon Sight of Danger, slip in through the Mouth of of the Parent *Shark*, and, when the Danger is over, issue forth to their proper Element again.

These Creatures are sometimes seen very numerous among the Ships in *Carlisle-Bay*; especially when there are many Vessels with Slaves from *Guinea:* For some Hundreds of these poor Wretches being often crouded together in one Bottom, a great many of them die with various Diseases; and being thrown over-board, bring together so great a Multitude of these voracious Animals, that it is not safe at such Times for the fatigued Sailors to refresh themselves by bathing in the Bay. And here let me take Occasion to testify my Esteem of these daring and profitable Adventurers of the Deep, by rescuing from Oblivion a memorable Atchievement of a common Sailor in the Destruction of a *Shark:* And when the Principle which prompted him to so very unequal and hazardous a Combat, and the Intrepidity of the Action itself are considered, abstractedly from the low and mean Circumstances of the Person, it will perhaps appear to be as heroic an Instance of disinterested Friendship and personal Bravery, as any recorded in History.

About

About the latter End of Queen *Anne*'s Wars, Captain *John Beams* Commander of the *York Merchant*, arrived at *Barbados* from *England*. Having disembark'd the last Part of his Loading, which was Coals, the Sailors, who had been employ'd in that dirty Work, ventured into the Sea to wash themselves; there they had not been long, before a Person on Board 'spyed a large *Shark* making towards them, and gave them Notice of their Danger; upon which they swam back and reach'd the Boat, all but one; him the Monster overtook almost within Reach of the Oars, and griping him by the Small of the Back, his devouring Jaws soon cut asunder, and as soon swallow'd the lower Part of his Body; the remaining Part was taken up and carried on Board, where his Comrade was. His Friendship with the deceased had been long distinguished by a reciprocal Discharge of all such endearing Offices, as imply'd an Union and Sympathy of Souls. When he saw the sever'd Trunk of his Friend, it was with an Horror and Emotion too great for Words to paint. During this affecting Scene, the insatiable *Shark* was seen traversing the bloody Surface in Search after the Remainder of his Prey; the rest of the Crew thought themselves happy in being on Board, he alone unhappy, that he was not within Reach of the Destroyer. Fired at the Sight, and vowing that he would make the Devourer disgorge, or be swallowed himself into the same Grave,—He plunges into the Deep, arm'd with a large sharp-pointed Knife. The *Shark* no sooner saw him, but he made furiously towards him,—both equally eager, the one of his Prey, the other of Revenge. The Moment the *Shark* open'd his rapacious Jaws, his Adversary dextrously diving, and grasping him with his left Hand somewhat below the upper Fins, successfully employs his Knife in his right Hand, giving him repeating Stabs in the Belly: the enraged *Shark*, after many unavailing Efforts, finding himself overmatch'd in his own Element, endeavours to disengage himself, sometimes plunging to the Bottom, then mad with Pain, rearing his uncouth Form (now stain'd with his own streaming Blood) above the foaming Waves. The Crews of the surrounding Vessels saw the unequal Combat, uncertain from which of the Combatants the Streams of Blood issued; till at length, the *Shark*, much weaken'd by the Loss of Blood, made towards the Shore, and with him his Conqueror; who, flush'd with an Assurance of Victory, pushes his Foe with redoubled Ardour, and, by the Help of an ebbing Tide, dragging him on Shore, rips up his Bowels; and unites and buries the sever'd Carcase of his Friend in one hospitable Grave.

The Story, I confess, is of so extraordinary a Nature, that I would not have dared to give it my Reader, had I not been authorised thereto by the Testimony of a very credible (1) Gentleman, who is ready to confirm by Oath, the Truth of what is here related. This Action, intrepid as it is, will unquestionably fall under the Censure of those, who are accustomed

(1) Lieut. Col. *Hillary Rowe*, of St. *Lucy*'s Parish, who was not far from the Place when this happened.

customed to judge by the Rules of moral or political Fitness; it not being prudent in any Man to expose himself to Danger, from which he must owe his Escape as much to Chance as Valour, nor consistent with the Value which ought to be set on the great Gift of Life, to risque it upon small and inadequate Occasions; the Exploit therefore had been more truly heroic, had it been performed for the Preservation of his Friend's Life, rather than the Recovery of his Body. But such Reflections are not the Sentiments of Sailors, a Class of Men to whom Courage is a Virtue, however madly or unseasonably exerted.

And yet if such an Action had been recorded of *Scipio* or *Alexander* in the Defence of *Lælius* or *Hephæstion*, whether it would not have been celebrated by their Admirers, among the most shining and magnanimous Atchievements of those renowned Heroes and Friends.

FINIS.

THE

EXPLANATORY

NOTES

OF ALL

Botanical and *Technical* Terms made uſe of in the foregoing WORK.

A.

ACINI, are ſmall Grains growing in Bunches, after the manner of Elder-tree Berries in *England*, and yellow Hercules Berries in this Iſland.

Apices or *Summits*, are thoſe Bodies which contain the *Farina fœcundans*. They generally hang upon ſlender Threads called Chives.

Aromatic Plants, are thoſe whoſe Leaves or Branches have an odoriferous Scent, as Roſemary, Lavender, &c.

B:

B*Acciferous* Trees, are thoſe which bear Berries. A *Berry* is a fleſhy Fruit incloſing ſeveral Seeds.

Bulbous-rooted Plants, are of two Sorts, *viz.* tunicated or coated Roots, which conſiſt of ſeveral Coats involving each other, as the Onion, Tulip, &c. The other ſquammoſe or ſcaly, which conſiſts of ſeveral Scales lying over each other, as the Lily, &c.

L l l l C. CAP-

C.

C*Apſula*, is the dry Huſk or Veſſel of Plants, in which the Seeds are contained.

Calyx, is the Flower-cup, or thoſe green Leaves which cover and embrace the Flower. Theſe in ſome Plants afterward incloſe the Seeds. Thoſe Plants whoſe Flowers have no Petals ſo incloſed, are termed *apetalous*. This is by ſome called the Empalement of the Flower.

Catkin or *Iulus*, is an Aggregate of Summits, collected into a long Body in form of a Rope. Theſe are the Male Flowers of Trees.

Chives, are thoſe ſlender Bodies which ſurround the *Ovarium* in the Centre of Flowers, and ſupport the Summits.

Clavicle or *Tendril (Capreolus)* is a Part of a Stalk curling and laying hold on any adjacent Body. As in the Vine, Briony, *&c.*

Clavicles, Claſpers, or *Tendrils*, are the young Shoots of creeping ſcandent Plants; it is by theſe that they take hold of the neighbouring Trees or Rocks; and ſometimes, if thus ſupported, they grow to a greater Height: Of this kind are the Claſpers of the Grape, Vine, and moſt others of that Tribe.

Coagulation, is the condenſing or thickening the Juices of any Tree or Plant, by the Heat of the Sun.

Coronated or umbilicated Fruits, are thoſe which have their Calyx or Empalement of the Flower growing on their Top, as the Medlar, Pear, Apple, *&c.*

Corrugated, or wrinkled, as the Leaves of Baum, Mint, *&c.*

Crenated, or notch'd Leaves, are thoſe, whoſe Edges are cut into ſeveral obtuſe Angles.

Corymbiferous Plants, are ſuch as have a compound or diſcous Flower; the Seeds having no Down ſticking to them.

D.

D*Entated* Leaves, are ſuch as have their Edges notch'd, ſomewhat like the Teeth of a Saw.

Digitated or finger'd Leaves, are thoſe which are divided into ſeveral Parts, which are connected together at the Tail, ſo as in ſome meaſure to reſemble a Hand, as in the Cinquefoil, Lupine, *&c.*

E.

E*Chinated* Pods, or Leaves, are thoſe that are ſet round with Prickles, ſuch as the Pods of Horſe-nickers in this Iſland, and Holly-leaves in *England*.

F. *Fiſtular*

F.

FIstular Flowers, are such as are compounded of many long hollow small Flowers like Pipes, all divided into many Jags at the End.

Falcated Pods, are those which bend in like an Hook or Semicircle.

Foliola, are those small Leaves which grow along the Mid-rib of the Leaf, in compound Leaves.

G.

GLaucous Leaves or Stalks, are of a whitish-green Colour, somewhat of a bluish Cast; such are the Leaves of the *French* Sorrel in *England*, and of the blue Edda in these Parts.

Galeated or hooded Flowers, are such, whose upper Part resembles a kind of Helmet or Hood, as in the Flower of Sage.

Granated, or that Fruit which is compos'd of many Grains or Kernels.

H.

HErbaceous Leaves, are such long small Leaves as come nearest in Shape and Colour to that of common Grass; such are the Leaves of Reeds.

I.

IULI, are Catkins, being Bunches of small dusty Flowers growing upon several Trees and Shrubs, such as Hassels and Walnuts.

Indented Leaves, are such as are slightly indented on their Edges.

L.

LAbiated Flowers, are difform monopetalous Flowers, divided usually into two Lips, as in Sage and Rosemary.

Laciniated Leaves, are those which are naturally jagg'd or notch'd to the Mid-rib by Vermin.

Ligulæ, or small Strings.

M. *Mem-*

M.

M*Embranaceous*, when taken in a Botanical Senſe, is generally applied to the ſeveral thin Filaments, which diſtinguiſh ſome partitional Cells from each other, as in the Pomegranate Fruit.

Monopetalous Flowers, are thoſe which have but one Leaf, or Petal, which, tho' it is ſeemingly cut into Four or Five ſmall Petals, or Flower-leaves, yet they are all one Piece.

Muricated, prickly Leaves.

N.

N*Ervous Leaves*, are thoſe which conſiſt of many Ribs or Fibres.

Nuciferous Trees, or Shrubs, are ſuch which bear Nuts.

O.

O*RÆ*, the Extremities of ſeveral Kinds of Flowers.

P.

P*Anicle*, is a Stalk diffuſed into ſeveral Pedicles, ſuſtaining the Flowers or Fruits, as in Oats, *&c.*

Papilionaceous Flowers, are thoſe which reſemble a Butterfly with the Wings extended, as in Peas and Beans in *England*, as well as in the Pigeon-Pea-Tree Bloſſoms in this Iſland.

Pappoſe Seeds, are ſuch as have a downy Subſtance ſitting on the Top of each Seed, as the Dandelion, Scorzonera, *&c.*

Paraſitical Shrubs or *Plants*, are thoſe which derive their Nouriſhment from other Plants; thus the Miſletoe from the Oak in *England*, and from the Orange Tree here.

Pennated or *Winged Leaves*, are ſuch as are compounded of many ſmall Leaves or Lobes, placed along the middle Rib, either alternately, or by Pairs. When the middle Rib is terminated by an odd Lobe, it is ſaid to be unequally *pennated*; and equally *pennated*, when it is not terminated by an odd Lobe.

Petals, are the fine-colour'd Leaves of the Flowers, to diſtinguiſh them from the Leaves of the Plant.

Pedicles, or the Foot-ſtalks of any Leaf, Fruit, or Flower.

Pen-

Pendulous Flowers, are those which hang downwards; such as the Aloe Flower.

Pistil, is that Body which arises in the Centre of Flowers, and generally supports the Ovary.

Pinnæ. The Number of these are reckoned from the Number of the several Pairs of winged Leaves upon a Branch.

Pomiferous Trees or *Shrubs*, are those which bear Apples, or some Fruit of the Apple-kind, such as Oranges and Lemons.

Pruniferous Trees, &c. are those which bear Plums.

R.

R*Eticulated Bark* or *Substance*, is that which is composed of a strong, Net-like Structure: Such is the most substantial Part of the Body of a Popais-tree.

Rugose. This is a Term chiefly made use of with regard to the Bark of such Trees as are wrinkled, or furrowed.

S.

S*Etæ*, are those stiff small Hairs which often cover the Leaves, or the Stalks, of Plants; such as Nettles, and the Foot-stalks of Belly-achs.

Scandent Plants, are those whose Stalks are too weak to support them upright; therefore by their Claspers or Tendrils they fasten unto, or climb up, the neighbouring Trees; by which means they often grow to a great Height; such is the Ivy in *England*, and all manner of wild Wyths in this Island.

Serrated, or *sawed Leaves*, are such as have acute Notches in their Edges, resembling the Teeth of a Saw.

Stellated Plants, are those whose Leaves grow round the Stalks at certain Intervals in the Form of a Star with Beams; as Crosswort in *England*, and Star-grass in this Island.

Siliquose. All Trees, Shrubs, and Plants which bear Pods, are of this Kind.

Spatha, is that Part of a Plant which incloses the Flowers of the Palms, *&c.*

Stamina (are the same as Chives) are those fine Threads, Capillaments, or Hairs, growing out of the Bottom of the Flower, as from Tulips in *England*, and from the Flowers of the Flower-fence in this Island. These *Stamina* are generally tip'd with *Apices*, or small Knobs, at their Points.

Squamminatim, in Scales.

Stamineous Flowers, are those imperfect ones which want the fine coloured Leaves called *Petala*, and consist only of the *Stylus* and *Stamina*; such as the Flowers of Ackatee.

T.

THrum, are those small setaceous *Villi*, or stiff Hairs, rising and forming an equal, even, plushy Bed, in the broad Bottom of a Flower, as in the Marygold and the Sunflower.

Transverse Ribs, are those smaller Ribs or Veins, which cross the middle Spine, or Ribs which run from the Stalk to the Extremity of the Leaf.

Tubular Shanks or *Stalks*, are those that are hollow like a Pipe; such as Popais-shanks in this Island, and the Stalks of Hemlock in *England*.

Tetrapetalous Flowers, are those that consist but of four single-coloured Leaves called *Petals*, set round the Stylus to compose the Flower.

V.

VErticillate Plants, are such as have their Flowers intermix'd with small Leaves, growing in a Kind of Whorles about the Joints of the Stalk, as Peny-royal and Harehound in *England*, and the small white Sage in this Island.

Vulnerary. All Plants that have a healing Quality in their Juices, or otherwise, are called *Vulnerary Plants*.

U.

UMbelliferous, signifies a Plant that bears many Flowers discompos'd, somewhat like an Umbrella, growing upon many Footstalks, like Fennel Angelica, in *England*.

Umbel, is the Extremity of the Stalk and Branches divided into several Pedicles, or Rays, beginning from the same Point, and opening in such a manner, as to form a kind of inverted Cone, as in a Parsnep. When these Pedicles, which surround the Stalks, are again divided

divided into others of the ſame Form, upon which the Flowers are diſpoſed, the firſt are called Rays, and the ſecond Pedicles.

W.

W*Inged Leaves*, are thoſe that are ſet oppoſite to one another on the Branches; ſuch are the Walnut-tree-leaves in *England*, and the Leaves of the yellow Sauders, with a great many others, in this Iſland.

N. B. *I owe moſt of the above* Explanatory Notes *to the accurate Mr.* Miller *of* Chelſea.

THE

INDEX.

INDEX

Crab

INDEX.

Ginger

INDEX.

Wild

INDEX.

O o o o Smalleft

INDEX.

Stopper

INDEX.

FINIS.

A D D E N D A.

To the Account of the MUREX, *ending in Page* 275. *add,*

THE ſeveral Authors who have treated of the Shell-fiſh which produces the Purple, have, in deſcribing it, promiſcuouſly made uſe of the Words *Conchylium*, *Purpura*, *Murex*, and *Oſtrum:* All theſe are here convertible Terms, or figuratively applied to expreſs either the Shell or Shell-fiſh, the Tincture it emits, or the Purple or Crimſon itſelf. The *Greeks* call'd all Shells *Conchs*; and that which afforded the Purple was, by way of Eminence, called the *Conch.* From hence *Plautus* calls purple Tapeſtry *conchyliata tapetia.* *Juvenal* and *Martial* have follow'd the *Greeks* in calling this the *Conch.*

Horum ego non fugiam conchylia.
JUVEN. III. 81.

Ebria Sidoniæ cum ſint de ſanguine conchæ.
MART.

There are two Sorts of Shells that are ſo term'd: The moſt generally ſo called, is that deſcribed and delineated by *Rondeletius* and others; this is ſtudded with ſeveral long Prickles, ſome of them chanell'd, and open on one Side. Thro' theſe Fiſſures it is ſaid that the Tongue of the *Murex* darts into the other leſſer Shell-fiſh, which theſe Prickles had before perforated, and thus feed upon them. The other is of the *Buccinum* Kind we have already deſcribed and delineated.

The learned *Fabius Columna*, as well as many antient Authors ſay, that this is the peculiar Characteriſtic of the true *Tyrian* Purple, and that the other *Conch*, or *Murex*, emits a Liquid which dyes a fine Violet Colour. That there were two Sorts of what is commonly called the Purple, is evident from a Paſſage in *Pliny* (1), where a Perſon is repreſented ſaying, *Me juvene violacea purpura vigebat, cujus libra* (2) *denariis centum venibat, nec multo poſt rubra Tarentina.* By

(1) *Lib.* 9. *Cap.* 29.

(2) *Denarius* was a *Roman* Peny, very near the Weight of the *Attic* Drachm, in our Money of the Value of Eight-pence Halfpeny Farthing; ſeven of them being made out of an Ounce Troy Weight in the Time of *Tiberius.* And we read of no higher Value either before or after that Time; which alſo continued legal Weight in the Time of *Veſpaſian.* After this they weigh'd leſs; eight of them being made out of an Ounce; which reduced them to Seven-pence Halfpeny of our Money. For whereas their *Libra* before contained but eighty-four, it now contained ninety-ſix of them. In the lower Empire they ſcarce weigh'd half ſo much in pure Silver. Tho' this was the Price of this Kind of Purple: Yet the ſo-much valuable *Tyrian* Scarlet was often ſold for One thouſand *Roman Denarii* a Pound; ſo that it was ten times as dear as the common Purple.

ADDENDA.

By the Help of this Distinction between these two Colours, we shall be able to throw Light upon several Passages in the Classic Authors, otherwise obscure; among these that of *Horace:*

- - - - Te bis Afro
Murice tinctæ
Vestiunt lanæ.
Lib. II. Od. 16.

The *bis tinctæ* hath been generally understood by Critics of Cloth double-dy'd in the same Materials, to give it a deeper Dye; but may not we be of Opinion, that this so much valued *Tyrian* Purple receiv'd first the Tincture of the Violet Colour mentioned as above? This being afterwards dyed in the deep *Tyrian* Red, the second Dying added a Redness to it, which could not have been done by being double-dyed in the Violet Colour.

That there was one Kind of the Purple of the Antients of a deep Red, like Blood, is evident from that beautiful Allusion of *Homer:*

τον δε
Ελαβε πορφυριος θανατος κ' μοιρα κραταιη.

From the above-mentioned Mixture of the Violet Colour, and the deep *Tyrian* Red, arose the most beautiful, and so valuable flaming Purple, varying in Degrees, as the Violet, or the *Tyrian* Red, prevail'd. These two Colours seem to be hinted at by *Catullus*, who, speaking of the Marriage of *Peleus* and *Thetis*, says,

Tincta tegit roseo conchylis purpura fuco.

And that the latter of these was more valuable, appears from its far greater Price, as well as from its being more frequently describ'd as the most distinguishing Mark of Royalty. Thus *Virgil*,

Et princeps Tyrio vestem prætexuit ostro.

And *Claudian*, speaking of the Marriage of *Honorius*, says,

Amplexu caleat purpura regio
Et vestes Tyrio sanguine fulgidas
Alter virgineus nobilitet cruor. Lib. 5. V. 114.

Many likewise are the Instances in Classic Authors, to prove that it is the peculiar Property of the *Tyrian Murex* to dye red, or of a crimson Colour.

Ovid

ADDENDA.

Ovid ſays,

Nec quæ de Tyrio murice lana rubet.

And *Virgil*,

Tyrioque ardebat murice læna. Æn. IV. 262.

The ſame Author elſewhere calls it the Purple of *Sarra* †.

Ut gemma bibat, et Sarrano dormiat oſtro.

From hence *Milton* takes occaſion to deſcribe the Dreſs of the Archangel deſcending to the Earth.

- - - - - O'er his lucid Arms
A military Veſt of Purple flow'd,
Livelier than Melibean, *or the Grain*
Of Sarra, *worn by Kings and Heroes old.*

Tarentum ſeems to be one of thoſe Places for preparing the Purple Dye, eſpecially, as *Pliny* obſerves, in the Time of *Auguſtus*; and the Ruins of the Buildings, as well as Heaps of broken Shells, remain'd there for ſeveral Ages after. As theſe Shells producing the Purple were found in the *Mediterranean*, this will perhaps help us to underſtand the Force of that Line of *Virgil*,

In mare purpureum violentior influit amnis.
Georg. IV. v. 373.

Virgil likewiſe alludes to this, when he ſays,

Lumenque juventæ
Purpureum, & lætos oculis afflârat honores.

The ſame Poet, deſcribing the Funeral of *Pallas*, gives us to underſtand that this Colour was of great Eſteem:

Tunc geminas veſtes auroque oſtroque rigentes
Extulit Æneas: quas illi læta laborum
Ipſa ſuis quondam manibus Sidonia Dido
Fecerat, & tenui telas diſcreverat auro. Æn. XI. 72.

In

† *Sar* was the Name of a *Phœnician* City, afterwards called *Tyre*

ADDENDA.

In After-ages it became the diſtinguiſhing Mark of Royalty, and made no ſmall Part of the Treaſures of the Eaſtern Monarchs. For when *Alexander* the Great had defeated *Darius*, and taken the City of *Suſa*, he found among other Spoils to the Value of five thouſand Talents of *Hermonic* Purple, that had been laid up in Store for near two hundred Years. In Proceſs of Time it became (as we have ſaid) the Mark of Grandeur and Dignity: Hence to put on the Purple, was to aſſume the Government: And ſometimes of Pride; thus the rich Man is repreſented in Scripture to be cloathed in fine Linen, and Purple. And as the *Romans* had a Law forbidding all private Perſons the Uſe of purple Garments, ſo likewiſe none but Emperors, or ſupreme Magiſtrates, were allow'd the Uſe of Purple, inſtead of Ink: Hence many of the Royal Grants and Laws were ſign'd with Purple; and *Pamphilus*, *Apelles*'s Maſter, made uſe of this Liquid in painting or enamelling upon Ivory, which was done by heating the Ivory to a great Degree, and then pouring into the engraved Lines this liquid Purple. Hence that Epigram of *Martial*,

Encauſtes Phaeton tabula tibi pictus in hac eſt.
Quid tibi vis, Dipyron qui Phaetonta facis?

This Colour was ſo admir'd by the Antients, that the Poets added the Epithet *Purple* to every thing that was rare, bright, or valuable. Hence, by a ſurpriſing poetic Licence, *Horace*, ſpeaking of a Swan, ſays that he was

Purpureis ales oloribus.

PAGE 195, *Line* 18, *add*, The Flowers are ſucceeded by Snuff-coloured Pods, theſe are deeply channell'd, and contain ſeveral thin ſhelled Nuts.

RESEARCH LIBRARY OF COLONIAL AMERICANA

An Arno Press Collection

Histories

Acrelius, Israel. **A History of New Sweden;** Or, The Settlements on the River Delaware . . . Translated with an Introduction and Notes by William M. Reynolds. Historical Society of Pennsylvania, MEMOIRS, XI, Philadelphia, 1874.

Belknap, Jeremy. **The History of New Hampshire.** 3 vols., Vol. 1—Philadelphia, 1784 (Reprinted Boston, 1792), Vol. 2—Boston, 1791, Vol. 3—Boston, 1792.

Browne, Patrick. **The Civil and Natural History of Jamaica.** In Three Parts . . . London, 1756. Includes 1789 edition Linnaean index.

[Burke, Edmund]. **An Account of the European Settlements in America.** In Six Parts . . . London, 1777. Two volumes in one.

Chalmers, George. **An Introduction to the History of the Revolt of the American Colonies:** Being a Comprehensive View of Its Origin, Derived From the State Papers Contained in the Public Offices of Great Britain. London, 1845. Two volumes in one.

Douglass, William. **A Summary, Historical and Political, of the First Planting, Progressive Improvements, and Present State of the British Settlements in North-America.** Boston, 1749–1752. Two volumes in one.

Edwards, Bryan. **The History, Civil and Commercial, of the British Colonies in the West Indies.** Dublin, 1793–1794. Two volumes in one.

Hughes, Griffith. **The Natural History of Barbados.** In Ten Books. London, 1750.

[Franklin, Benjamin]. **An Historical Review of the Constitution and Government of Pennsylvania, From Its Origin . . .** London, 1759.

Hubbard, William. **A General History of New England, From the Discovery to MDCLXXX.** (*In* Massachusetts Historical Society, COLLECTIONS, Series 2, vol. 5, 6, 1815. Reprinted 1848.)

Hutchinson, Thomas. **The History of the Colony of Massachusetts Bay** . . . 3 vols., Boston, 1764–1828.

Keith, Sir William. **The History of the British Plantations in America** . . . London, 1738.

Long, Edward. **The History of Jamaica:** Or, General Survey of the Antient and Modern State of that Island . . . 3 vols., London, 1774.

Mather, Cotton. **Magnalia Christi Americana;** Or, The Ecclesiastical History of New-England From . . . the Year 1620, Unto the Year . . . 1698. In Seven Books. London, 1702.

Mather, Increase. **A Relation of the Troubles Which Have Hapned in New-England, By Reason of the Indians There From the Year 1614 to the Year 1675** . . . Boston, 1677.

Smith, Samuel. **The History of the Colony of Nova-Caesaria, Or New-Jersey . . . to the Year 1721** . . . Burlington, N.J., 1765.

Thomas, Sir Dalby. **An Historical Account of the Rise and Growth of the West-India Collonies,** and of the Great Advantages They are to England, in Respect to Trade. London, 1690.

Trumbull, Benjamin. **A Complete History of Connecticut,** Civil and Ecclesiastical, From the Emigration of Its First Planters, From England, in the Year 1630, to the Year 1764; and to the Close of the Indian Wars . . . New Haven, 1818. Two volumes in one.

Personal Narratives and Promotional Literature

Byrd, William. **The Secret Diary of William Byrd of Westover, 1709–1712,** edited by Louis B. Wright and Marion Tinling. Richmond, Va., 1941.

Byrd, William. **The London Diary (1717–1721) and Other Writings,** edited by Louis B. Wright and Marion Tinling. New York, 1958.

A Genuine Narrative of the Intended Conspiracy of the Negroes at Antigua. Extracted From an Authentic Copy of a Report, Made to the Chief Governor of the Carabee Islands, by the Commissioners, or Judges Appointed to Try the Conspirators. Dublin, 1737.

Gookin, Daniel. **An Historical Account of the Doings and Sufferings of the Christian Indians in New England in the Years 1675, 1676, 1677** . . . (*In* American Antiquarian Society, Worcester, Mass. ARCHAEOLOGIA AMERICANA. TRANSACTIONS AND COLLECTIONS. Cambridge, 1836. vol. 2.)

Gookin, Daniel. **Historical Collections of the Indians in New England.** Of Their Several Nations, Numbers, Customs, Manners, Religion and Government, Before the English Planted There . . . Boston, 1792.

Morton, Thomas. **New English Canaan or New Canaan.** Containing an Abstract of New England, Composed in Three Books . . . Amsterdam, 1637.

Sewall, Samuel. **Diary of Samuel Sewall, 1674–1729.** (*In* Massachusetts Historical Society. COLLECTIONS, 5th Series, V–VII, 1878–1882.) Three volumes.

Virginia: Four Personal Narratives. (Hamor, Ralph. *A True Discourse on the Present Estate of Virginia . . . Till the 18 of June 1614* . . . London, 1615/Hariot, Thomas. *A Briefe and True Report of the New Found Land of Virginia . . .* London, 1588/Percy, George. *A Trewe Relacyon of the Proceedings and Ocurrentes of Momente Which Have Happened in Virginia From . . . 1609, Until . . . 1612.* (In *Tyler's Quarterly Historical and Genealogical Magazine,* Vol. III, 1922.)/Rolf, John. *Virginia in 1616.* (In *Virginia Historical Register and Literary Advertiser,* Vol. I, No. III, July, 1848.) New York, 1972.

Winthrop, John. **The History of New England From 1630–1649.** Edited by James Savage. Boston, 1825–1826. Two volumes in one.

New England Puritan Tracts of the Seventeenth Century

Cobbett, Thomas. **The Civil Magistrate's Power in Matters of Religion Modestly Debated** . . . London, 1653.

Cotton, John. **The Bloudy Tenent, Washed, and Made White in the Bloud of the Lambe** . . . London, 1647.

Cotton, John. **A Brief Exposition with Practical Observations Upon the Whole Book of Canticles.** London, 1655.

Cotton, John. **Christ the Fountaine of Life:** Or, Sundry Choyce Sermons on Part of the Fift Chapter of the First Epistle of St. John. London, 1651.

Cotton, John. **Two Sermons.** (*Gods Mercie Mixed with His Justice* . . . London, 1641/*The True Constitution of a Particular Visible Church, Proved by Scripture* . . . London, 1642.) New York, 1972.

Eliot, John. **The Christian Commonwealth:** Or, The Civil Policy of the Rising Kingdom of Jesus Christ. London, 1659.

Hooker, Thomas. **The Application of Redemption,** By the Effectual Work of the Word, and Spirit of Christ, for the Bringing Home of Lost Sinners to God. London, 1657.

H[ooker], T[homas]. **The Christian's Two Chiefe Lessons,** Viz. Selfe Deniall, and Selfe Tryall . . . London, 1640.

Hooker, Thomas. **A Survey of the Summe of Church-Discipline** Wherein the Way of the Churches of New England is Warranted Out of the Word, and All Exceptions of Weight, Which Are Made Against It, Answered . . . London, 1648.

Increase Mather Vs. Solomon Stoddard: Two Puritan Tracts. (Mather, Increase. *The Order of the Gospel, Professed and Practised by the Churches of Christ in New-England* . . . Boston, 1700/Stoddard, Solomon. *The Doctrine of Instituted Churches Explained, and Proved From the Word of God.* London, 1700.) New York, 1972.

Mather, Cotton. **Ratio Disciplinae Fratrum Nov-Anglorum.** A Faithful Account of the Discipline Professed and Practised, in the Churches of New England. Boston, 1726.

Mather, Richard. **Church Covenant:** Two Tracts. (*Church-Government and Church-Covenant Discussed, in an Answer to the Elders of the Severall Churches in New-England* . . . London, 1643/*An Apologie of the Churches in New-England for Church-Covenant, Or, A Discourse Touching the Covenant Between God and Men, and Especially Concerning Church-Covenant* . . . London, 1643.) New York, 1972.

The Imperial System

[Blenman, Jonathan]. **Remarks on Several Acts of Parliament Relating More Especially to the Colonies Abroad . . .** London, 1742.

British Imperialism: Three Documents. (Berkeley, George. *A Proposal for the Better Supplying of Churches in our Foreign Plantations, and for Converting the Savage Americans to Christianity by a College to be Erected in the Summer Islands, Otherwise Called the Isles of Bermuda . . .* London, 1724/[Fothergill, John]. *Considerations Relative to the North American Colonies.* London, 1765/*A Letter to a Member of Parliament Concerning the Naval-Store Bill . . .* London, 1720.) New York, 1972.

Coke, Roger. **A Discourse of Trade** . . . London, 1670.

[D'Avenant, Charles]. **An Essay Upon the Government of the English Plantations on the Continent of America** (1701). An Anonymous Virginian's Proposals for Liberty Under the British Crown, With Two Memoranda by William Byrd. Edited by Louis B. Wright. San Marino, Calif., 1945.

Dummer, Jeremiah. **A Defence of the New-England Charters . . .** London, 1721.

Gee, Joshua. **The Trade and Navigation of Great Britain Considered:** Shewing that Surest Way for a Nation to Increase in Riches, is to Prevent the Importation of Such Foreign Commodities as May Be Rais'd at Home. London, 1729.

[Little, Otis]. **The State of Trade in the Northern Colonies Considered;** With an Account of Their Produce, and a Particular Description of Nova Scotia . . . London, 1748.

Tucker, Jos[iah]. **The True Interest of Britain, Set Forth in Regard to the Colonies:** And the Only Means of Living in Peace and Harmony With Them, Including Five Different Plans for Effecting this Desirable Event . . . Philadelphia, 1776.